Early Modern Philosophers and the Renaissance Legacy

INTERNATIONAL ARCHIVES OF THE HISTORY OF IDEAS

ARCHIVES INTERNATIONALES D'HISTOIRE DES IDÉES

220

EARLY MODERN PHILOSOPHERS AND THE RENAISSANCE LEGACY

Cecilia Muratori • Gianni Paganini

More information about this series at http://www.springer.com/series/5640

Cecilia Muratori • Gianni Paganini
Editors

Early Modern Philosophers and the Renaissance Legacy

Springer

Editors
Cecilia Muratori
School of Modern Languages and Cultures
University of Warwick
Coventry, UK

Gianni Paganini
Dipartimento di Studi Umanistici
Università del Piemonte Orientale (Vercelli)
Vercelli, Italy

Centro di ricerca della Accademia dei Lincei
Rome, Italy

ISSN 0066-6610 ISSN 2215-0307 (electronic)
International Archives of the History of Ideas Archives internationales d'histoire des idées
ISBN 978-3-319-32602-3 ISBN 978-3-319-32604-7 (eBook)
DOI 10.1007/978-3-319-32604-7

Library of Congress Control Number: 2016947951

Printed on acid-free paper

This Springer imprint is published by Springer Nature
The registered company is Springer International Publishing AG Switzerland

Contents

1 **Renaissance and Early Modern Philosophy: Mobile Frontiers and Established Outposts** 1
Cecilia Muratori and Gianni Paganini

Part I The Endurance of Tradition

2 **What's Wrong with Doing History of Renaissance Philosophy? Rudolph Goclenius and the Canon of Early Modern Philosophy** 21
Guido Giglioni

3 **Italian Renaissance Love Theory and the General Scholar in the Seventeenth Century** 41
Stephen Clucas

4 **The Critique of Scholastic Language in Renaissance Humanism and Early Modern Philosophy** 59
Lodi Nauta

5 **Henry More and Girolamo Cardano** 81
Sarah Hutton

Part II Natural Philosophy

6 **From *Attractio* and *Impulsus* to Motion of Liberty: Rarefaction and Condensation, Nature and Violence, in Cardano, Francis Bacon, Glisson and Hale** 99
Silvia Manzo

7 **Telesio Among the Novatores: Telesio's Reception in the Seventeenth Century** 119
Daniel Garber

8 Looking at an Earth-Like Moon and Living on a Moon-Like Earth in Renaissance and Early Modern Thought 135
Natacha Fabbri

Part III Changing Conceptions of the Human

9 Descartes, the Humanists, and the Perfection of the Human Being 155
Emmanuel Faye

10 The Return of Campanella: La Forge versus Cureau de la Chambre 169
Emanuela Scribano

11 From Animal Happiness to Human Unhappiness: Cardano, Vanini, *Theophrastus Redivivus* (1659) 185
Cecilia Muratori

Part IV Moral and Political Theory

12 Ethics, Politics, and Friendship in Bacon's *Essays* (1625): Between Past and Future 203
Annalisa Ceron

13 Thomas Hobbes Against the Aristotelian Account of the Virtues and His Renaissance Source Lorenzo Valla 221
Gianni Paganini

14 Debating "Greatness" from Machiavelli to Burton 239
Sara Miglietti

15 John Upton from Political Liberty to Critical Liberty: The Moral and Political Implications of Ancient and Renaissance Studies in the Enlightenment 259
John Christian Laursen

Part V Epilogue

16 A Story in the History of Scholarship: The Rediscovery of Tommaso Campanella 277
Germana Ernst

Index 293

Chapter 1
Renaissance and Early Modern Philosophy: Mobile Frontiers and Established Outposts

Cecilia Muratori and Gianni Paganini

Abstract Difficulties with periodization are often symptoms of internal diseases affecting the history of philosophy. Renaissance scholars and historians of early modern philosophy represent two scholarly communities that do not communicate with each other, as if an abrupt change of scenery had taken place from the sixteenth to the seventeenth century, from the age of Campanella to the age of Descartes. The assumption of an arbitrary division between these two periods continues to have unfortunate effects on the study of the history of philosophy. This chapter provides a diagnosis of this problem by looking at the way in which periodization crystallized in the history of philosophy. It then lays a foundation for attempting a new approach to this issue, which consists in mapping direct connections and conceptual links of seventeenth-century philosophers with the philosophies of the Renaissance. We intend to shift the weight from the problem of assessing the 'modernity' of Renaissance philosophers to the creation of a space of interaction between Renaissance and early modern thinkers in the spirit of 'conversation', with special attention to tracing sources, direct allusions, confutations and continuities.

Renaissance Thinkers as "Conversation Partners"

Difficulties with periodization are often symptoms of internal diseases affecting the history of philosophy. Nowadays, Renaissance scholars and historians of early modern philosophy represent two scholarly communities that do not communicate with each other, as if an abrupt change of scenery had taken place from the sixteenth to the seventeenth century, from the age of Campanella to the age of Descartes. This would be understandable if one could locate a clear watershed between the end of the Renaissance and the beginning of early modern philosophy. But marking a clear

C. Muratori (✉)
School of Modern Languages and Cultures, University of Warwick, Coventry, UK
e-mail: c.muratori@warwick.ac.uk

G. Paganini
Dipartimento di Studi Umanistici, Università del Piemonte Orientale (Vercelli), Vercelli, Italy

Centro di ricerca della Accademia dei Lincei, Rome, Italy
e-mail: gianenrico.paganini@uniupo.it

C. Muratori, G. Paganini (eds.), *Early Modern Philosophers and the Renaissance Legacy*, International Archives of the History of Ideas Archives internationales d'histoire des idées 220, DOI 10.1007/978-3-319-32604-7_1

line between the two periods appears rather to be an impossible task, since, to give just one example, philosophers like Tommaso Campanella (1568–1639), Pierre Gassendi (1592–1655), and René Descartes (1596–1650) were contemporaries, and their lives and works largely overlapped. Nonetheless, the first is considered to be in all respects a Renaissance philosopher, the second is supposed to have had one foot in humanism with his *Epicurea anastasis* and the other in scientific revolution with his atomism ("le plus excellent Philosophe qui fût parmi les Humanistes, et le plus savant Humaniste qui fût parmi les Philosophes", according to Bayle's astute description[1]), and the last is considered without doubt as the father of early modern philosophy.

The assumption of an arbitrary division between these two periods continues to have unfortunate effects on the study of the history of philosophy. Early modern scholars, for instance, tend to perceive scholars of the Renaissance as belonging to an entirely different group, not precisely defined as to its disciplinary focus.[2] Indeed, Renaissance philosophical studies are often considered to be related thematically to the literature and arts of the fifteenth and sixteenth centuries, rather than as lying within a time-frame that would connect them to the great tradition of modern philosophy. Such lack of symmetry is the result of long-established presuppositions, which have been integral to the structure of modern philosophical historiography since the discipline took its first steps. This is reflected most notably in the terminology used to talk about the line to be drawn between the Renaissance and the beginning of modern philosophy. As Guido Ruggiero fittingly puts it, the Renaissance seems to have remained under siege for a long time: "It has been reincorporated into the Middle Ages, dissolved into the early modern period, obliterated by the premodern, and largely ignored by history done from a local or a world perspective".[3]

[1] Bayle (1740), vol. I, art. "Catius", rem E, 102b. This poignant description of Gassendi follows from rem. D, 1o2a, where Bayle famously defines the "Rèpublique des Lettres" as a free state in which truth and reason reign, a state characterized by a freedom similar to the indipendence of the natural state. All these elements contribute to the concept of *libertas philosophandi*, which, as we will discuss below, plays a key role in drawing the boundaries between Renaissance and early modern period, from the origins of modern historiography (Brucker) onwards. On the importance of this conception of modern philosophy in Bayle, especially in the *Dictionnaire*, see Paganini (1980), 331–348. Bayle viewed modernity in philosophy as a 'culture of evidence' and thus distinguished it from the age of the Renaissance. Yet, because of his sympathy for scepticism he was far from attributing an absolute supremacy to one single position among those of the 'modern' thinkers: from this point of view Brucker's eclecticism has its roots in Bayle's sceptical approach to the history of philosophy.

[2] As C.B. Schmitt has remarked, it was not "until the first quarter or so of the twentieth century that the history of Renaissance philosophy emerged as a subject in its own right – a subject distinguishable form medieval philosophy [...] and from 'modern' philosophy which had been initiated by Bacon, Descartes and their contemporaries" (Schmitt (1989), 11). Yet, the word 'contemporaries' implies using chronology as a watershed: if applied practically, this would imply, for instance, placing Campanella, Gassendi and Descartes in the same 'group'.

[3] Ruggiero (2002), 3. Luca Molà has discussed the fragility of the historiographical category of 'Renaissance' in Molà (2008), where he also considers the impact of revealing the shaky historical foundation of Burckhardt's idealised construction. Cf also Martin (2003), Part I ("The Renaissance Paradigm in Crisis").

National background and historiographical traditions play a significant role, if one compares the Italian distinction between *Rinascimento* and *Età moderna* with the English concepts of Renaissance and early modern period (sometimes even used as synonyms, even if with some discomfort), and with the German definition of *Frühe Neuzeit* (where the word *früh*, just as in the case of *early* modern, preludes the fulfilment of the following 'new' period, *Neuzeit*). In the aftermath of Burckhardt, the word 'Renaissance' tended to be geographically focused on Italy, in such a way that defining the Renaissance was in fact equivalent to tracing the contours of a national culture in Italy, taking shape against the background of the Middle Ages. The Renaissance thus became nothing other than the defining mark of the "Italian spirit", a specific phenomenon with precise geographical as well as historical borders.[4]

In recent years a conception of a 'long Renaissance' has emerged, according to which the end point of the Renaissance period might be extended to include the whole first half of the seventeenth century. This conception is for instance foundational for the new *Encyclopaedia of Renaissance Philosophy*, in preparation for Springer at the time of writing. The border between the two epochs thus appears to be, at least with regard to philosophy, a mobile one, with the tendency of the former (the Renaissance) to invade the latter (the early modern period), probably as a reaction to the contrary tendency in past historiography. Other scholars, notably James Hankins, have argued in favour of the opportunity and indeed the necessity of establishing continuities and connections between Renaissance philosophy and early modern philosophy. In this sense Hankins spoke of "continuities" and "similarities" extending from Renaissance to seventeenth- and eighteenth-century philosophy: his aim was to claim the right of Renaissance philosophers to be "conversational partners" in current philosophical discourse ("partners who can join in modern conversations", in the famous phrase of Richard Rorty to which Hankins referred), despite the peculiarities for which they were too often dismissed out of hand, primarily for their methodology and their approach to constructing a theory of knowledge.[5]

It might be objected that such an attempt to open a space of dialogue is based on a precise idea of methodology: Renaissance thinkers could be brought into dialogue with later philosophical developments *despite* the fact that their stand on philosophical method seemed to set them apart from seventeenth-century philosophy. It is especially the identification of philosophy with natural philosophy, or more precisely with epistemology and theory of knowledge (closely related to scientific method) which was, retrospectively, responsible for the reinterpretation of what philosophy's aims are, and, ultimately of what philosophy proper is,[6] as distinguished

[4] See Walther Rehm's useful introduction to Burckhardt (2014), especially 8–9.

[5] Hankins (2007), 339. See also 2–3 for further "continuities" between "the thought of the fourteenth through sixteenth centuries, often labelled late medieval or Renaissance or premodern or transitional, and that of the seventeenth and eighteenth centuries, generally regarded as modern or early modern".

[6] For a discussion of the meaning of philosophy, and to the role of historiography in defining this, see Kristeller (1985). Kraye and Stone (2000), xiii, have pointed to the continuity of humanism and seventeenth-century scientific developments: "Various features of the new 'mechanical philosophy' also benefit from being seen against the backdrop of the ideals and achievements of the

from other disciplines.[7] Yet the interpretation of seventeenth-century philosophy, too, has been subject to constant revision, up to the point where the issue of methodology is no longer the primary object of historians, even though it still occupies an important place in their narratives. This is not just matter of either chronology or academic organisation: behind the fact that Renaissance studies on the one hand, and early modern philosophy and history on the other, function *de facto* as two separate fields in the scholarly division of labour (as also in the case of ancient and medieval philosophy) lies a precise idea of what modernity was or should have been.[8] Historians of philosophy have thus until now based the reconstruction of continuity between the Renaissance and early modern period exclusively upon an emphasis of the 'modernity' of Renaissance thinkers, rather than by highlighting the Renaissance legacy in early modern thought.

While this is the trajectory of the approach that seeks to connect the Renaissance with the (early-)modern period in philosophy, there have been fewer attempts in the converse direction, to map direct connections and conceptual links of seventeenth-century philosophers with the philosophies of the Renaissance. This is precisely what the essays in this book set out to do. The present volume thus takes further the challenge posited by Hankins to think in terms of continuities and similarities, but proposes to invert the direction, going beyond fixed periodization by considering the 'Renaissance legacy' in early modern philosophy. We thus intend to shift the weight from the problem of assessing the 'modernity' of Renaissance philosophers to the creation of a space of interaction between Renaissance and early modern thinkers in the spirit of 'conversation', with special attention to tracing sources, direct allusions and confutations within a frame of continuity.

Back to the Founders: Brucker's Sense of Continuity and the Rise of True Modernity

For a professional historian it is always tempting to go back to the founders of the discipline in order to understand when and why the two periods were conceived as separate ages and in what ways they could possibly be connected to each other. Looking at this kind of Ur-history of the history of philosophy, it emerges that the division is certainly older than the organisation of the university into departments

humanist movement, which, contrary to conventional wisdom, remained a powerful force throughout most of the seventeenth century."

[7] See Celenza (2013), 368–369. Celenza reconstructs in detail how this judgement on fifteenth-century thinkers, not recognized as philosophers in the proper sense of the word, became dominant in philosophical historiography, and contrasts this with an analysis of what philosophy actually meant in the fifteenth century.

[8] James Hankins has studied twentieth-century interpretations of the role and meaning of Renaissance philosophy. See on this aspect Hankins (2002), especially 274–275 and 290 on the problem of 'modernity'.

and faculties during the nineteenth century, and even older than the birth of the notion of the Renaissance itself with Burckhardt.[9] If one considers the 'founding fathers' of philosophical historiography – Brucker in the eighteenth century, Hegel in the nineteenth – it is clear that the emergence of the modern conception of a history of philosophy developed in parallel with the theorization of a sharp distinction between these two periods – a theorization which was practically applied as well.

The German early Enlightenment is the birthplace of modern philosophical history, with Brucker's *Historia critica philosophiae*, which had a deep influence on the articles dealing with the history of philosophy in Diderot and d'Alembert's *Encyclopédie*.[10] In Brucker's work the early modern age (with Bacon, Descartes, Hobbes, Locke, Leibniz, and so forth) and the Renaissance were tightly associated into the same "period", the third one, which stretched "from the restoration of letters to our times".[11] In line with this sense of unity, Brucker emphasised continuities from the seventeenth century and the Renaissance, and even more with Luther and the Reformation, due to the importance he attributed to the battle against Scholasticism that was common both to Renaissance philosophers and to theologians and philologists of the Reformation. Yet, at the same time, the author of *Historia critica* was convinced that a sharp line could be drawn between the first part of this period, i. e. Humanism and the Renaissance,[12] and the latter part ("Pars altera. De studio philosophiae eclecticae post renatas litteras").[13] For Brucker, the watershed is located between philosophy which was still "sectarian" (like the several 'renaissances' of the old school) and modern philosophy proper, which he labelled "eclectic". Eclecticism meant for Brucker not to be enslaved to one single school, as instead, he claimed, had happened during the Renaissance, and to practice by contrast full philosophical liberty, as in seventeenth-century philosophy. In Brucker's own words, modern philosophy "as it does not swear on the words of the masters, chooses out of all things that which is proven to the highest degree, and discovers the truth by means of accurate rational reflection on the very nature of things".[14]

Even if he saw continuities from one period to another, Brucker was nevertheless convinced that a new kind of philosophical research began in the seventeenth cen-

[9] On the status of Renaissance philosophy within Burckhardt's interpretation of the Renaissance (also with attention to the influence of Hegel's own treatment of Renaissance philosophy) see Hankins (2002), 273–274.

[10] For a pre-history of philosophical historiography in Germany see Santinello et al. (1993), 371 ff.

[11] See Piaia and Santinello (2011), 512–513.

[12] For Brucker's opinion of Ficino, for instance, whom he considers to have been very learned but philosophically feeble, see Celenza (2013), 367–368, and 373–374.

[13] This part occupies the whole of Brucker (1744), vol. 5, and it is also divided into three books: "De restauratoribus philosophiae universae", i. e. philosophers who attempted an "eclectic" reform of all philosophy, like Bruno, Cardano, Bacon, Hobbes, Descartes, Leibniz, Ch. Tomasius (V, 3–543); "De emendatione philosophiae in singulis partibus" (V, 544–803); "De philosophia exotica" (V, 804–923), devoted to non-European philosophy (Chinese, Indo-Chinese, Canadian, Japanese).

[14] Brucker (1742–1744), I, 44. Quoted and trans. in Piaia and Santinello (2011), 518.

tury, "due mainly to the emergence of eclectic philosophy", which realised in one single century more advancements than in the previous ages.[15] At the same time, "love of Antiquity was harmful for the men of the Renaissance", who were driven by the study of the texts to "a state of veneration and absolute respect for Classical civilization", which prevented them from being true innovators. In Brucker's view, continuities and discontinuities coexist;[16] nonetheless he was able to point out a criterion, thus identifying a precise defining feature of modern philosophy: eclecticism.[17]

In Brucker's case the notion of *libertas philosophandi* has a twofold function: it allows him not only to distinguish between two different conceptions of modernity, but also to see them at work within the same period. When used as a polemical instrument against the dominance of *auctoritas*, Brucker considers it as forming part of the the initial phase (a phase more destructive than constructive, more linked to the renaissance of antiquity than open to novelties). But if it is instead viewed as full freedom within the frame of eclecticism, going beyond positive as well as negative prejudices, then it represents the *pars altera*, that is true modernity. In any case, continuity and discontinuity are never radically separate: Brucker maintains a strong sense of the unity of modernity, and he attributes the key role in achieving this to the Reformation.

Conversely, the development of a separation can be seen at work in the most important German histories of philosophy of the eighteenth and nineteenth centuries, which commonly refer to what we now call the Renaissance as the "more recent time" ("neuere Zeit")[18] and to early modern philosophy as the "most recent time" ("neueste Zeit"). For instance, Rixner's *Handbuch der Geschichte der Philosophie* (published in several volumes in the first half of the nineteenth century: 1823–1850) lists a series of events that in his opinion concurred in lending the sense of an epoch to the centuries from the fourteenth to the mid-seventeenth. He calls these events 'symptoms', thus implying an acknowledgement of the fact that precise demarcation lines are always artificial: these include political events, as well as literary changes connected to the rediscovery of antiquity. At the same time he mentions the aspects that mark the beginning of something different in the seventeenth century, primarily the emergence of "systems of philosophical doctrine" ("philosophische Lehrgebäude"),[19] such as those of Bacon and Descartes. Despite the flexi-

[15] Piaia and Santinello (2011), 533. (The whole section on Brucker was written by Mario Longo: see ibid., 479–577; this is a monograph on Brucker in its own right).

[16] See Piaia and Santinello (2011), 532.

[17] Leo Catana (2008) has drawn attention to Brucker's treatment of Bruno as "an innovator of eclectic philosophy" (35). C. Schmitt has emphasized the connection between Brucker's understanding of the history of philosophy and his interpretation of the modern revival of ancient scepticism, which appears as a key example of the way in which an ancient philosophical tradition reached modern times through Renaissance interpreters (Schmitt (1989), 193).

[18] On the emergence of the literary category of "history of the more recent philosophy" [*Geschichte der neueren Philosophie*] see also Piaia (1998), 167–180, 169.

[19] Rixner (1823), 3–4. See Catana (2008), 3–4 for a brief overview of the meaning of 'system' in philosophical historiography.

bility of Rixner's definition of the two periods, he nevertheless considers the Renaissance a largely literary rather than philosophical period, and this is why Bacon and Descartes mark in his eyes a definite change.

Tennemann argues in a similar way in his *Geschichte der Philosophie* when he identifies as a coherent period that from Descartes and Bacon to Kant, a period markedly different from the previous ones because of the "new attemps to ground a system of philosophical knowledge".[20] This is, in his opinion, the phase during which "the interest in philosophy emerges more strongly and spreads further and further. [...] It begins a new and indeed a very interesting period for the advancement of philosophy".[21] Tennemann thus frames Renaissance philosophies (he deals directly with Telesio and Patrizi, among others, and proceeds as far as Bruno) as "attempts" ("Versuche"): "particular philosophical attempts and combinations", as he terms them, placed somewhere in between the rediscovery of the ancients and the systems of modernity.[22]

The Renaissance as Intermezzo: Hegel

The other great forefather of this discipline beside Brucker – Hegel – posited so neat and even dramatic a separation that in order to depict it comprehensively he even reinterpreted historical facts with a certain liberty, especially from the point of view of chronological order. The Reformation, in Hegel's case, appears to be fully 'on the side' of the moderns, thus ceasing to work as a possible *trait-d'union* with the Renaissance. The crucial watershed is indeed not the conception of *libertas* but that of self-consciousness. Famously, Hegel's *History of Philosophy* pinpoints a radical change in the emergence of Cartesian philosophy. Descartes is described as "a hero who restarted again from the beginning, and reconstituted the foundation of philosophy anew, to which it now came back after a 1000 years."[23]

When turning to the discussion of Descartes, Hegel, like a new Columbus, exclaims: "land!".[24] This newly discovered territory – which Hegel nevertheless calls a place in which "we are home" – is for him the solid ground of self-consciousness, even though in the form, not yet mature, of "the thinking under-

[20] Tennemann (1817), vol. 10, 1: "neue[] Versuche[], ein System philosophischer Erkenntnisse zu gründen."

[21] Ibid., 1–2: "Das Interesse für Philosophie tritt kräftiger hervor und verbreitet sich immer weiter. [...] Es beginnt eine neue und zwar sehr interessante Periode für das Fortschreiten der Philosophie."

[22] Tennemann (1829), 320.

[23] Hegel (1836), 331: "Er ist so ein Heros, der die Sache wieder einmal ganz von vorne angefangen, und den Boden der Philosophie erst von Neuern konstituirt hat, auf den sie nun erst nach dem Verlauf von tausend Jahren zurückgekehrt ist."

[24] Ibid., 328: "Hier, können wir sagen, sind wir zu Hause, und können, wie der Schiffer nach langer Umherfahrt auf der ungestümen See 'Land' rufen; Cartesius ist einer von den Menschen, die wieder mit Allem von vorn angefangen haben; und mit ihm lebt die Bildung, das Denken der neueren Zeit an."

standing". For this reason, Hegel regards Descartes as the truly pivotal figure who pulls Renaissance and modernity apart.

The background and impact of Hegel's sharp separation of modern philosophy from the preceding period is well known. What is less widely acknowledged is the fact that Hegel consciously turned historical sequence upside down in order to strengthen this view. As a good Lutheran and exactly like Brucker, the author of the lectures on *Geschichte der Philosophie* connects Descartes to the father of the Reformation, even though he does so not under the auspices of the struggle against Scholasticism but rather in a more sophisticated form, emphasising rather moment in which the spirit comes to itself.

In sharp contrast with Brucker's approach, Hegel's strategy led him to what could be called a subversion of chronology, in order to unearth the full value of a crucial development in the history of philosophy. Descartes established a new terrain for philosophy itself, and indeed the conjunction of "exact sciences" (also called "sciences of the determinate intellect") and philosophy began in this period, with Descartes.[25] As a result of this, Renaissance philosophy was pushed into the background, disconnecting it decisively from modernity, while at the same time Hegel emphasized the role of the Lutheran reformation.[26] As in Brucker, periodisation is the key, but Hegel instead opts for a different order – one that pays more attention to *conceptual* development – rather than simply following historical succession.

The fact that most Renaissance authors were little known in Germany at the time partly explained their treatment in these histories of philosophy. Hegel's *History of Philosophy* is a case in point: he mentions certain authors only via second-hand material, often without direct knowledge of the texts. Moreover, many authors are consciously left out of the history of philosophy – a selection which is paramount to the specific understanding of what philosophy is, and thus of what can be included in a history of philosophy and what should be left out of it. This is notably the case with Montaigne, as well as with Machiavelli, who do not feature in Hegel's *History of Philosophy* simply because from Hegel's point of view they rather belong to the history of general learning: for him, they are not true philosophers.[27]

Hegel briefly discusses a series of authors from the fifteenth to the early seventeenth century under the heading "Revival of the Sciences" ("Wiederaufleben der Wissenschaften"). They are divided into three main groups. To the first – labelled

---

[25] Ibid., 331. Cf. also ibid., 332 on "Wissenschaften des bestimmten Verstandes".

[26] Hegel (1969–1981), I, 99–100 (Hegel to Voss, April 1805): "Luther hat die Bibel, Sie den Homer deutsch reden gemacht, – das größte Geschenk, das einem Volke gemacht werden kann [...] [W]enn Sie diese beiden Beispiele vergessen wollen, so will ich von meinem Bestreben sagen, daß ich die Philosophie versuchen will, deutsch sprechen zu lehren."

[27] Hegel (1836), 252: "Noch viele andere merkwürdige Männer fallen in diese Zeit, die auch in der Geschichte der Philosophie aufgeführt zu werden pflegen, als Michael de Montaigne, Charron, *Machiavell* u.s.f. Dergleichen Männer werden genannt; aber sie gehören nicht eigentlich der Philosophie, sondern der allgemeinen Bildung an."

"The Study of the Ancients" – belong thinkers who rediscovered ancient philosophy in its original form, without adding anything new. The border dividing philosophy and literature is here particularly subtle, and Hegel underlines that during this phase philosophy did not achieve anything original, as engagement with philosophical texts (especially Greek ones) consisted merely in their rediscovery through the learning of ancient languages.[28] Hegel dedicates a few lines to the thinkers of this period who studied strands of ancient philosophical: Pomponazzi as an example of Aristotelianism, Ficino for the Platonic tradition, followed by Gassendi (Epicureanism), Lipsius (Stoicism) and Reuchlin (the Kabbalah).

It is true that Cardano, Vanini, Bruno, and Campanella – who are included in the next section, "Idiosyncratic Endeavours of Philosophy" ("Eigentümliche Bestrebungen der Philosophie") – receive friendly treatment from Hegel (Campanella is the most neglected figure of the three, receiving in the Michelet edition one third of a page, less than Cardano and Vanini, much less than Bruno).[29] But what is most relevant for the purpose of our investigation in this volume is that this section dedicated to peculiar and rather wild philosophical attempts is a subsection of the part on "Philosophy of the Middle Ages". This might appear paradoxical, given that these philosophers were all "novatores", that is fierce opponents of medieval and scholastic philosophy.

If we consider a particular course of Hegel's history of philosophy, that held in the year 1825–1826, rather than Michelet's edition (in which materials related to various courses were merged), we still find the same structure but with a notable awareness on Hegel's part of the implications of his shaping procedure. Renaissance thinkers are still discussed before the Reformation, and the link between the Reformation and modern philosophy features just as prominently. Nevertheless, according to the structure of this course, which can be reconstructed from the *Nachschriften* prepared by Hegel's students, *Renaissance* and *Reformation* are here placed together under one heading. Just as in Michelet's version, here too the general section within which the Renaissance and the Reformation are included is that of "medieval philosophy": Renaissance and Reformation are considered as one section, following three previous ones: (1) the Church fathers, (2) the philosophy of the Arabs, and (3) the Scholastics.[30] Again, the section in which Renaissance and Reformation are included is in itself divided into three subheadings. The short section dedicated to the rediscovery of ancient languages and philosophies is entitled here "The Interest in Ancient Philosophy", and Hegel affirms firmly that "free phi-

[28] Ibid., 213: "Die Wiedererweckung der Wissenschaften und Künste, besonders des Studiums der alten Literatur in Beziehung auf Philosophie war aber zuerst eines Theils eine Wiedererweckung bloß der alten Philosophie in ihrer früheren ursprünglichen Gestalt; Neues ist noch nicht aufgekommen."

[29] Ibid., 224.

[30] Hegel (1986), 45.

losophy, systems that were initiated by thought, have not yet emerged, but ancient philosophical systems were now merely renewed and revived."[31]

This is followed by the presentation of the same selection of Renaissance thinkers as in Michelet's edition. Hegel underlines the fact that this period – that is the Renaissance – was "full of such individuals, who experimented in the most genial but then in the most corrupt way", stressing the contrast between the confused energy of these figures and the calmer and less original character of those who occupied themselves only with the rediscovery of ancient thought.[32] When Hegel comes to the section on the Reformation, leading directly to the third period ("Die neuere Philosophie"), it emerges clearly that he is well aware of the tension in his unfolding of the history of philosophy – the tension between the use of chronology and the requirements of different principles, inherent in philosophical content rather than mere chronological succession. In the course of 1825–1826 we read: "Here a transition must be mentioned, which interests us because of the universal principle that is recognized here at a higher level and it is recognized in its justification. Giordano Bruno, Vanini and others fall in the age of Reformation and later. The Reformation therefore commenced in this time."[33] This comment shows both Hegel's attempt to trace the unfolding of universal principles, such as the principle that human "activity, reason, imagination"[34] gradually develop, while also acknowledging actual historical patterns.

Given this outline, it seems that Hegel's history pf philosophy could allow no *tertium* between the Middle Ages and Modernity: the Renaissance must be merely a descendent of medieval philosophy, a transition towards the achievements of philosophy in the proper sense of the word. Luther, meanwhile, is discussed not only in the third and last part of "Renaissance of science", titled "The Reformation", but also, and at great length, in the third introductory explanation to the third period: "Modern Philosophy" ("Neuere Philosophie"). Here Hegel mentions again the group of Renaissance authors (Bruno, Vanini, Ramus; Campanella is no longer mentioned) who lived after Luther, adding a telling explanation for this chronological twist: "With the Reformation we actually enter the *third period*, regardless of the fact that Bruno, Vanini and Ramus, who lived later, still belong to the Middle Ages. A point of inversion occurred."[35]

---

[31] Ibid.: "Freie Philosophie, Systeme, die vom Denken ausgegangen wären, sind noch nicht aufgekommen, sondern nur die alten philosophischen Systeme wurden jetzt erneuert und erweckt" (translation by Cecilia Muratori, but see also Hegel (1990), 71).

[32] Hegel (1986), 50: "Die Zeit war reich an solchen Individuen, die sich auf die genialste und dann aber auf die korrupteste Weise herumtrieben [...]."

[33] Ibid., 61: "Hier ist nun ein Übergang zu erwähnen, der uns angeht des allgemeinen Prinzips wegen, das darin höher erkannt und in seiner Berechtigung erkannt ist. Jordanus Bruno, Vanini und andere fallen in die Zeit der Reformation und später. Die Reformation ist also in diese Zeit eingetreten." (Translation by Cecilia Muratori, but see Hegel (1990), 94.) Cf. the same passage, differently formulated, in Hegel (1836), 255.

[34] Hegel (1986), 62.

[35] Hegel (1836), 265: "Mit der Reformation treten wir so auch eigentlich in die *dritte Periode* hinüber, ungeachtet Bruno, Vanini und Ramus, die später lebten, noch zum Mittelalter gehören. Ein Punkt der Umkehrung trat ein" (translation by Cecilia Muratori).

In the 1825–1826 course the introduction to modern philosophy is short, and does not directly mention Luther and the Reformation. Yet, Hegel continues to posit a direct link between the guiding principle of the Lutheran Reformation – the turn inward, towards subjectivity – and the beginnings of modern philosophy. Indeed, here too Renaissance thinkers are discussed before the Reformation, which thus concludes the section on medieval philosophy, leading to the new period of modern philosophy. Moreover, Hegel stresses the fact that thought makes its appearance now as "something subjective, with the reflection of its being-in-itself"[36]: once again, the role of subjectivity plays the role of a bridge directly from Luther to modernity, while the Renaissance is left in the background, placed logically – and thus chronologically – before the Reformation. The essential tenets of Hegel's position are thus clearly evident in this course, too, but so is his awareness of the tension between chronology and logical unfolding. From Hegel's perspective there are two overriding reasons for discussing Renaissance thinkers *before*, rather than *after* Luther, contrary to what history and chronology would require. The first is a conceptual one: these Renaissance authors did not have the modern notion of self-consciousness. The second reason is, so to speak, a strategic one. If he had respected and not inverted the real sequence, the Renaissance intermezzo, made of such irregular thinkers, would have disturbed the direct connection between Luther and Descartes he aimed to establish.

History of philosophy, even of the most speculative kind, is still history, which means that it depends on knowledge or ignorance of matters of fact, and not just on the choice of "conversation partners" – and this principle is valid even for Hegel.[37] For instance, the better treatment Bruno received from Hegel in comparison to Campanella is clearly due to Bruno's renown in Germany from Lessing to Schelling. This is understandable, but also a little paradoxical, because Campanella would fit better than Bruno into Hegel's historical scheme. If the author of the *Geschichte der Philosophie* had known Campanella, he could have appreciated in the latter a strong supporter of his own metaphysical thesis that modernity is fundamentally the era of self-consciousness.[38] Campanella thought that any kind of consciousness, even if it is diffused throughout nature, still remains some sort of self-consciousness, according to his theory of the three 'primalities': besides power (pon = posse) and love (mor = amor), every kind of being exhibits another fundamental quality ('primality'), which is "sap" (sapere = to feel, to know), and most of all to have a certain knowledge or feeling of itself.

This is a clear example of the fact that the transmission and reception of certain texts, to the detriment of others, had a direct impact on the discipline of philosophi-

---

[36] Hegel (1986), 71: "Dies [das Denken] tritt wesentlich jetzt auf als ein Subjektives, mit der Reflexion seines Insichseins".

[37] C. Schmitt has pointed to the 'imbalances' that emerged from the tendency to select certain data to form a 'history of Renaissance philosophy', for instance privileging certain geographical areas, such as Italy, over others (Schmitt (1988), 10).

[38] See Paganini (2008a), 11–29. See also Paganini (2008b), chapter III, which focuses on the intellectual relations between Campanella, Descartes, and Mersenne.

cal historiography. In Campanella's case this is blatant: from the chronological point of view Campanella was a true contemporary of Descartes (with whom he was even indirectly in contact through Mersenne). Further, his thought is in harmony with the trajectory of Hegel's section on modern philosophy, to the extent that the idea of self-consciousness, so crucial in Campanella's thought, is the criterion for tracing a line to divide the new, proper philosophy from previous, pseudophilosophical attempts. Of course, his idea of self-awareness is profoundly different from Luther's and from idealistic conceptions, because Campanella attributes self-knowledge to every kind of being, and not only to spiritual ones (taking word 'spiritual' in the meaning of the German word *geistig*, that is not referring to spiritual beings as those endowed with a higher form of life, but rather to all living beings). If one takes into serious consideration this idea of extended self-awareness, which has several different levels, up to the highest one of the pure *mens*, even the emphasis put by German historiography on the central place of consciousness could point towards a different theory for the birth of modernity: it would imply going beyond the connection Luther-Descartes, and moving back in the direction of the Italian Renaissance. The case of Campanella and of his treatment within Hegel's history of philosophy is thus an instance of the fact that a different assessment of philosophical texts, or simply access to them, could have changed the path of thought about the history of philosophy, or at least its fundamental contours.

Philosophical Periodization: The Issues at Stake

This historiographical survey, and the examples of Hegel and Brucker in particular, highlights the crucial issues at stake in dealing with the Renaissance legacy in early modern philosophy: were Renaissance thinkers philosophers in their own right, or do they rather belong to the history of literature and the arts? Were they essentially polemical, anti-scholastic thinkers, or did they instead contribute innovatively to the birth of modernity? Were they mainly philosophers (as Garin argued), or mainly philologists (Kristeller's thesis), that is 'humanists' in the technical, narrowest sense of the word (scholars of ancient languages and texts)?[39] Did their work consist mainly in the reappraisal of antiquity, or what were the (other) elements that contributed to their inclusion in a separate group apart from that of modern philosophers? And especially, how did the legacy of their philosophical approaches persist in the following centuries through the direct encounters of subsequent generations with their texts?

By asking such questions, this volume deals collectively with the broad historiographical problem of bridging the distance between phases fixed by subsequent historiography, considering in particular the role of the Renaissance between the

[39] On the terminological juxtaposition of humanism and Renaissance seen as a historiographical phenomenon, see Hankins (2005), 73–96. See further the clear overview of the terms Renaissance and Humanism and the history of their uses in Black (2005).

Middle Ages and the Early Modern period.[40] More specifically, it addresses the question regarding the legitimacy of setting milestones to delimit important changes of scenery and conceptual shifts. One of the crucial ones is the year 1650: for instance, Jonathan Israel has argued that after this date, "a general process of rationalization and secularization set in which rapidly overthrew theology's age-old hegemony in the world of study, slowly but surely eradicated magic and belief in the supernatural from Europe's intellectual culture, and led a few openly to challenge everything inherited from the past".[41] The contrast between the adjective "rapidly" and the following phrase "slowly but surely" exemplifies the complexity of the task of marking boundaries while at the same time acknowledging the persistence of certain conceptual problems. Indeed, the intertwining of 'rapid' and 'slow' changes characterizes the fabric of the case studies presented in this collection: while highlighting the continuous transmission of texts and sources, these studies attempt to pinpoint what changed and how problems were reinterpreted and framed in new contexts, not only strictly philosophical but also religious.

While the legacy of philosophical historiography sets a critical frame for the present volume, the studies included here collectively question the assumption of an abrupt border dividing Renaissance and early modern philosophies. They thus represent concrete alternatives to a division into two periods which only by a convention (which scholars often employ without either justifying or questioning it) has come to be identified with the year 1650, or more traditionally some decades before, with the publication of *Discours de la méthode* by Descartes (1637) and Campanella's death (1639). In so doing, the chapters in the volume set aside this conventional border in order to explore in detail how thinkers of the fifteenth to the seventeenth centuries underwent a varied philosophical afterlife, comprising influence as well as reaction, through the engagement of later philosophers with their work: thus it is early modern philosophers, in this case, who are viewed as joining their predecessors as 'conversation partners'. In this way the volume aims to establish a new methodological approach to study permanencies, modifications to and new interpretations of philosophical theories from the period usually labelled as 'Renaissance' to that termed 'early modern'. We wish to point in the direction of reconstructing the sources known to early modern philosophers, in order to restore the missing link between the Renaissance and the early modern period, thus attaining the sense of a continuity, in which each philosopher's approach to his immediate predecessors marks at the same time a certain change of perspective. We aim to highlight elements of continuity without losing sight of the various points of difference and of

[40] Of course the problem of the continuity of Renaissance and the Middle Ages can also be considered – as it was by Ferguson (1948) – from the point of view of the scholars of the latter period as well (see the chapter 'The Revolt of the Medievalists: The Renaissance Interpreted as Continuation of the Middle Ages', in Ferguson (1948), 329–397. See also Burnett et al. (2008).

[41] Israel (2001), 4.

change, creating a frame of investigation which allows for a plurality of chronological as well as geographical and ideological viewpoints.[42]

Case Studies of 'Conversation'

The case studies presented in this volume of course do not aim at exhaustive coverage of the Renaissance legacy in early modern thought: rather, they attempt to inaugurate a new way of studying the history of philosophy of this period. Core topics and key authors running through the volume give a sense of the continuous engagement with certain philosophical approaches and texts, while the timeframe spans from the Platonic revival in the fifteenth century, to the late eighteenth century. The chapters are divided into four main areas, bringing attention specifically to the endurance and change of philosophical theories with regard to (1) the topic of creating and maintaining a philosophical tradition; (2) issues in natural philosophy; (3) the reception and reinterpretation of political and moral theory; and finally (4) the changes in the anthropological conception of the human being, also considering the difference between man and the animals. The volume culminates in the Epilogue, 'A Story in the History of Scholarship: The Rediscovery of Tommaso Campanella' by Germana Ernst. This is the story of a personal encounter with Campanella's writings that contributed significantly to the reappraisal of his philosophy.

I. The first section addresses a crucial question for the entire book: how is a philosophical tradition constructed? The essays deal with the concepts of eclecticism and philosophical systematicity, investigating their connections and oppositions in selected historiographical cases. Guido Giglioni's essay analyses the case of Rudolph Goclenius's *Lexica* in order to answer precisely this question, by looking at the focal point on which a philosophical tradition is constructed – that is, language. Dealing with the afterlives of Renaissance philosophical terminology, Goclenius presents a practical example of the fact that continuities persist despite any historiographical attempt to draw a boundary line. A principal example of this endurance, apparent in Goclenius's *Lexica*, is the 'construction' of a Platonic tradition from the Renaissance into the seventeenth century. Stephen Clucas's starting point is the conception of the 'general scholar' in the seventeenth century: focussing in particular on Burton's *Anatomy of Melancholy*, Clucas shows how Renaissance love theory was received and reinterpreted in the new context of

[42] As Black (2010) has aptly put it: "Renaissance humanism [...] may have been a movement, but as such it was far from uniform, and so it is arguably more appropriate to speak in the plural rather than in the singular of Renaissance chronologies, ideologies and geographies" (44). Black also interestingly stresses the fact that at the roots of attempts at periodization (even while using apparently flexible formulations, such as the concept of 'movement') there is often a tendency to give primacy to continuity, coherence and uniformity within one selected 'period', such as the Renaissance, at the expense of the many differences and changes that of course also need to be taken into account.

what can be called "general scholarship". Also considering the persistence of Humanism in later centuries, Lodi Nauta brings attention to the topic of language: the criticism of Scholastic language – Nauta argues – has a rich afterlife which connects Humanists like Lorenzo Valla to early modern philosophers like Hobbes and Gassendi. Sarah Hutton sets out to investigate a particular, and often forgotten, case of reception history: Henry More's engagement with the writings of Girolamo Cardano. She employs this case study to pose a series of crucial questions: does the engagement with the past make a philosopher less 'modern'? In what sense is More 'anachronistic' in his interests in Humanistic and Renaissance philosophers, and what does his way of employing arguments drawn from Renaissance sources tell us about early modern philosophical methodologies?

II. The second section focuses on Renaissance and early modern natural philosophy, analysing especially the conception of the living being that formed through the legacy of authors like Cardano, Telesio and Bacon. Silvia Manzo considers ways of describing and interpreting the changes of matter – especially rarefaction and condensation – from Cardano to his seventeenth-century readers, especially Bacon. By reconstructing the history of specific termini (*impulsus* and *attractio* feature prominently), Manzo offers a detailed textual basis for framing the philosophical debate over the qualities of matter in the *longue durée* from the Renaissance to the early modern period. Daniel Garber's essay retraces the varied afterlives of a Renaissance thinker whose philosophical legacy often seems to have been almost entirely forgotten: Bernardino Telesio. From Bacon's engagement with Telesio to Sorel's inclusion of the Italian philosopher among the *novatores*, Garber shows how Telesio's fame was established, while his philosophy was nevertheless left more and more in the background. The last essay of this session, by Natacha Fabbri, considers the changes that one of the most lively debates in the Renaissance – on the idea of an earth-like moon – developed and changed, especially after the introduction of the telescope. Indeed the intertwining of astronomical theories with ontological ones, regarding the similarity of the earth and the moon, explains on the one hand the necessary changes in the frame of this debate, while on the other linking firmly together Renaissance approaches and early modern reinterpretations.

III. The conception of man, and the distinction between man and the other animals, is an exemplary topic with regard to the long afterlives of Renaissance philosophy.[43] Emmanuel Faye's essay looks at a pivotal conception in philosophical historiography – that of the perfection of man, in relation to Descartes' position – in order to reassess its position in relation to the debt owned to humanism. Emanuela Scribano continues the analysis of the relation and indebtedness of the Cartesian tradition to the previous centuries by considering the specific case of the debate on the mechanistic versus vitalistic view of nature: she shows how the legacy of Campanella's vitalism was at the heart of discussions about the

[43] Descartes' automatism theory has been often used as a watershed marking the beginning of a new, modern era in thinking about the difference between man and the animals. This view is discussed and methodologically challenged in Dohm and Muratori (2013).

explanation of the animals' capabilities. Cecilia Muratori reconstructs a parallel reception history which also pivots around the man-animal distinction in the mid-seventeenth century. The anonymous *Theophrastus redivivus* – she argues – weaves together selected Renaissance sources (especially Cardano, Vanini and Campanella) in order to intensify the sense of a continuity of all beings – and yet, in so doing, gives to these sources a different frame and a different tone.

IV. The four chapters in this section explore the Renaissance sources in early modern moral and political theory. Annalisa Ceron reveals in Bacon's *Essays* a whole series of echoes to the moral and political literature of the fifteenth century (with particular reference to the so-called 'mirrors for princes'), up to Machiavelli's prince. She shows how Bacon's treatment of the topic of friendship is ultimately deeply linked to this background: ultimately, Bacon frames his personal view of friendship by drawing ideas from the past and by reframing them so as to make them applicable to the present. Gianni Paganini's essay casts light on often forgotten Renaissance sources of Hobbes's account of virtue. Intervening in the contemporary debate about the appropriateness of considering Hobbes a 'virtue ethicist', Paganini shows that in order to answer this question it is essential to think in terms of continuity, bringing the Renaissance thinkers in dialogue with 'modernity'. Paganini demonstrates that the legacy of Lorenzo Valla is crucial for understanding Hobbes's conceptions of virtue, equality, self-preservation, alongside the critique directed to the concept of glory. From virtue to the greatness of states: Sara Miglietti selects the concept of *greatness* – one that is considered the corner stone of political thought on modern state building – to retrace changes of perspective from Machiavelli to Burton, but also reveal what might appear surprising points of agreement (for instance between Botero and Machiavelli) on the topic of how to recognize the greatness of a state. The final essay of this section, by John Christian Laursen, expands the area of investigation of dis-continuities between Renaissance and early modern philosophy by reaching the eighteenth century: John Upton (1707–1760) serves here as a case study for evaluating the practical effects of Renaissance studies on changing political and moral views. By including in the analysis literary Renaissance sources as well – from Shakespeare to Spenser – Laursen points to the necessity of dealing yet with another border: that which often artificially divides literary and philosophical approaches.

The volume ends with an epilogue which functions as exemplary case study for our topic and as homage to a study of Campanella that has been carried out indefatigably despite the prejudices that often still surround this writer. As Giglioni puts it in his essay, there is still a sense that "there is something wrong with the history of Renaissance philosophy", and Campanella is one of the thinkers who most suffered the consequences of such a feeling. Germana Ernst's presentation of her approach to Campanella – a contemporary of Descartes so often misjudged against the backdrop of fixed conceptions of 'modernity' – is a plaidoyer for a new approach to Renaissance philosophy and to its relation to the early modern period.

***

This volume is the result of a collaboration between the two editors, generously sponsored by the Center for Advanced Studies, Ludwig Maximilian University, Munich. We wish to acknowledge the Center's support in the organisation of the conference from which this book originates. The conference was sponsored by LMU Munich (programme 'Research Fellowships') and the Università del Piemonte Orientale (Vercelli): we thank both institutions for having made the conference, and thus the book, possible. We also wish to thank the Research Centre of the Accademia dei Lincei (Rome) for supporting our research collaboration. We are also very grateful for editorial support financed by the LMU Research Fellowships Programme. Finally, we thank the anonymous reviewers of our book manuscript for their useful suggestions.

Germana Ernst passed away during the production of this book. We dedicate this volume to her memory.

Bibliography

Bayle, Pierre. 1740. *Dictionnaire Historique et Critique*, 5th ed. Amsterdam-Leiden-La Haye-Utrecht: P. Brunel and others.

Black, Robert. 2005. The Renaissance and Humanism: Definitions and Origins. In *Palgrave Advances in Renaissance Historiography*, ed. Jonathan Woolfson, 97–117. Basingstoke/New York: Palgrave Macmillan.

Black, Robert. 2010. The Renaissance and the Middle Ages: Chronologies, Ideologies, Geographies. In *Renaissance? Perceptions of Continuity and Discontinuity in Europe, c.1300–c.1550*, ed. Alexander Lee, Pit Péporté, and Harry Schnitker, 27–44. Leiden: Brill.

Brucker, Johann Jakob.1742–1744. *Historia critica philosophiae a mundi incunabulis ad nostram usque aetatem deducta*, 5 vols. Leipzig: Breitkopf.

Burckhardt, Jacob. 2014 (first edition 1960). *Die Kultur der Renaissance in Italien. Ein Versuch*, ed. Walther Rehm. Stuttgart: Reclam.

Burnett, Charles, Meirinhos, José and Hamesse, Jacqueline (ed.). 2008. *Continuities and Disruptions Between the Middle Ages and the Renaissance*. Proceedings of the colloquium held at the Warburg Institute, 15–16 June 2007, jointly organised by the Warburg Institute and the Gabinete de Filosofia Medieval. Louvain-la-Neuve: Fédération Internationale des Instituts d'Études Médiévales (*Textes et Études du Moyen Âge*, 48).

Catana, Leo. 2008. *The Historiographical Concept 'System of Philosophy': Its Origin, Nature, Influence and Legitimacy*. Leiden: Brill (Brill's Studies in Intellectual History, volume 165).

Celenza, Christopher. 2013. What Counted as Philosophy in the Italian Renaissance? The History of Philosophy, the History of Science, and Styles of Life. *Critical Inquiry* 39(2) 367–401.

Dohm, Burkhard, and Cecilia Muratori (eds.). 2013. *Ethical Perspectives on Animals in the Renaissance and Early Modern Period.* Galluzzo: SISMEL (Series: Micrologus' Library, volume 55).

Ferguson, Wallace K. 1948. *The Renaissance in Historical Thought*. Riverside Press: Cambridge, MA.

Hankins, James. 2002. Renaissance Philosophy Between God and the Devil. In *The Italian Renaissance in the Twentieth Century. Acts of an International Conference (Florence, Villa I Tatti, June 9–11, 1999)*, ed. Allen J. Greco, Michael Rocke, and Fiorella Gioffredi Superbi, 269–293. Florence: Olschki.

Hankins, James. 2005. Renaissance Humanism and Historiography Today. In *Renaissance Historiography*, ed. Jonathan Woolfson, 73–96. Basingstoke/New York: Palgrave Macmillan.

Hankins, James. 2007. The Significance of Renaissance Philosophy. In *The Cambridge Companion to Renaissance Philosophy*, ed. Hankins James. Cambridge: Cambridge University Press.

Hegel, Georg Wilhelm Friedrich. 1836. *Vorlesungen über die Geschichte der Philosophie*, vol. 3, ed. Karl Ludwig Michelet. In Hegel, G.W.F. 1832–1845; 1887. *Werke*, ed. by *Verein von Freunden des Verewigten*, 17 vols. Berlin/Leipzig: Duncker and Humblot.

Hegel, Georg Wilhelm Friedrich. 1969–1981. *Briefe von und an Hegel*, ed. Johannes Hoffmeister and Friedhelm Nicolin, 4 vols, 3th ed. Hamburg: Meiner.

Hegel, Georg Wilhelm Friedrich. 1986. *Vorlesungen. Ausgewählte Nachschriften und Manuskripte, vol 9: Vorlesungen über die Geschichte der Philosophie, Part 4: Philosophie des Mittelalters und der neueren Zeit*, ed. Pierre, Garniron and Walter, Jaeschke. Hamburg: Meiner.

Hegel, Georg Wilhelm Friedrich. 1990. Lectures on the History of Philosophy. The Lectures of 1825–1826. In *Medieval and Modern Philosophy*, vol. 3, ed. Robert F. Brown. Berkeley: University of California Press.

Israel, Jonathan. 2001. *Radical Enlightenment: Philosophy and the Making of Modernity 1650–1750*. Oxford: Oxford University Press.

Kraye, Jill, and M. W. F. Stone (eds.). 2000. *Humanism and Early Modern Philosophy*. New York: Routledge.

Kristeller, Paul Oskar. 1985. Philosophy and Its Historiography. *The Journal of Philosophy* 82(11): 618–625.

Martin, John Jeffries. 2003. *The Renaissance: Italy and Abroad*, ed. John Jeffries Martin. London/ New York: Routledge.

Molà, Luca. 2008. Rinascimento. In *Le parole che noi usiamo. Categorie storiografiche e interpretative dell'Europa moderna*, ed. Marcello Fantoni and Amedeo Quondam, 11–31. Rome: Bulzoni.

Paganini, Gianni. 1980. *Analisi della fede e critica della ragione nella filosofia di Pierre Bayle*. Florence: La Nuova Italia.

Paganini, Gianni. 2008a. Le cogito et l'âme qui "se sent". Descartes lecteur de Campanella. *Bruniana & Campanelliana* 14(1): 11–29.

Paganini, Gianni. 2008b. *Skepsis. Le débat des modernes sur le scepticisme. Montaigne – Le Vayer – Campanella – Hobbes – Descartes – Bayle*. Paris: Vrin.

Piaia, Gregorio. 1998. Cassirer, Historiker der Renaissancephilosophie. In *Die Renaissance und ihr Bild in der Geschichte. Die Renaissance als erste Ausklärung III*, ed. Enno Rudolph. Tübingen: Mohr Siebeck.

Piaia, Gregorio, and Santinello, Giovanni. 2011. *Models of the History of Philosophy*, vol. 2: *From the Cartesian Age to Brucker*, ed. Gregorio Piaia and Giovanni Santinello. Dordrecht: Springer, 2011 (Series International Archives of the History of Ideas, volume 204; trans. from the Italian: Piaia Gregorio and Santinello, Giovanni. 1979. *Storia delle storie generali della filosofia*, vol. 2, Brescia: La Scuola).

Rixner, Thaddä Anselm. 1823. *Handbuch der Geschichte der Philosophie*, vol. 3. Sulzbach: Seidel.

Ruggiero, Guido (ed.). 2002. *A Companion to the Worlds of the Renaissance*. Oxford: Wiley.

Santinello, Giovanni et al. 1993. *Models of the History of Philosophy*, vol. 1, ed. Giovanni Santinello. Dordrecht: Kluwer.

Schmitt, Charles B. 1987. The Development of the Historiography of Scepticism: From the Renaissance to Brucker. In *Scepticism from the Renaissance to the Enlightenment*, ed. Charles B. Schmitt and Richard H. Popkin, 185–200. Wiesbaden: Harrassowitz (Wolfenbütteler Forschungen, volume 35), republished in Schmitt, Charles B. *Reappraisals in Renaissance Thought*, ed. by Charles Webster. London: Variorum Reprints.

Schmitt, Charles B. 1988. Towards a History of Renaissance Philosophy. In *Aristotelismus und Renaisssance in memoriam Charles Schmitt*, ed. Eckhard Kessler, Charles Lohr and Walter Sparn, 9–16. Wiesbaden: Harrassowitz (Wolfenbüttler Forschungen, volume 40), republished in Schmitt, Charles B. *Reappraisals in Renaissance Thought*, ed. by Charles Webster. London: Variorum Reprints.

Schmitt, Charles B. 1989. *Reappraisals in Renaissance Thought*, ed. by Charles Webster. London: Variorum Reprints.

Tennemann, Wilhelm Gottlieb. 1817. *Geschichte der Philosophie*, vol. 10. Leipzig: Barth.

Tennemann, Wilhelm Gottlieb. 1829. *Grundriss der Geschichte der Philosophie für den akademischen Unterricht*, 5th ed. Leipzig: Barth.

Part I
The Endurance of Tradition

Chapter 2
What's Wrong with Doing History of Renaissance Philosophy? Rudolph Goclenius and the Canon of Early Modern Philosophy

Guido Giglioni

Abstract The chapter is divided into two main parts. In the first, I offer some general remarks on the elusive place of Renaissance philosophy within the larger disciplines of philosophy, philosophy of history and history of philosophy. In the second part, I rely on a specific case study – Rudolph Goclenius's dictionaries of philosophy (published in 1613 and 1615) – to emphasize the value and importance of the philosophical production during the fifteenth and sixteenth centuries. As a thinker straddling two centuries, Goclenius demonstrates how the contribution of seventeenth-century philosophers, with their innovative ideas about language, science and religion, cannot be properly understood without taking into account the philosophical work elaborated during the fifteenth and sixteenth centuries. Rather than perpetuating the image of these two centuries as impoverished and unoriginal in terms of ideas and commitments, Goclenius helps us to have a more historicized and positive consideration of eclectic contaminations among philosophical trends, the influence exercised by the classical tradition, the persistence of scholastic ways of arguing and the decisive impact of philological methods.

Introduction

Between 2004 and 2007, Anthony Kenny published a new history of Western philosophy in three volumes (reissued as one volume in 2010). In Volume 3, devoted to early modern thought, 32 out of 331 pages are devoted to the sixteenth century (the fifteenth century, with Renaissance Platonism, Renaissance Aristotelianism and a few pages on Lorenzo Valla and Nicholas of Cusa, had already been dealt with in Volume 2, dedicated to medieval philosophy). The authors examined in Volume 3 are Giordano Bruno, Galileo Galilei and Francis Bacon. Two pages are on Montaigne

G. Giglioni (✉)
The Warburg Institute, University of London, London WC1H OAB, UK
e-mail: Guido.Giglioni@sas.ac.uk

C. Muratori, G. Paganini (eds.), *Early Modern Philosophers and the Renaissance Legacy*, International Archives of the History of Ideas Archives internationales d'histoire des idées 220, DOI 10.1007/978-3-319-32604-7_2

(117–118), three on Suarez's metaphysics (181–184), three on issues of religious casuistry (247–250) and eight on Machiavelli's *Prince* and More's *Utopia* (273–281). Kenny's survey is certainly better than Bertrand Russell's ("Until the seventeenth century, there was nothing of importance in philosophy. The moral and political anarchy of fifteenth-century Italy was appalling, and gave rise to the doctrines of Machiavelli") or D. W. Hamlyin's ("It may seem a paradox that a period that saw the flowering of much else – of science, of art and of literature – was a period in which philosophy was at a low ebb. It is nevertheless a fact").[1] Kenny's *The Rise of Modern Philosophy* remains, however, disappointing, for it ignores all the efforts to legitimate the philosophical production of the Renaissance undertaken in the past by such historians as Giovanni Gentile, J.-Roger Charbonnel, Henri Busson, Eugenio Garin, Paul Oskar Kristeller, Charles Schmitt, Brian Copenhaver and Cesare Vasoli.[2] It is nonetheless evidence that Renaissance philosophy continues to have a precarious place in the history of philosophy, and this for a number of reasons such as the bad publicity that the word and notion of eclecticism has received among philosophers from the seventeenth century on, the allegedly non-philosophical nature of Renaissance humanism, a lingering uneasiness about early modern theories of universal animation and, finally, a certain tendency to regard theological debates from the heretical movements of the fifteenth century to the Reformation and Counter-Reformation as philosophically spurious. For all these reasons, Renaissance philosophy continues to be relegated in a limbo of pseudo-philosophy placed between the genuinely and gloriously philosophical epochs of the Middle Ages and the seventeenth century. The story goes that, in the early modern period, serious philosophy resumed its course with Descartes, Hobbes and Spinoza, after an interlude of literary experiments and slavish imitations of classical authors. Although there may be an element of historical truth in this view (for the seventeenth century was indeed a golden age for philosophy), the problem behind a blasé and patronizing attitude towards Renaissance philosophy boils down – very prosaically – to issues of monolingualism, nationalism, educational and institutional settings, publishing marketing plans and, more recently, criteria and strategies through which research funds are allocated to historians and institutions of higher education.

This may sound obvious, but it is fair to say that every age elaborates its own philosophical consideration of the surrounding reality and the major events that shape such reality. Depending, however, on philosophical preferences and tastes, a sort of tacit assumption has established itself among historians of philosophy according to which some ages are more philosophical than others. To cut a long story short, classical antiquity (Plato and Aristotle), scholastic philosophy (from

[1] Kenny (2006); Russell (2004 [1946]), 453; Hamlyn (1987), 123. Recent attempts to present a more conciliatory view of the philosophical relationship between the sixteenth and the seventeenth centuries can be found in Sorell (1993) and Rogers, Sorell and Kraye (2010).

[2] Gentile (1968 [1920]); Gentile (1968 [1923]); Charbonnel (1919); Busson (1957 [1922]); Kristeller (1964); Kristeller (1979); Garin (1978 [1966]); Schmitt *et al.* (1988); Copenhaver and Schmitt (1992); Vasoli (2002).

Abelard to William of Ockham), the seventeenth century (Descartes, Hobbes, Locke, Spinoza and Leibniz) and German idealism (from Kant to Hegel) are seen as authentic philosophical ages. In different ways and with different aims, this view has been perpetuated by both continental and analytical philosophers. It has also been sanctioned by a growing industry of handbooks and companions to philosophy and history of philosophy, which helps reinforce the stereotype that there are in fact serious and less serious periods of philosophy in human history.

To this situation, which is specific to the discipline of history of philosophy,[3] one should add the question of the narrow and contested space left to the history of Renaissance philosophy within the broader field of Renaissance studies, a field that is still dominated by Burckhardtian prejudices. Jacob Burckhardt (1818–1897) famously ignored Renaissance philosophy in his *Civilization of the Renaissance in Italy* (1860). Censors of Renaissance philosophy can therefore be found both within and without the field of the history of philosophy. The ones from within are often historians of seventeenth-century philosophy and analytically-trained historians of philosophy. The censors from without are historians who are simply uninterested in intellectual history, either for militant reasons (and therefore actively and aggressively uninterested) or because they prefer to devote their energies to investigating various aspects of material history (they are passively uninterested).

Finally, I should at least hint at a general philosophical question – a question pertaining to philosophy of history. This has to do with the uncomfortable relationship which has always characterized the two intellectual activities of philosophy and history. In an article published in 1994, Paola Zambelli noted that "[h]istory of philosophy is now a part of history, no longer a part of philosophy, nor its completion".[4] This is an important point to be borne in mind, which is relevant for the definition of the history of philosophy as a discipline. In the second half of the nineteenth century and the first half of the twentieth, the history of Renaissance philosophy emerged from a glorious historiographical past and from a speculative setting that was as glorious. Within the traditions of both German idealism and historicism Renaissance philosophy was indeed a privileged field of both scholarship and inspiration. One should only think of G. W. F. Hegel (1770–1831) and Wilhelm Dilthey (1833–1911). In Italy, the shaping of the field of Renaissance philosophy coincided with the activity of Bertrando Spaventa (1817–1883), Francesco De Sanctis (1817–1883), Francesco Fiorentino (1834–1884), Benedetto Croce (1866–1952) and Giovanni Gentile (1875–1944). Within this context, the history of Renaissance philosophy was indeed meant to be an integral part of the philosophical investigation of reality, in some cases a crucial stage in its very development. Even Ernst Cassirer (1874–1945) demonstrated a certain penchant for this speculative tendency in doing history of Renaissance philosophy. His *Individuum und Cosmos* (1927), however, remains one of the most fascinating contributions to the study of Renaissance philosophy.

---

[3] Rorty, Schneewind and Skinner (1984).

[4] Zambelli (2012 [1994]), 384.

Today history of Renaissance philosophy has definitely become part of the history of the fifteenth and sixteenth centuries, while philosophical assumptions on the part of the historian of philosophy are seen – and probably rightly so – with suspicion and scepticism. We all should be ready to acknowledge, though, that in this field philosophy of history has often been replaced by practical instructions in historiographical methodology, the subject of many seminars in graduate programmes at university. Historiography is a bland version of philosophy of history and gives hope for method in an area where in fact there is no method. The result of this very recent development is that history of Renaissance philosophy has inadvertently mutated into history of philosophical historiography, and historians are often keen to maintain their allegiances to hoary historiographical traditions.[5] This, however, should not come as too much of a surprise, for, by its very nature, history of philosophy is strongly opinionated and often judgmental.

The bad publicity that Renaissance philosophy is receiving at the moment dates back, in fact, to the Renaissance itself. This cannot be denied, for already at the time, a significant chunk of philosophical production – the so-called scholastic philosophy – was strongly criticized by two main fronts: the humanists (Lorenzo Valla, Desiderius Erasmus, Luis Vives, Thomas More) and a number of vanguard philosophers (Marsilio Ficino, Baldassarre Castiglione, Leone Ebreo, Giordano Bruno, to mention only a few). Both groups shrank from the technicalities of scholastic philosophy. On the other hand, both Neo-Hegelianism and Neo-Kantianism during the nineteenth and early twentieth centuries imparted an aura of speculative dogmatism on Renaissance thinkers. This fact has certainly not helped the cause of Renaissance philosophy. In a way, sixteenth-century philosophers find themselves in an infelicitous situation, even more so than fifteenth-century philosophers. It must be, I suspect, the proximity of the seventeenth century, for doing history of Renaissance philosophy with one's eyes turned towards the seventeenth century transforms some of the most tantalizing and original philosophers of that period into disquieting hybrids. Girolamo Cardano (1501–1576), Bernardino Telesio (1509–1588), Tommaso Campanella (1568–1639) and Francis Bacon (1561–1526), to mention a few names, look like spooked centaurs, suddenly caught in the light of a better, more rational century. Sometimes they are reluctantly included in accounts of history of philosophy which I would call "history in the optative", that is, *if-only* history. If only Telesio hadn't maintained that everything is sentient in nature; if only Cardano had not been so prone to astrology and demons; if only Campanella had not got lost chasing theocratic dreams; if only Bacon hadn't assumed that appetite rules nature, including inanimate nature. If only indeed: we could have had modern physics, algebra, global studies and the scientific revolution already in the sixteenth century! Instead we had to wait for later developments in the seventeenth century to see the longed-for *telos* fulfilled.

We all know that such things as fifteenth, sixteenth and seventeenth centuries are in fact figments of the historiographical imagination. They fit extremely well in long-tested and official periodizations and categories; however, they simply have no

[5] Celenza (2004).

ontological consistence apart from being useful conventions. The view of Renaissance philosophy that is being legitimized by august companions and handbooks from institutions which are as august is that of a squeezed middle, squeezed between medieval and seventeenth century. Let us instead resort to the anamorphic resources of the imagination and think of a "long" philosophical Renaissance, stretching as it were from Anselm of Canterbury (c. 1033–1109) to Spinoza (1631–1677). The great historian of Renaissance philosophy, Edward Cranz (1914–1998), who managed to combine philosophical ingenuity with historical rigour, argued that towards the end of the eleventh and the beginning of the twelfth century a momentous "reorientiation" in thinking took place in the major European centres of philosophical investigations.[6] This is the path I would like to follow, for here I can see a pattern, emerging from the matrix of scholastic philosophy, with extraordinarily valuable accretions coming from the Latin translations of Avicenna, Averroes and Jewish thinkers and with decisive influences from cutting-edge vernacular thinking.

In this chapter, in order to practise this exercise in anamorphic vision, I will use Rudolph Goclenius's philosophical dictionaries as barometers to test the climate of pre-Cartesian philosophical endeavours in Europe. This decision is certainly open to methodological objections. They concern, firstly, the extent to which Goclenius's dictionaries can be said to be representative of Renaissance philosophy; secondly, whether they transcended the limits imposed by the so-called national styles of thought; and, finally, whether they were in fact prone to individual philosophical preferences. I will address these points briefly in my conclusion.

An Entire Library in One Book: Goclenius's *Lexicon Philosophicum*

To begin with, I summarily describe the physiognomy of Goclenius's *Lexica*, starting with those traits that we expect to find in them judging from what we know about its author. Rudolph Goclenius (1547–1628), father of another illustrious philosophical Rudolph (1572–1621) better known for his place in the history of magnetic therapies, wrote two seminal dictionaries of philosophy: the *Lexicon philosophicum, quo tanquam clave philosophiae fores aperiuntur* ("Philosophical Lexicon, Which Opens the Doors of Philosophy Like a Key"), published in 1613, and the *Lexicon philosophicum Graecum* ("Greek Philosophical Lexicon"), which came out two years later. After having studied in Erfurt, Marburg and Wittenberg following the Philippist line within the Lutheran fold, he taught physics, logic and ethics at the universities of Kassel and Marburg. He wrote a number of metaphysical writings which clearly reflect the theological debates between Lutherans and Calvinists at the time. His *Analyses in exercitationes aliquot J. C. Scaligeri de*

[6] Cranz (2006).

subtilitate (1599) and *Adversaria ad exotericas aliquot Julii Caesari Scaligeri acutissimi philosophi exercitationes* (1606) testify to the pervasive influence of Julius Caesar Scaliger (1484–1558) in Protestant philosophical circles. When the University of Marburg turned into a Calvinist institution in 1605, Goclenius's sympathies towards Calvinism became more evident.[7]

I will focus on his two dictionaries as philosophical texts that are both rhetorically and theologically savvy: these for me are two very important reasons why this books can be said to be representative of the philosophical climate of the Renaissance. Part of Goclenius's rhetorical awareness lies in his constant attention to the linguistic aspects of the philosophical problems. Goclenius was clearly a philo-Hellenist in his approach to philosophy, so much so that he felt the need to complement his lexicon of philosophical Latin terms with one devoted to the principal Greek concepts. He remained nonetheless a strong advocate of the importance of the Latin tradition – both scholastic and humanist – in Western philosophy. And while he was in principle against the use of barbarisms in philosophy (especially scholastic barbarisms), the painstaking care – almost of an entomological kind – with which he collected, scrutinized and dissected all sorts of inappropriate and incorrect terms reveals in him a passion for the domain of the philosophically inarticulate or the barely articulable, i.e., for those awkward protrusions of clotted meaning (the barbarisms, that is) which especially accrue on the technical terms dividing language from the process of thought. Here it is significant to recall that by 1615 Goclenius had added to the Latin dictionary an appendix entirely devoted to a meticulous analysis of the inappropriate ways of expressing philosophical concepts in Latin, a "Collection of Words and Phrases that are Obsolete, Less Ordinary, Recently Born, Improper, Impure, Uncouth, including Barbarisms, Solecisms and Slight Solecisms" (*Sylloge vocum et phrasium quarumdam obsoletarum, minus usu receptarum, nuper natarum, ineptarum, lutulentarum, subrusticarum, barmibarbararum, soloecismorum et ὑποσολοίκων*).[8]

One way of shedding further light on the characteristic physiognomy of the work is by looking at the imposing array of sources used by Goclenius and considering in particular the authors whom he refers to with more frequency. At the top, I would put the already mentioned Scaliger, who is cited and quoted in almost every single page. The principal text by Scaliger to be referenced by Goclenius is the *Exotericarum exercitationum liber quintus decimus de subtilitate ad Hieronymum Cardanum* ("The Fifteenth Book of Exoteric Exercises about Cardano's *On Subtlety*", 1557), but other works are also well represented, such as *De causis linguae Latinae* ("The Principles of the Latin Language", 1540), *Poetices libri VII* ("Seven Books of Poetics", 1561) and his dialogues on the pseudo-Aristotelian *De plantis* (1556). Then, in descending order of frequency, we encounter Jacopo Zabarella (1533–1589), Jakob Schegk (1511–1587) and a large number of Reformed metaphysicians and theologians such as Philipp Melanchthon (1497–1560), Girolamo Zanchi

[7] On Goclenius, see Ashworth (1967); Jensen (1990), 32–36; De Angelis (2010), 158–192; Lamanna (2013); Stiening (2014). On early modern philosophical lexicons, see Canone (1988).

[8] Goclenius (1980 [1613]), 282. On Goclenius's "Sylloge", see Giglioni (2015).

(1516–1590), Joachim Mörlin (1514–1571), Amandus Polanus von Polansdorf (1561–1610) and Daniel Tillen (1563–1633). In keeping with the philosophical and theological guidelines of Lutheran and Calvinist debates, Goclenius manifested a clear preference for Scholastic Aristotelianism.[9] For him, this tradition warranted a fundamentally rational understanding of reality. Relying on Amandus Polanus, he stated with confidence that truth was "besides, below and above reason", and reason was perceived "through the intellect, through the senses and through faith", and as a result truth could never be "against reason".[10] After all, it is worth remembering here that Goclenius came up with quite a notable philosophical term: "ὀντολογία", ontology, understood as the philosophical inquiry about being and its more general properties.[11]

The classics, of course, are well represented in both dictionaries (Plato, Aristotle, Cicero and Seneca), together with the Fathers of the Church (Augustine and Boethius). Diogenes Laertius ("non ignobilis rerum philosophicarum rapsodus") is often mentioned as a reliable source of philosophical information.[12] Albert the Great, Thomas Aquinas and John Scotus are among the most important sources. The presence of Gabriel Biel (1425–1495) is evidence of Goclenius's attention to the contribution of nominalism and the Ockhamist *via moderna*. Averroes's point of view is frequently consulted. The same is true of Thomas Cajetan (1469–1534) and the Coimbra Commentators.[13] Among the authors with no affiliation to the university system, Scaliger is not the only one to be referred to by Goclenius. We also have Bessarion (1403–1472), Giovanni and Gianfrancesco Pico della Mirandola (1463–1494 and 1470–1533, respectively) and, above all, Ficino (1433–1499). The latter is a key source in Goclenius's *Lexica* also for the authors he translated and commented upon, first among them Dionysius the Areopagite, who is well represented in both dictionaries. Finally, it is worth noting how Goclenius drew profusely on a number of encyclopaedic accounts of Renaissance learning, such as Conrad Gessner's *Bibliotheca universalis* (1545–1549), Theodor Zwinger's *Theatrum vitae humanae* (1565) and Paul Skalić de Lika's *Encyclopedia, seu orbis disciplinarum tam sacrarum quam prophanarum epistemon* ("Encyclopedia or Knowledge of the World of Disciplines, both Sacred and Profane"), published in Basel in 1559.

Like Campanella and Bacon, whom I mentioned earlier, Goclenius is a philosopher who inhabited both the sixteenth and the seventeenth centuries. The main reason why I decided to focus my chapter on the *Lexicon philosophicum* and the

---

[9] On the relationship – mainly of a pragmatic nature – between metaphysics and theology in Lutheran contexts, see Jensen (1990), 25: "In the late sixteenth century, for Lutherans in particular, metaphysics became subordinate to theology in a far more direct way [than it used to be in the thirteenth century], and no secret was made of this subordination. The principles of metaphysics were derived from theology and proved *a posteriori*".

[10] Goclenius (1980 [1613]), 24ab.

[11] Ibid., 16.

[12] Ibid., 210b.

[13] On the influence of scholastic philosophy in Germany during the sixteenth and seventeenth centuries, the following studies remain fundamental: Weber (1907); Petersen (1921); Lewalter (1967 [1935]); Wundt (1939); Leinsle (1985). See also Lohr (1988).

Lexicon philosophicum Graecum is that they provide – in the shape of a microcosm as it were – a relatively faithful image of what doing philosophy was like during the Renaissance. This is made clear in an elegy written by Raphael Eglinus (1559–1622), professor of theology at the University of Marburg with strong interests in alchemy, who composed the poem placed at the beginning of the Latin *Lexicon*: "we don't need shelves any longer; no more thousands books. This book alone is like a whole library".[14] In using Goclenius's dictionaries as one coherent book of philosophy, I intend to concentrate on four aspects that, while they may give the impression of being philosophically meagre or illegitimate from a post-Cartesian point of view, represent in fact a most original contribution to philosophical inquiry between the thirteenth and sixteenth century. These aspects are: a reassessment of the virtues of eclecticism, which means that the philosophical past – and therefore history – could be used as legitimate matter for philosophical investigations (otherwise said: doing history of philosophy is part of the philosophical exercise); a background ontology based on the idea that substances of different nature are nevertheless able to interact and that life in particular acts as the principal mediator between physical and cognitive reality; a sophisticated understanding of the role played by language in articulating human thought; and, finally, a positive consideration of theology seen as a source of philosophical knowledge, in which the cognitive faculty of the imagination plays a key role. In this respect, I feel entitled to adopt the term with which Johann Wirz, a professor of theology active in Zurich during the 1650s, described Goclenius's endeavour: *theiosophia*.[15] If we consider these four points carefully, they can also be seen as a way of taking *history*, *life*, *language* and *imagination* seriously from a philosophical point of view – which for me it's another way of stressing the specific contribution of Renaissance philosophy to the definition of early modern thought.

History, Life, Language and the Imagination: A Précis of Renaissance Philosophy

Like many philosophers at the time, Goclenius explained the way in which the human mind understands reality as a result of a continuous adjustment between cognitive, linguistic and natural factors. From a strictly metaphysical point of view, his dictionaries chronicle the emergence of what can be seen as the great conundrum of the modern age: the contested relationship between *res* ("reality"), *ideae* (the underlying template of reality, both in cognitive and operative terms) and a number of devices left to both nature and the human minds to bridge the gap dividing reality from its ideal underpinnings. These devices correspond to a cluster of philosophical notions that denote the human ability to reflect and represent being:

---

[14] Quoted in Goclenius (1980 [1613]), sig.)( )(1r. On Eglinus see Moran (1994).

[15] Quoted in Goclenius (1980 [1613]), sig.)()(1v.

imago ("appearance"), *species sensibilis* and *species intellegibilis* ("liknesses", both sensible and intelligible), *figura* ("shape"), *repraesentatio* ("representation") and *signum* ("sign").[16] Goclenius held that *imago* could be understood as either archetype or ectype. He distinguished between σωματικαί and πνευματικαί images (the latter also termed *species immateriales et intelligibiles*).[17] Ontologically speaking, he regarded *ideae*, *imagines* and *species* as all instances of *formae*, "forms", that is, constitutive principles of both nature and the mind. While *formae mentales* were key in adjusting representations to reality, *formae reales* were further subdivided by Goclenius into forms that governed things from without while remaining separate from them (*formae assistentes*), and forms that shaped things from within, becoming one single entity with them (*formae informantes*). In addition, real forms were also divided into "natural" and "artificial". Among natural forms, "substantial" forms were certainly the more problematic, for they remained the mainstay of scholastic metaphysics despite being increasingly exposed to objections coming from different fronts of the European republic of philosophical letters. They were either "separable" (such as the rational soul) or "inseparable" (celestial or sublunary, the latter further subdivided into inanimate and animate). The rational soul, because of its dependence on the body, was characterized by a limited degree of self-sufficiency (*subsistentia incompleta*).[18]

Goclenius's loyalty to the Aristotelian notion of life as ἐντελέχεια helped accelerate the crisis of the Latin *forma* when this was associated to matters of soul and identity. With his *De immortalitate animae* (1516), Pomponazzi had demonstrated how the Aristotelian ἐντελέχεια remained perilously too intimate with the structure of the body. As is often the case with his *Lexica*, Goclenius smudged the boundaries between Platonism and Aristotelianism adding the Platonic meaning of form to complete the picture of traditional scholastic accounts. In these circumstances, Goclenius showed that he was an Aristotelian who leant towards an irenically Thomist interpretation of being, while being eclectically open to Platonism through the mediation of Ficino's philosophy. Forms could therefore be characterized as either *ideae* (i.e., *exemplaria*, patterns "devoid of matter and participating in intelligence") or *imagines*, understood as reflections of ideas "joined to matter", that is, a "σκιαγραφία, sketch and rough outline (*adumbratio et rudis delineatio*), to which colours are yet to be applied (*vivis coloribus nondum adhibitis*)", also described as τύποι, "characters".[19] Judging from the amount of lexicographic attention paid to such lemmas as *ratio*, *idea*, *archetypus*, *similitudo* and *species*, it seems evident that in Goclenius's framework the notion of form continued to be the ontological backbone of the whole universe. Another recurrent way used by Goclenius to express the relationship between *res* and *idea* was to assume a correspondence between *esse*

---

[16] On the history of the adjustment of intellectual knowledge to sensible reality through the category of "representation" (*species*), see Spruit (1994–1995).

[17] Goclenius (1980 [1613]), 215b-216a.

[18] Ibid., 588-593b; Goclenius (1980 [1615]), 104a. On the early modern evolution of the Aristotelian notion of form, see Des Chene (1996); Des Chene (2000).

[19] Goclenius (1980 [1613], 593a; Goclenius (1980 [1615]), 244a.

reale (or *formale*) and *esse ideale*. The *esse ideale* – said Goclenius while acknowledging that he was using the way of speaking of the "barbarian schoolmen" – was reality in its being fashioned by mental forms, "the being of a thing, as this is in the mind according to the species through which, as in a representational principle (*obiectivum principium*), a thing is known".[20]

In Goclenius's reconstruction of Western metaphysics, the ultimate foundation of the congruity between *esse reale* and *esse ideale* was, however, of a theological nature. The two levels corresponded – and indeed interacted – because they had been originally created by God in such a way that they mirrored each other and to a certain extent allowed forms of mutual dependency and interaction (for, as causal entities, ideas were productive and reproductive entities, besides establishing the link through which knowledge connected to reality). A further consequence coming from this metaphysical setting was Goclenius's belief in a fundamental unity or harmony between the mind and the material universe. Citing Augustine, he defined ideas as *species*, *formae* or *rationes* of an external reality, and in this sense they were "outside" reality (*extra rem*). As such, *ratio* was not *res*. And yet, as Goclenius explained while progressing in the argument, "some likeness (*similitudo*) between ideas and things" needed to be assumed. As in the realm of physical generation, this *similitudo* was either "univocal" or "equivocal".[21] Envisaging concepts in terms of fertile seeds, Goclenius defined ideas as principles of activity (*principia operationis*), in which operations and implementations followed the instructions included in an original pattern (*per modum exemplaris*).[22] This model of causality predicated upon notions of likeness and archetypal productivity secured a level of interaction between the world of ideas and physical reality. In discussing the entry "Reactio", for instance, Goclenius confirmed that a world of "spiritual" responses (*in potentia cognoscente*, that is, in the field of knowledge) was running parallel to the material universe organized by networks of physical actions and reactions. It was certainly not by accident that a key notion in Goclenius's dictionaries was the power of being affected (*vis recipiendi*).[23]

However influential in bridging the gaps between reality and appearance, being and activity, Platonism was kept carefully at bay when the issue under scrutiny concerned theological matters. As the creator of both ideas and things, Goclenius's God was no constrained by any pre-existing and extra-mental ideal reality, for, in opposition to the Platonic notion of idea, he unambiguously stated that "nothing outside God is eternal" (*nulla enim res extra Deum est aeterna*). As all things were deemed to be in God beyond any degree and measure (*eminenter*) and as objects of His infinite thinking power (*secundum esse cognitum*), Goclenius looked at God's mind as the boundless repository of all that could be actually thought: "the whole realm of possible things (*tota multitudo rerum possibilium*) is not in nature in actuality (*actu*),

---

[20] Goclenius (1980 [1613]), 209a.

[21] Ibid., 208ab.

[22] Ibid., 209a.

[23] Ibid., 960ab. On *dispositio recipiendi*, see 565b. See also Goclenius (1980 [1615]), 161b-162a, s.v. "Πάθος".

but in God's knowledge, for there is no possible thing that God ignores in actuality (*actu*)".[24] In addition, as the creator of physical, intelligible and linguistic objects, and the guarantor that nature, knowledge and language were different expressions of the same reality, God was also the foundation upon which moral certainty rested: "the right judgment of the mind is the judgment of right reason, which is congruent with the eternal and immovable norm in God's mind, as revealed in the Decalogue".[25]

Despite rejecting the most radical assumptions underlying Plato's exemplarism, Goclenius showed nevertheless a favourable disposition towards the "double world" (*duplex mundus*) of the Platonists, and not simply for its tendency to smooth the asperities of Aristotelian naturalism.[26] Although, as already said, Aristotelian scholasticism remained the great argumentative platform of Goclenius's metaphysics (and, unsurprisingly, Zabarella was often cited and quoted), in his *Lexica* the Platonic tradition worked as the speculative glue that could cement theology with ontology. The theologians' distinction between *imago increata* and *imago creata*, for example, was Platonic in kind. It was yet another way for Goclenius to underline an original congruity between reality and appearance, with the difference that while a divine "image" was "essential and immutable", a created "image" could only be "accidental and mutable". When seen along these lines, it's easy to understand why the theological matters most debated by Goclenius concerned the divinity of Christ, His humanity, the difference between essence and person within the Trinity, the nature of divine presence in the Eucharist, the effect of Grace and the importance of biblical hermeneutics. Considering himself primarily a philosopher, however, he left to contemporary divines the task of discussing with caution (*sobrie disputanda*) the most controversial issues in theology.[27]

For all its philosophical significance, the pivotal juncture created by the many relationships between *idea*, *res*, *verbum* and *imago* has momentous theological reverberations throughout the dictionary. Innumerable entries are organized in such a way that they often end with a significant Christological coda. For instance, in discussing the meaning of *regressus*, that is, the logical procedure from effects to causes and then back from causes to effects, Goclenius found a way of further expanding on how to interpret the body of Christ in the Eucharist. He argued against the "corporeal presence" of Christ in the bread in favour of the Calvinist thesis that the Lord's Supper signified a real participation in a ritual of divine transformation: "Corpus Christi est ubicunque est Ecclesia".[28] The entry "Repraesento" sheds more light on this crucial point. Goclenius explained that the verb "to represent" could be understood in two principal senses: as *significare* in a purely denotative way ("The breaking of bread in the Lord's Supper represents [*repraesentat*] the passion and sacrifice of the body of Christ on the cross") and as *rem praesentem facere*, that is,

---

[24] Goclenius (1980 [1613]), 209b; 989b.

[25] Ibid., 963a.

[26] Ibid., 209ab.

[27] Ibid., 206b.

[28] Ibid., 974ab.

as a way of re-enacting and reproducing the event in question.[29] The relationship between *idea*, *res*, *verbum* and *imago* was also a crucial issue with respect to the meaning and effectiveness of sacraments and rituals. While discussing the concept of *identica praedicatio*, Goclenius used the statement "Hoc est corpus meum" of the Eucharist as an opportunity to discuss the ontological and causal status of religious signs. While Thomas Aquinas had interpreted that which was "hidden under the *species* of bread" as the body of Christ, for the Lutherans, His body was that which was hidden "under the *substantia* of bread".[30]

As already said, a large number of entries in Goclenius's Latin and Greek dictionaries demonstrate a clear willingness on his part to engage with theological issues. Above all, he showed a great deal of exegetical subtlety and philosophical acumen in discussing the power of signs and rituals. In the entry "Relatio", while examining the different meanings of the words *patronus* and *cliens*, he argued that the proposition according to which Christ was "our priest" and "the victim who sacrificed himself for our sins" was to be understood in a spiritual sense.[31] Also, the reason why the devil would be warded off and sick people were healed by invoking the name of Christ did not depend on the name as such, but on the intention with which one invoked Him.[32] Regarding the symbol of the cross, Goclenius quoted John Chrysostom (c. 347–407) to reinforce the thesis that religious symbols were effective in triggering the inner development of the soul: "One should not simply make the sign of the cross with his finger, but shape the cross mentally with intense faith".[33] Closely related to this point is Goclenius's way of addressing the interplay of *idea* and *res* by relying on the rhetorical tradition. An example of the many rhetorical and theological intertwinements that run through Goclenius's dictionaries is his discussion of the difference between "clarity" (*perspicuitas*) and "certainty" (*certitudo*). While he rested on Cicero's authority to argue that the meaning of the adjective "apparent" (*evidens*, ἐναργές in Greek) was the same as the adjective "clear" (*perspicuus*), he thought, however, that certainty had a different status in theological matters: "many of the foundations of our faith", Goclenius went on to explain, "are not apparent (*evidentia*); they all are, however, said to be certain and stable within us, for certainty (*certitudo*) refers to the infallibility of the believed object (πιστοῦ)". Goclenius defined certainty as the *firmitas*, i.e., the strong urge that led the intellect to adhere to the known object from within; *evidentia*, on the other hand, indicated "the way in which" the truth manifested itself and forced the intellect to give its assent.[34]

---

29 Ibid., 981b.

30 Ibid., 212b. On the meaning of Eucharist among Protestant theologians, see Wandel (2005); Wandel (2014).

31 Goclenius (1980 [1613]), 978b.

32 Ibid., 966a.

33 Ibid., 966b.

34 Ibid., 206b. See also Goclenius (1980 [1615]), 73b-74a, under "Ἐνάργεια". For Cicero on *evidentia*, see *Academicae quaestiones*, II, vi, 17.

Goclenius's discussion of the religious power of symbolical meanings presupposes the ongoing debate among Reformed theologians about the value of sacraments, saints, images and rites, but it is also firmly grounded on both his rhetorical expertise and his appropriation of the Renaissance notion of life (with special attention to the scholastic doctrine of substantial forms and the analogical meanings of life refracted through the prism of Aristotelian metaphysics). It is especially the all-encompassing nature of the question of life in Goclenius's dictionaries that constantly allows notions to migrate from the metaphysical to the physical, from the moral to the political, from the logical to the theological. Indeed, migration of meanings is one of the most riveting experiences that today's readers of the Latin and Greek *Lexica* may undergo while perusing the various entries, and this is certainly a tribute to the richness of the philosophical experience of the Renaissance. Trained as a linguist and a rhetorician, Goclenius revealed the extent to which synecdochic and metonymic transfers of meanings could shape one's philosophical inquiries.[35] In doing so, he proved to be a typical product of Renaissance culture, for he showed how philosophy intersected various domains of knowledge and spanned many levels of abstraction. Precisely because he duly recognized and recorded all possible metaphorical, metaleptic and catachrestic shifts every time they occurred while scrutinizing the content of a given concept, almost every entry can be read as a forum in which ontological, epistemological, ethical, political and theological meanings are held together in the most productive of hermeneutic tensions. A telling illustration (*pulchrum exemplum*) of the analogical correspondences among the natural, artificial, ethical and political aspects of reality is the one that, in Goclenius's opinion, brings to the fore the "similarities (*convenientiae*) between physical qualities and human wills":

> just as in compound substances (*mixta*) the primary qualities are weakened or as it were blunted, so that another quality may arise or emerge, in the same way, in society, the wills of the individuals are weakened, and from there a common will emerges.[36]

To add further examples, the entry "Recidiva" ("Relapse") prompted Goclenius to expand on various transfers of meaning concerning the domains of medicine and theology ("Transfertur a Scholasticis Theologis ad vitia cum dicitur, Recidiva peccati"), and so did the entries "Facies" and "Fames".[37]

As a thesaurus of both linguistic and rhetorical wisdom, Goclenius's work marks in quite remarkable terms the conceptual evolution of philosophical Latin during the Long Renaissance, from around mid-twelfth century to mid-seventeenth century. In reflecting a number of dramatic changes in interests, topics and priorities, it epitomizes a characteristic Renaissance way of doing philosophy, that is, philosophy through a dictionary. The Renaissance was a time of intense experimentation in literary genres, and this also applies to philosophy. Dialogues, treatises, essays,

---

[35] For some examples of synecdochic predications, see Goclenius (1980 [1613]), 963a, 970b; Goclenius (1980 [1615]), 222b.

[36] Goclenius (1980 [1613]), 955a.

[37] Respectively, Goclenius (1980 [1613]), 962a; 565a; 569ab.

commentaries and supercommentaries, plays, letters and poems were often used to convey philosophical arguments and to debate the most urgent issues of the time. Goclenius's dictionaries can be said to be a typical product of Renaissance culture in that literary creativity is a constitutive element of the speculative exercise. While a modern philosophical eye would consider the stylistic unevenness of his *Lexica* as a sign that their author was unable to organize the material in a proper way, creating a sort of philosophical pastiche or cento, or a didactic tool that more or often than not seems to verge on the pedantic and the pedestrian, the diversity and multiplicity of writing practices put to fruition by Goclenius are in fact another attempt to compress a whole library into one book, to repeat Eglinus's phrase. His *Lexica* are systematic and yet sufficiently loose to allow all sorts of digressions and detours (*paululum saltare extra chorum*).[38] They contain discussions about proverbs, responses to specific queries addressed through private letters, grammatical debates about the proper use of terms and even their correct spelling. See, for instance, the discussion about which of the two Latin words *redarguitio* and *redargutio* is the correct one.[39] As demonstrated by the important appendix on philosophical barbarisms at the end of the Greek *Lexicon*, Goclenius was as much a humanist almost as he was a scholastic thinker.[40] He did not refrain, to give some other examples, from referring to Euripides to make a philosophical point[41] or from inserting epigrams while discussing philosophical matters; indeed, we even find an epigram in the middle of a discussion about whether asses *rudunt* or *rudiunt*.[42] Sometimes he also recorded vernacular terms (in German) corresponding to their Latin equivalents.[43] In full agreement with the rhetorical spirit of Renaissance philosophy, copiousness and accumulation should therefore be seen as resources, not limits or defects.[44]

Between the fifteenth and the sixteenth centuries, linguistic *copia* and *congeries* were seen as manifestations of the vital power of nature and inner creativity. As such, they could be used to foster the power of thinking. This was, after all, one of the most significant legacies of Erasmus in philosophy. The transfers of meaning recorded by Goclenius in his dictionaries paralleled the overlapping of semantic exchanges within the domain of life. Goclenius distinguished between three principal meanings of life (*vita*) – physical, political and theological – which intersected with the Aristotelian division into animal (*pecuina seu voluptaria*), civic (*civilis*) and contemplative (*contemplativa*) existence. In a physical sense, "life" meant the natural power to assimilate food (*vis alendi*), to grow (*vis augendi*) and to perform elementary vital and cognitive operations (*motus vitalis* and *sensus*). While in a broader, less technical sense (*improprie*) "life" coincided with the meaning of soul understood as a principle of life (*essentia rei viventis*), in a political sense, *vita*

[38] Ibid., 174a.

[39] Ibid., 964ab.

[40] Goclenius (1980 [1615]), 282-371b.

[41] Goclenius (1980 [1613]),172b.

[42] Ibid., 173a.

[43] Ibid., 172b.

[44] On copiousness in early modern culture, see Shinn and Vine (2014).

denoted one's way of life: *ratio, modus, genus agendi seu vivendi*.[45] Referring to Ficino's commentary on *De divinis nominibus* by Dionysius the Areopagite, Goclenius was also willing to mention the Platonists' contribution to the definition of *vita*, seen as a spiritual force pervading the universe in its entirety.[46] It is an important acknowledgment, which once again testifies to the influence of Ficino's translations while signalling the persistence of Dionysian words and tropes.[47] By the time Goclenius had published his *Lexicon philosophicum* in 1613 and his *Lexicon philosophicum Graecum* in 1615, the Latinization of Dionysius had reached its end, both conceptually and linguistically. Goclenius did not hesitate to acknowledge the presence of this legacy in the philosophical armoury of contemporary theologians, but he added significant scholastic qualifications of a distinctively Reformed kind. It wasn't therefore by chance that Goclenius reported Girolamo Zanchi's definition of life as "the unremitting movement (*agitatio*) of the soul in the body (ἐντελέχεια) through which the body nourishes itself and grows", and, relying on the Bible, he identified this *agitatio* with the life pervading the blood.[48] An Italian Protestant, Zanchi (1516–1590) shared with Goclenius a basically irenic and eclectic position between the Lutheran and Calvinist fronts, especially on matters pertaining to the interpretation of the Eucharist.[49]

Conclusion

As a thinker straddling two centuries, Goclenius shows how seventeenth-century philosophy, with its innovative aspects in the fields of language, science and religion, cannot be properly understood without taking into account the philosophical background of the fifteenth and sixteenth centuries. Rather than looking at these two centuries as an unusually barren and unoriginal age in terms of speculative ideas and commitments, Goclenius helps us embrace a more historicized and positive consideration of such cultural trends as eclecticism, the reception of the classical tradition and the role of philological inquiry. In Goclenius's dictionaries, the eclectic layering that forms the texture of each entry reveals how old traditions interweaved with new ideas. For this reason, as I stated at the beginning of this chapter, Cranz's notion of a Long philosophical Renaissance seems to me a perspective that, from a historical point of view, is much more convincing and stimulating than explaining away the fifteenth and sixteenth centuries as philosophically trite and derivative. There are certainly risks in describing the period of time between the twelfth and the seventeenth century as a Long philosophical Renaissance, for in doing so some crucial differences may disappear (such as a more prominent sense of the self, a more

---

45 Goclenius (1980 [1613]), 324–328.

46 Ibid., 328b.

47 On Ficino's commentaries on Dionysius, see now Allen (2015).

48 Goclenius (1980 [1613]), 326.

49 On Zanchi, see Gründler (1963); Goris (2001).

nuanced sense of the historicity of human experience, a closer attention to the linguistic conditions of thinking and a growing awareness of the role played by economic factors in shaping the world of human beings). These differences, however, would not be perceived unless they were set against the background of long-term, tectonic shifts in the domain of metaphysics and theology.

If we read Goclenius's lexicographic accomplishments as a book of philosophy, a sinuous but seamless narrative emerges. To cut a very long story short, Goclenius attempted to adjust the Aristotelian ontology in its latest scholastic versions to the principles of Reformed theology *via* a reinterpretation of Platonic metaphysics. In this chapter, I decided to limit my analysis to a few specific remarks concerning the fields of metaphysics and the practice of writing about philosophy. The more than 1500 pages of the Latin and Greek *Lexica* make the work too vast in scope for the limited amount of words of this chapter, but my aim was to provide a case study that could testify to the healthy state of Renaissance philosophy. It may sound like a hackneyed commonplace, but the printing press, the discovery of new worlds and an astounding proliferation of political and religious conflicts had immeasurably expanded, within the space of a century, the boundaries of knowledge. Renaissance philosophy was inextricably related to these technological, anthropological and bellicose developments. It coincided with a momentous linguistic turn in that conflicts, controversies and commerce fostered the emerging of a plural and quarrelling multilingualism. Goclenius's rich and articulate account of contemporary philosophy thus provides historical evidence that there was diffuse awareness of these rapid changes among fifteenth- and sixteenth-century philosophers, which is yet another instance of early modern intellectuals coping with information overload.[50]

I began my chapter by criticizing the limits of the contemporary hand-bookish view of Renaissance philosophy, and I ended up using a Renaissance dictionary of philosophy to defend the value of Renaissance philosophy in the history of modern thought. This is not a contradiction on my part. The fact is that, for a historian, handbooks are good material to probe the perceptions, preconceptions and expectations of a particular age. A contemporary text-book of history of philosophy speaks volumes about the philosophical concerns of our age. Even more so, a companion to Renaissance philosophy of our time will tell future historians what view of Renaissance thought was predominant among twenty-first-century scholars.[51] For the same reason, therefore, Goclenius's dictionaries can be used now to assess the state of philosophical experience around the 1610s.

Of course, these dictionaries express a particular point of view and cannot be used to represent the totality of philosophical endeavours occurring during the Renaissance. While I look at them as a distillation of Renaissance thought, I am fully aware that this material remains one synthesis of a kind, which can be employed for tentative generalizations only if these are made with a good dose of caution and scepticism. Another important reason why I decided to use Goclenius's dictionaries

[50] See now Blair (2010).

[51] Some examples of recent text-books of Renaissance philosophy are: Ernst (2003); Hankins (2007) and Blum (2010 [1999]).

as reliable specimens of thinking practice is that they provided me with what I earlier called an anamorphic resolution of long-term developments in Western philosophy. Each philosophical notion is there, safely located in its corresponding entry, in the historically conditioned setting of a particular dictionary written between the sixteenth and the seventeenth century. And yet those philosophical notions also cut through centuries, disciplinary fields and the synchronic assessment of foundational notions. This, too, can be safely taken as evidence of Goclenius's historicist and humanist attitude. Together with his reliance on a metaphysics of interactive substances and his speculatively creative use of theological commonplaces, his attitude brought to the fore the philosophical legacy of the Renaissance while highlighting the differences with later centuries.

Bibliography

Allen, Michael J.B. 2015. Introduction. In Marsilio Ficino, *On Dionysius the Areopagite*, 2 vols. Cambridge: Harvard University Press.

Ashworth, E.J. 1967. Joachim Jungius (1587–1657) and the Logic of Relations. *Archiv für Geschichte der Philosophie* 49: 72–85.

Blair, Ann M. 2010. *Too Much to Know: Managing Scholarly Information Before the Modern Age*. New Haven: Yale University Press.

Blum, Paul Richard, ed. 2010 [1999]. *Philosophers of the Renaissance*. Trans. Brian McNeil. Washington: Catholic University of America Press.

Busson, Henri. 1957 [1922]. *Le rationalisme dans la littérature française de la Renaissance, 1533–1601*. Paris: Vrin.

Canone, Eugenio. 1988. *Phantasia/imaginatio* come problema terminologico nella lessicografia filosofica tra Sei-Settecento. In *Phantasia Imaginatio*, ed. Marta Fattori, Massimo Bianchi, 221–257. Rome: Edizioni dell' Ateneo

Celenza, Christopher S. 2004. *The Lost Italian Renaissance: Humanists, Historians, and Latin's Legacy*. Baltimore: The Johns Hopkins University Press.

Charbonnel, J.-Roger. 1919. *La pensée italienne au XVIe siècle et la courant libertin*. Paris: Champion.

Copenhaver, Brian P., and Charles Schmitt. 1992. *Renaissance Philosophy*. Oxford: Oxford University Press.

Cranz, F. Edward. 2006. Augustine and Anselm of Canterbury. In *Reorientations of Western Thought from Antiquity to the Renaissance*, ed. Nancy Struever, 1–87. Aldershot and Burlington: Ashgate.

De Angelis, Simone. 2010. *Anthropologien: Genese und Konfiguration einer 'Wissenschaft vom Menschen' in der Frühen Neuzeit*. Berlin: De Gruyter.

Des Chene, Dennis. 1996. *Physiologia: Natural Philosophy in Late Aristotelian and Cartesian Philosophy*. Ithaca: Cornell University Press.

Des Chene, Dennis. 2000. *Life's Form: Late Aristotelian Theories of the Soul*. Ithaca: Cornell University Press.

Ernst, Germana (ed.). 2003. *La filosofia del Rinascimento*. Rome: Carocci.

Garin, Eugenio. 1978 [1966]. *Dal Rinascimento alla Controriforma* [Volume 2 of *Storia della filosofia italiana*]. Turin: Einaudi.

Gentile, Giovanni. 1968 [1920]. *Il pensiero italiano del Rinascimento* [originally published as Giordano Bruno e il pensiero del Rinascimento]. Florence: Sansoni.

Gentile, Giovanni. 1968 [1923]. *Studi sul Rinascimento*. Florence: Sansoni.

Giglioni, Guido. 2015. Philosophy. In *The Oxford Handbook of Neo-Latin*, ed. Sarah Knight and Stefan Tilg, 249–262. Oxford: Oxford University Press.
Goclenius, Rudolph. 1980 [1613]. *Lexicon philosophicum quo tanquam clave philosophiae fores aperiuntur*. Hildesheim: Olms.
Goclenius, Rudolph. 1980 [1615]. *Lexicon philosophicum Graecum*. Hildesheim: Olms.
Goris, Harm. 2001. Thomism in Zanchi's Doctrine of God. In *Reformation and Scholasticism: An Ecumentical Enterprise*, ed. Willem J. van Asselt and Eef Dekker, 121–139. Grand Rapids: Baker Book House.
Gründler, Otto. 1963. *Thomism and Calvinism in the Theology of Girolamo Zanchi*, 1516–1590. Ph.D. dissertation: Princeton Theological Seminary.
Hamlyn, D.W. 1987. *The Penguin History of Western Philosophy*. London: Penguin.
Hankins, James (ed.). 2007. *The Cambridge Companion to Renaissance Philosophy*. Cambridge: Cambridge University Press.
Jensen, Kristian. 1990. Protestant Rivalry – Metaphysics and Rhetoric in Germany c. 1590–1620. *Journal of Ecclesiastical History* 41: 24–43.
Kenny, Anthony. 2006. *The Rise of Modern Philosophy* [Volume 3 of *A New History of Western Philosophy*]. Oxford: Clarendon Press.
Kristeller, Paul Oskar. 1964. *Eight Philosophers of the Renaissance*. Stanford: Stanford University Press.
Kristeller, Paul Oskar. 1979. *Renaissance Thought and Its Sources*, ed. Michael Mooney. New York: Columbia University Press.
Lamanna, Marco. 2013. *La nascita dell'ontologia nella metafisica di Rudolph Göckel (1547–1628)*. Hildesheim: Olms.
Leinsle, Ulrich Gottfried. 1985. *Das Ding und die Methode: Methodische Konstitution und Gegenstand der frühen protestantischen Metaphysik*, vol. 2. Augsburg: Maro.
Lewalter, Ernst. 1967 [1935]. *Spanisch-jesuitische und deutsch-lutherische Metaphysik des 17. Jahrhunderts: Ein Beitrag zur Geschichte der iberisch-deutschen Kulturbeziehungen und zur Vorgeschichte des deutschen Idealismus*. Darmstadt: Wissenschaftliche Buchgesellschaft.
Lohr, Charles. 1988. Metaphysics. In *Cambridge History of Renaissance Philosophy*, ed. Charles B. Schmitt, Quentin Skinner, Eckhard Keßler, and Jill Kraye, 537–638. Cambridge: Cambridge University Press.
Moran, Bruce T. 1994. Alchemy, Prophecy, and the Rosicrucians: Raphael Eglinus and Mystical Currents of the Early Seventeenth Century. In *Alchemy and Chemistry in the 16th and 17th Centuries*, ed. Piyo Rattansi and Antonio Clericuzio, 103–119. Dordrecht: Kluwer.
Palmer, Wandel Lee (ed.). 2014. *A Companion to the Eucharist in the Reformation*. Leiden: Brill.
Petersen, Peter. 1921. *Geschichte der Aristotelischen Philosophie im Protestantischen Deutschland*. Leipzig: Meiner.
Rogers, G.A.J., Tom Sorell, and Jill Kraye (eds.). 2010. *Insiders and Outsiders in Seventeenth-Century Philosophy*. London: Routledge.
Rorty, Richard, J.B. Schneewind, and Quentin Skinner (eds.). 1984. *Philosophy in History: Essays on the Historiography of Philosophy*. Cambridge: Cambridge University Press.
Russell, Bertrand. 2004 [1946]. *History of Western Philosophy*. London: Routledge.
Schmitt, Charles B. *et al.* 1988. *Cambridge History of Renaissance Philosophy*, ed. Charles B. Schmitt, Quentin Skinner, Eckhard Keßler, Jill Kraye. Cambridge: Cambridge University Press.
Shinn, Abigail, and Angus Vine, ed. 2014. *The Copious Text: Encyclopaedic Books in Early Modern England*. Special issue of *Renaissance Studies* 28: 167–332.
Sorell, Tom (ed.). 1993. *The Rise of Modern Philosophy: The Tension Between the New and Traditional Philosophies from Machiavelli to Leibniz*. Oxford: Clarendon.
Spruit, Leen. 1994–1995. Species Intelligibilis *from Perception to Knowledge*, 2 vols. Leiden: Brill.
Stiening, Gideon. 2014. Goclenius (Gockel, Göckel), Rudolph d. Ä. In *Frühe Neuzeit in Deutschland 1520–1620: Literaturwissenschaftliches Verfasserlexikon*, vol. 3, ed. Kühlmann Wilhelm et al., 31–38. Berlin: De Gruyter.

Vasoli, Cesare. 2002. In *Le filosofie del Rinascimento*, ed. Paolo Costantino Pissavino. Milan: Bruno Mondadori.

Wandel, Lee Palmer. 2005. *The Eucharist in the Reformation*. Cambridge: Cambridge University Press.

Weber, Hans Emil. 1907. *Die philosophische Scholastik des deutschen Protestantismus im Zeitalter der Orthodoxie*. Leipzig: Quelle und Meyer.

Wundt, Max. 1939. *Die deutsche Schulmetaphysik des 17. Jahrhunderts*. Tübingen: Mohr.

Zambelli, Paola. 2012 [1994]. From the *Quaestiones* to the *Essais*: On the Autonomy and Methods of the History of Philosophy. In P. Zambelli, *Astrology and Magic from the Medieval Latin and Islamic World to Renaissance Europe: Theories and Approaches* 373–390. Farnham (Surrey) and Burlington, VT: Ashgate (originally published in: *Science, Politics and Social Practice: Essays on Marxism and Science, Philosophy of Culture and the Social Sciences*, ed. Kostas Gavroglu, John J. Stachel and Marx W. Wartofsky, 373–390. Dordrecht: Kluwer, 1994).

Chapter 3
Italian Renaissance Love Theory and the General Scholar in the Seventeenth Century

Stephen Clucas

Abstract This essay considers the uses made of Renaissance love theory by the seventeenth-century English scholar Robert Burton in his *Anatomy of Melancholy* (first published in 1621). It is argued that Burton's approach is that of a 'general scholar', and a close examination of his sources reveal that he made use not only of the primary texts of Renaissance love theory such as the works of Marsilio Ficino and Leone Ebreo, but also the compendious works of later scholars working in medicine and law, as well as philosophy. Drawing on sources as diverse as Francesco Piccolomini's weighty philosophical tome on civil science, *Vniversa Philosophia de Moribus* to a diminutive collection of Platonic commonplaces by Niccolò Liburnio, Burton's work makes it clear that a history of the reception of Platonism in the seventeenth century needs to consider the various milieux of European general scholarship.

Many scholars in seventeenth-century Europe, working in a variety of disciplines, re-visited the themes of late-fifteenth- and early sixteenth-century Italian Platonic love theory. In this essay, I will be looking at the uses made of Renaissance love theory in the work of the English author Robert Burton (1577–1640). Burton's generically complex work *The Anatomy of Melancholy* was first published in 1621, but was re-published (and expanded) several times before the author's death in 1640. Drawing together religious and medical concerns, Burton's voluminous work is a typical (although highly self-conscious and sometimes playful) product of late humanism in that it proceeds by the compilation and collection of passages from a bewildering variety of authors, both Classical and modern. Although John Charles Nelson asserted in his influential 1958 study *Renaissance Theory of Love*, that by the time that Giordano Bruno published his *Degli eroici furori* in 1583, "the courtly

S. Clucas (✉)
English and Humanities, Birkbeck College, University of London, London WC1E 7HU, UK
e-mail: s.clucas@bbk.ac.uk

C. Muratori, G. Paganini (eds.), *Early Modern Philosophers and the Renaissance Legacy*, International Archives of the History of Ideas Archives internationales d'histoire des idées 220, DOI 10.1007/978-3-319-32604-7_3

tradition of superficially learned Platonizing comment upon love, deriving from Pietro Bembo, had become hackneyed",[1] in this essay I will be considering the persistence of the philosophical themes of Platonic love commentary in the milieu of "general scholarship" in the sixteenth century through to the first two decades of the seventeenth century.

General Scholarship in the Seventeenth Century

What exactly do I mean by "General Scholarship"? It is a term which could perhaps be replaced by the terms *Philologia* or *critice* as they were understood in the sixteenth and seventeenth century.[2] A useful definition is provided in the manuscript treatise *Generall Learning* written in 1668 by the Anglican scholar and divine Meric Casaubon (1599–1671) son of the great humanist scholar Isaac Casaubon, and – like Burton – a scholar of Christ Church, Oxford. "Generall Learning" or "that learning which make's à GENERALL SCHOLLER", was the kind of humanistic scholarship pursued by jurists, physicians and divines in pursuit of their professions. "There be, and haue beene", Casaubon wrote,

> of all professions, Lawyers, physitians, & others, who (whateuer theire aime hath beene) have deserved the title of generall schollers, as their learned labours, dealing with, & in all kinde of learninge, with good choyce & judgment will beare them testimony.[3]

Divines in particular, had need of this kind of learning, for how could any man "doe any good" in this field, Casaubon asks, "without competent knowledge of the originall tongues; a good stocke of human learninge, some insight in all sciences, good knowledge of former tymes, of actions & events; rites and customes, sacred and civill […]?"[4] The scholar needed to be able to make intelligent and judicious use of the works he read, be "able to make good use with choyce and variety of what he reades, and […] in such multiplicitie, & contrarietie sometymes, of interpretations & opinions, to judge what [is] most probable, or warrantable."[5] In 1645, in a letter concerning *De Methodo Studiorum*, Casaubon defined "Studia Philologica" as including "the whole ἐγκυκλοπαιδείαν, all liberal Arts & Sciences",[6] while in *A Treatise of Vse and Custome* (published anonymously in 1638), he defined "a true Philologist" etymologically as a "a lover of learning in generall".[7]

---

[1] Nelson (1958), 257. All translations from Latin are my own unless otherwise stated.

[2] See Bravo (2006). See also Ligota and Quantin's "Introduction" to the same volume, 1–38.

[3] Casaubon, *Generall Learning*, 89 and 92. See also Serjeantson's "Introduction", especially "General Learning and the Encyclopaedia" and "The Ideal General Scholar", 13–21.

[4] Ibid., 95.

[5] Ibid., 96.

[6] Meric Casaubon to Oliver Withers, 24 February 1645, Lambeth Palace Library, Lambeth MS 595, 101–103, printed as an 'Appendix' by Serjeantson in Casaubon, *Generall Learning*, 194.

[7] Casaubon (1638), cit. Serjeantson in Casaubon, *Generall Learning*, 20.

Robert Burton, writing in the 1620s also promoted the ideal of the general scholar. Burton published the *Anatomy of Melancholy* under the pseudonym of "Democritus Junior", and in a preface entitled "Democritus Junior to The Reader", he explains why he has chosen this *nom-de-plume*. Democritus, he says,

> was a litle wearish old man, verie melancholy by nature [...] and much giuen to solitarinesse, a famous Philosopher in his age [...] wholly addicted to his studies at the last, and to a priuate life [...] A great Diuine, according to the Diuinitie of those times, an expert Physitian, a Politician, an excellent Mathematician [...] He was much delighted with the studies of Husbandry, saieth Columella [...] [and] He knew the natures, [and] differences of all Beasts, Plants, Fishes [and] Birds [...] In a word he was *omnifarium doctus*, a generall Schollar [...].[8]

It is on the basis of their shared pursuit of general scholarship, that Burton claims an affinity with the illustrious Greek philosopher, comparing his own life as a fellow of Christ Church, and as an Anglican divine (he was appointed as preacher at St Thomas Church Oxford in 1616, and was also rector of Seagrave in Leicestershire from 1630 onwards) to that of Democritus. Like the melancholy Greek he "liu'd a silent, sedentarie, solitarie, priuate life [...] penned vp most part in my Study."[9] As Serjeantson has noted, "The vast majority of statutorially sanctioned scholarship pursued in an early modern English university beyond the level of Master of Arts was directed towards divinity."[10] The role of general scholar as later outlined by Casaubon, was therefore an important part of the ecclesiastical life, and more so for those who were also university scholars. The philological ideal of being *omnifarium doctus*, codified by the great humanist scholars of the previous century such as Erasmus in his *De ratione studii*,[11] had perhaps become a little jaded by the 1620s, and Burton steers a fine line between self-mockery of the general scholar ideal, and promotion of it as a still-vibrant tradition. He cites Lipsius's *Manuductio ad Stoicam Philosophiam* in praise of the scholar who is not a "slave" to any single discipline, and Montaigne's praise of the early sixteenth-century French humanist Adrien Turnèbe as a general scholar, but he also lampoons his own humanistic procedures: "I haue confusedly tumbled ouer diuers Authors in our Libraries, with small profit, for want of Art, Order, Memorie, Iudgement".[12] While he concedes – in a rhetorical move designed to pre-empt criticism – that his work is nothing more than "Apish imitation, a Rapsody of Rags gathered together from seuerall Dung-hills, excrements of Authors [...] confusedly tumbled out without [...] Iudgement, Wit [or] Learning", he nonetheless defends the "*Cento*" which he has patched together out of the works of others: "I haue wronged no Authors", he insists, "but giuen euerie man his owne. [...] I cite and quote mine Authors, *sumpsi non surripui* [I have taken, not stolen]."[13]

---

[8] Burton (1628), 2 (hereafter cited as *Anatomy*). All quotations are from this edition, together with a cross-reference to the modern edition, Burton (1989–2000) (hereafter cited as AOM).

[9] Burton, *Anatomy*, 2 [AOM, vol. 1, 3].

[10] Serjeantson, 'Introduction' in Casaubon, *Generall Learning*, 18.

[11] See Erasmus (1511). For the mediaeval background of this ideal see Rohling (2012).

[12] Burton, *Anatomy*, 3 [AOM, vol. 1, 4].

[13] Burton, *Anatomy*, 8 [AOM, vol. 1, 11].

He quotes a Terentian tag out of a medical author, Johann Jacob Wecker's *Medicae Syntaxes* (1562), that "*nihil dictum* [*est*] *quod non dictum prius; methodus sola artificem ostendit*, wee can say nothing but what hath been said, the composition & method is ours onely, and shewes a Schollar."[14] This is the "choyce" and judgement which Casaubon insisted that the general scholar bring to his reading, and while Burton satirically scoffs at the composition of such works of general scholarship, he is also tacitly defending its values, and cites Seneca, the "painfull omniscious Philosopher" as a type of the general scholar, defending him against some of the harsher judgments of Lipsius in the introduction to his 1605 edition of Seneca's *Opera omnia*. "If Seneca be thus lashed", asks Burton, "what shall I expect? I that am *vix vmbra santi* [sic=*tanti*] *Philosophi* [scarcely the shadow of so great a philosopher] [...]?"[15] Shade or shadow he might be, but Burton clearly emulated the idea of the "omniscious" scholar.

The General Scholar and Renaissance Love Theory

So – what happens to the Renaissance philosophy of love in the hands of a seventeenth-century general scholar? Burton cites as a singular precedent for his own work Antonio Zara's *Anatomia ingeniorum et scientiarum*, published in Venice in 1615.[16] A divine, like Burton, Zara presents in his work a compendium of classical quotations, together with judicious selections of omnivorous sixteenth-century general scholars, such as Girolamo Cardano and Julius Caesar Scaliger. Like Burton, Zara divides his *Anatomia* into *membra* and *sectiones*, playfully alluding to the processes of anatomical dissection, and like Burton he uses his anatomy as a pretext for promiscuously ranging across a whole variety of topics, from astronomy, optics and architecture, to the magical arts, law and theology. Like Casaubon, both Burton and Zara cite modern authors alongside ancient ones,[17] and this process is seen vividly at work in Burton's treatment of two sub-varieties of melancholy which can be found in the final part of his work: "love melancholy" and "religious melancholy". It is in these sections of the *Anatomy*, that we find Burton gathering and selecting from the work of Marsilio Ficino and Leone Ebreo.

So why did neoplatonic love theory feature in a book which sought to provide a comprehensive account of a range of early modern mental illnesses and their

[14] Burton, *Anatomy*, 8 [AOM, vol. 1, 11]. Cf. Wecker (1562), 'Pro Lectore', sig. α 1 verso. I am indebted here (and throughout this paper) to the annotations of Burton in AOM, although I have occasionally amended them slightly, in the light of checking their references against the originals.

[15] Burton, *Anatomy*, 11 [AOM, vol. 1, 15].

[16] Burton, *Anatomy*, 5 [AOM, vol. 1, 6]: "I haue honourable Presidents for this which I haue done: I will cite one for all, Anthony Zara Pap. Episc. his Anatomie of Wit." See Zara (1615).

[17] On Casaubon's use of ancient and modern sources, see Serjeantson, 'Introduction', in Casaubon, *Generall Learning*, 26–27.

"causes, symptoms, prognostickes & severall cures"? The inclusion of Ficino is not as anomalous as it might seem, given the profoundly medical orientation of his work. In his commentary on Plato's *Symposium*, the *Commentarium in Convivium Platonis de amore*, first published in Florence in 1484, Ficino emphasized the physiological causes of "vulgar love" (*Amor uulgaris*), which he presents in Oratio VII.4 as a contagious disease comparable to venereal itch, mange, leprosy, dysentery, the plague and "other diseases which attack through contagion".[18] Burton cites this passage, and others of a medical nature from Ficino's work. He cites Oratio VII.10, for example, where Ficino explains how the disease (*morbus*) of love is a kind of fascination or enchantment (*fascinatio*) which has a physiological foundation. "The beginning of this disease", Burton says, translating Ficino, "is in the eye" (*causa* [...] & *origo* [*morbi*] *est oculus*), and he cites in support of this a demonological work on *fascinatio* by the Neapolitan Benedictine Leonardus Vairus [Leonardo Vairo] (1540–1603), Bishop of Pozzuoli, who says that the spirits are infected by rays emitted by the eye.[19] Burton concedes that his contemporaries may baulk at the extromissive theory of sight (which had largely been discredited by the 1620s) but "*Ficinus* proues it by bleare eyes" (*lippus*) in Oratio VII.4, which Ficino explains by rays of light leaving the eyes, carrying with them "a vapour of corrupt blood" (*cum radio unà uaporem corrupti sanguinis emanare*).[20] Ficino, in fact, believed that melancholic blood was the cause of irrational love. As Burton notes, "*Ficinus* [...] *in Convivium Platonis, will haue the blood to be the part affected.*"[21] Burton cites Ficino's definition of vulgar love as "a species of madnesse", and when we consult the chapter in Oratio VII.12 on the noxiousness of earthly love, we find a thoroughgoing humoral definition of this madness: "First in the course of their love they are kindled by bile [choler]; then they are afflicted by the turning brown of the black bile [i.e., melancholy] and thence they rush into madness and raging passion."[22]

However, Burton's interest in Renaissance love theory does not end with this medicalized view of vulgar love as a contagious disease caused by blood infected by adust melancholy. Burton's reasons for digressing into the area of love are more complex, and – as we shall see – are intimately connected with the following treatment of religious melancholy. Part of the complexity of Burton's work, and interpretations of it, is due to its satirical framework. As the title of the work announces, the preface of Democritus Junior is "Satyricall". This preface is described as

---

[18] Ficino, *Commentarium in Convivium Platonis de Amore*, VII.4. In Ficino (1576), 1358 (hereafter cited as *Opera*). English edition: Ficino (1944), 224 (hereafter cited as Jayne).

[19] Burton, *Anatomy*, 431 [AOM, vol. 3, 88]. See Ficino, *Opera*, 1360 (Jayne, 113) and Vairus (1583), I.3, 13. On Vairo's *De fascino* see Brann (2002), 213–214.

[20] Burton, *Anatomy*, 431 (AOM, vol. 3, 88). Ficino, *Opera*, 1357 (Jayne, 108).

[21] Burton, *Anatomy*, 413 [AOM, vol. 3, 58]. See Oratio VII.7 "Vulgaris amor est sanguinis perturbatio", *Opera*, 1359: "In sanguine igitur illam meritò collocamus. In sanguine uidelicet melancholico". "In the blood, therefore, we rightly place the fever of love; that is to say, in the melancholic blood" (Jayne, 226).

[22] Burton, *Anatomy*, 412 [AOM, vol. 3, 57]. See Ficino, *Opera*, 1361: "qui amore durante bilis incendio primum, deinde atrae bilis adustione afflicti, in furias, ignemque ruunt." (Jayne, 114).

"conducing to the following Discourse". Interpreters have been divided about whether this implies that the whole work is satirical in intent, or whether "conduce" should be construed in a more neutral sense. While some of the sections of the *Anatomy* are clearly satirical in intent – the "Digression on Air", for example, which scoffs at the bewildering variety of new natural philosophical ideas emerging at this time – others seem to have more serious intentions, even though they are sometimes underpinned by a certain wry scepticism (see, for example, the deliberately inconclusive conclusions of the section where Burton outlines the many conflicting medical accounts of the location of melancholy in the human body). The introduction of the section on love melancholy certainly seems to suggest the possibility that it will be a satirical, or at least comical interlude. "Tis a Comicall subiect", Burton says,

> in sober sadnesse I craue pardon of what is amisse [...] I am resolued howsoeuer, *velis, nolis*, in this Trage-comedy of Loue, to Act seuerall parts, some Satyrically, some Comically, some in mixt Tone, as the subiect I haue in hand giues occasion.[23]

This would certainly seem to suggest a rather light-hearted approach to the topic, and yet, I would argue, his handling of Renaissance love theory is in earnest. Burton, echoing one of his sources, the sixteenth-century French jurist Pierre Godoffroy [Petrus Godefredus] (who tells his readers that his *Dialogus de amoribus*, was written as a way of relaxing his mind from his legal studies),[24] presents his discussion of love as a diversion from the more serious business of discussing melancholy. To discourse of love, he says, will allow him to "recreate himself after laborious studies" on melancholy which have been "harsh and vnpleasing". "Giue me leaue then", he opines "to refresh my muse a little, and my weary readers, to expatiate in this delightsome field."[25]

Despite this suggestion that his treatment of love will be a harmless and light-hearted diversion, he also mounts a serious defence of love as a topic for scholarship. Some may think that love is a topic more suited to "a wanton Poet [...] an effeminate Courtier, or some such idle person", and that discussing love is to use one's time badly. "I am not perswaded it is [...] so ill spent", he says, addressing a subject on which

> many graue and worthy men haue written whole volumes, *Plato, Plutarch, Plotinus, Maximus Tyrius, Alcinous, Avicenna, Leon: Hebraeus*, in three large dialogues [...] *Picus Mirandula, Marius Equicola*, both in Italian [...] [and] almost euery Physician [...] haue treated of a part.[26]

---

[23] Burton, *Anatomy*, 379 [AOM, vol. 3, p. 8].

[24] Godeffroy (1552), "Intentio Autoris summaque operis", 1: "Et quî non inquam de amoribus nos etiam agemus vel relaxandi animi gratia, laborissimis omnium istius iuris studijs fatigati, quando & poëtarum genus dudum, otiosissimum his se iuuari ac iuuare illaesis moribus vult?" ("And who, I say, would not wish us to occupy ourselves with love or to alleviate our mind wearied with all its legal studies, when but a short time ago the idlest kind of poets have delighted, and are delighting themselves with this subject without any harm to morals?").

[25] Burton, *Anatomy*, 377 [AOM, vol. 3, 4].

[26] Burton, *Anatomy*, 376 [AOM, vol. 3, 2].

The Platonic orientation of this list is striking, as is its close attention to love treatises of the Italian Renaissance, including the works of Pico della Mirandola, Leone Ebreo and Mario Equicola.[27] Burton also cites Marsilio Ficino in his defence: "They reproue *Plato* then, but without cause (as *Ficinus* pleads) for all loue is honest and good, and they are worthy to bee loued that speake well of loue."[28] Here Burton cites in the margin phrases from Oratio I.4 where Ficino defends Plato's philosophy of love against his critics such as Dicaearchus of Messana (c. 350 – c. 285 BCE). Love and the desire for physical union are two separate things says Ficino, as both ancient and Christian theologians agree. Therefore we should be careful to use the word "love" correctly, and refrain from applying it to irrational emotions (*insanias perturbationes*):

> Let Dicaearchus, and whoever else makes bold to accuse the Platonic majesty of indulging too much in love, blush with shame, for we can never indulge too much, or even enough, the proper, pure, and divine passions. From this it follows that all true love is honorable, and every lover virtuous [...] But the turbulent passion by which men are seduced to wantonness [...] is considered the opposite of love.[29]

Burton's treatment of love is very much concerned with "*Plato*'s majesty", and the "pure and divine passions" of man in relation to God promoted by Ficino and Ebreo, and it is my contention that in his treatment of Platonic love theory Burton is very much in earnest. Love, he says, is not "scurrile, but chast, honest, most part serious and euen of religion itselfe. *Incensed* (as he said) with the loue of finding loue, *we haue sought it, found it*."[30] The allusion here is to the exultant closing chapter of Ficino's *Commentarium* (Oratio VII.17) where he is celebrating the fact that he has reached the true definition of love: that is to say, the love of God himself.[31]

Burton's Compositional Process

But let us take a closer look at Burton's compositional process, to bring us a little closer to the way in which Platonic love theory is integrated into the work of a seventeenth-century general scholar. Following Plato's emphasis on beauty as a cause of love, Burton echoes *Phaedrus* 250C-D:

---

[27] On this tradition of love treatises see Zonta (1975) [1910].

[28] Burton, *Anatomy*, 377 [AOM, vol. 3, 4].

[29] Ficino, *Opera*, Oratio, I.4, 1323: "Erubescat Dicaearchus, & si quis alius Platonicam maiestatem quod amori nimium indulserit, carpe non ueretur. Nam decoris, honestis, diuinis affectibus, nec nimium, nec satis unquam possumus indulgere. Hinc efficitur, ut omnis amor honestus sit & omnis amator iustus. Pulcher enim est omnis atque decorus, & decorum propriè diligit. Turbulentus autem ardor, quo ad lasciuiam rapimur [...] amor contrarius iudicatur." (Jayne, 41).

[30] Burton, *Anatomy*, 378 [AOM, vol. 3, 7].

[31] Ficino, *Opera*, 1361 (Jayne, 235). See, in particular, "inueniendis amoris accensi, amorem quaesiuimus & inueniemus." ("aroused by love of finding love, so to speak, we have sought and found love").

> Beauty shines, Plato saieth, and by reason of its splendor and shining causeth admiration, and the fairer the obiect is, the more eagerly it is sought. For as the same Plato defines it *Beauty is a liuely shining or glittering brightnesse resulting from effused good, By Ideas, seeds, reasons, shadowes, stirring vp our mindes, that by this good they may be vnited and made one.*[32]

In the "Preface", Burton says that Greek authors have been "cited out of their Interpreters [i.e. Latin translations], because the Originall was not so ready".[33] However, here the "Plato" passage cited is not from Marsilio Ficino, or Jean de Serres, but is taken from another work of general scholarship, Francesco Piccolomini's *Vniversa Philosophia de Moribus*, first published in 1583.[34] Piccolomini (1520–1604) was a Jesuit and the first ordinary Professor of natural philosophy at the University of Padua. In this work Piccolomini aims to give the outlines of a "universal civil philosophy", which attempts to synthesise and harmonize the moral philosophies of Plato, Aristotle and the Stoics, all of which are corrected against the norms of Christianity (albeit within what is essentially an Aristotelian framework).[35] The passage cited by Burton comes from the eighth *gradus*, which deals with "The instruments of the virtues and the greatest good, which are called the gifts of nature and fortune".[36] One of the "gifts of nature" discussed is beauty, and the passage Burton cites is from Chap. 35, "The opinion of Plato concerning Beauty" (*De Pulchro Opinio Platonis*). However, the cited passage is not a second-hand citation from a translation of Plato, but rather a summary of Plato's doctrine in Piccolomini's own words. The phrase quoted by Burton is preceded by the words "In my opinion beauty, according to the view of Plato is to be defined thus", and continues, much as Burton translated it:

> Beauty is a vital brightness emanating from the good itself, poured forth through Ideas, Reasons, seeds, and shadows, exciting our minds so that by the good they are reduced to unity. From this definition every kind of cause shines forth.[37]

Burton continues from here to cite the opinion of "Others" who think beauty is "the perfection of the whole composition, *caused out of the congruous symmetry, measure, order and manner of parts* [etc]."[38] The Latin passage in the margin is also

---

[32] Burton, *Anatomy*, 381 [AOM, vol. 3, 10].

[33] Burton, *Anatomy*, 14 [AOM, vol. 1, 19].

[34] Piccolomini (1583). In this essay I refer to the Geneva edition of 1596. On Piccolomini's moral philosophy see Kraye (2002), and Poppi (1997), 59–78 and 206–213.

[35] Kraye (2002), 59–60. On "universal civil philosophy" see Piccolomini, *Vniversa Philosophia*, 4: "Dum quaeritur subiectum Ciuilis Scientiae, nomine Ciuilis Scientiae vniuersam Philosophiam Ciuiliem denoto, non partem eius […]."

[36] Piccolomini (1596), 531: "De instrumentis virtutum et summi boni quae naturae et fortunae munera dicuntur."

[37] Piccolomini (1596), VIII.35, 595: "Pulchritudinem ex sententia Platonis ita definiendam censeo: Pulchritudo est vitalis fulgor ex ipso bono manans, per Ideas, Rationes, semina, & vmbras effusus, animos excitans vt per bonum in vnum redigantur. Per hanc definitionem omne genus causae elucescit."

[38] Burton, *Anatomy*, 381 [AOM, vol. 3, 10].

taken from Piccolomini, this time from Chap. 36, “Concerning Aristotle’s opinion of beauty” (*De Pulchro Opinio Aristotelis*). Once again Burton has taken the quotation out of context (here Piccolomini is exploring the various ways in which Aristotelians define beauty),[39] and the definition which Burton cites (without naming Aristotle), is in actual fact a summary in Piccolomini’s own words:

> In my opinion according to Aristotle’s view in book 13 of the *Metaphysics* it should be defined thus: Beauty is the perfection of a composition, arising from the congruous order, measure and proportion of the parts. The explanation of this definition is obvious from what is said. This in my opinion was the view of Aristotle concerning beauty.[40]

Burton clearly has no interest in Piccolomini’s central concern – that of reconciling the Aristotelian and Platonic understandings of beauty (to which Piccolomini devotes part of Chap. 39: “An Aristoteles cum Platone conciliari possit & dubia nonnulla diluuntur”).[41] The original controversial nature of Piccolomini’s text is elided, and he segues seamlessly into a Platonic account of “Grace” (*Gratia*) culled selectively from the opening passage of Piccolomini’s Chap. 38, “Quid sit Gratia”, which reports the opinions of Plato and the Academics (*Academicorum plurimi*), which compares beauty and grace to the rays and beams of the divine sun shining in various ways in various things,[42] which (Burton says), “are diverse, as they proceed from the diverse obiects, to please & affect our seuerall senses”.[43] This is followed by a quotation about the “species of beauty” (*Species Pulchritudinis*) which are received by the senses and “conceiued in our inner soul” (*concipiuntur interna mente*), an idea which, Burton says, “*Plato disputes at large in his Dialogue de Pulchro, Phaedo, Hyppias* [etc.]”.[44] Burton has moved here from Piccolomini’s *Vniversa Philosophia* to the *Controversiarum medicarum et Philosophicarum* of the Spanish medical humanist Francisco Vallès [Franciscus Valesius] (1524–1592), first published in 1556.[45] A professor of Medicine at Madrid, and commentator of Galen and Hippocrates, Vallès’s *Controversiarum*, as its title suggests, was a compendium of medical controversies. His remarks about beauty appear, oddly, in Book III of the work, which is dedicated to the use of the pulse as a method of diagnosing illnesses. In Chap. 15, “Whether there is a lover’s pulse” (*Vtrum sit aliquis pulsus amatorius*), Vallès departs from his stated object to dilate more generally on the subject of love.

---

[39] Piccolomini (1596), VIII.36, 597: “Definitur Pulchritudo à Peripateticis vario modo.”

[40] Piccolomini (1596), VIII.36, 597: “Censerem ego ex sententia Arist. 13 Metaphysicae ita esse definiendam, Pulchritudo est perfectio compositi, ex congruente ordine, mensura, & ratione partium consurgens. Cuius definitionis explicatio ex dictis satis est conspicua. Hanc censeo fuisse sententiam Aristotelis de Pulchritudine.” The reference is to *Metaphysics*, XIII.3.10, 1078b.

[41] Piccolomini (1596), VIII.37, 600–604.

[42] Piccolomini (1596), VIII.38, 599: “Pulchritudo & Gratia sint, tanquam radij & splendores Diuini Solis, in rebus variis vario modo fulgentes”.

[43] Burton, *Anatomy*, 381 [AOM, vol. 3, 10].

[44] Burton, *Anatomy*, 381 [AOM, vol. 3, 11].

[45] I am using the third edition, Vallès (1591).

> If all the things which have been written concerning love by Plato and other most virtuous philosophers were to be surveyed now, it would take a whole book of its own. But since now we are not undertaking to deal with a single part of philosophy, but exercise our pen with many and varied questions: we will only deal with that kind of love called "tempestuous", which seems to be completely necessary to resolve the question which has been proposed.[46]

Like Burton, Vallès finds opportunities in what is essentially a medical enterprise, to "exercise his pen" as a general scholar, exhibiting his knowledge of a wide range of philosophical as well as medical topics. For Burton, the Spanish humanist is simply another source of the Platonic doctrines which interest him, substituted for – or supplementing – the works of Plato himself. Vallès goes on to discuss the difference between rational and irrational appetites and loves, and in a passage Burton will return to later in his work, situates rational love in the brain and irrational love in the liver.[47]

Platonism and *amor Dei*

Burton then breaks away from his mosaic of sources, to reflect on the "seuerall kindes of loue", dividing it into two basic kinds: the love of God (*amor Dei*) of which he says "many fathers and Neotericks haue written iust volumes [...] many paranetical discourses", and the love of God's creatures – which seeks both spiritual beauty which is discerned with "the eyes of our minde", and physical beauty which is "discerne[d] with [...] corporall eyes".[48] This "twofold Diuision" is, he says, advanced by Vallès, Scaliger and Melanchthon "out of Plato φιλειν & ἐραν, from that speech of Pausanias belike, that makes two *Veneres* and two loues."[49] This is followed by a quote, given in Latin the margin: "*One Venus is antient without a mother, and descended from heauen, whom we call caelestiall; The Younger, begotten of Iupiter and Dione, whom commonly we call Venus.*"[50] The quotation from Plato is, once again, not from a translation but a secondary source, in this case, a

46 Vallès (1591), III. xiiii, 361: "Si omnia quae de amore à Platone, & aliis probatissimis philosophis scripta sunt, forent modò recensenda, iustus liber in sola hac tractatione consumeretur. Sed, *cum* modò non vnicam philosophiae partem susceperimus pertractandam, sed in multiplicibus & variis quaestionibus, stylem exerceamus: illud tantum de amore dicere erat tempestiuum, quod ad quaestionis modò propositae dissolutionem videbitur esse omnino necessarium."

47 Vallès (1591), III. xiiii, 362: "Amor igitur generatur in appetitu nonnumquam rationali, qui in cerebro residet: nonnumquam irrationali qui in hepate." Cf. Burton, *Anatomy*, 385 [AOM, vol. 3, 16], where the passage is paraphrased "Affectus nunc appetetiuae potentiae, nunc rationalis, alter cerebro residet, alter epate, cor &c."

48 Burton, *Anatomy*, 381 [AOM, vol. 3, 11].

49 Burton, *Anatomy*, 382 [AOM, vol. 3, 11]. The passage referred to is *Symposium* 180D-E.

50 Burton, *Anatomy*, 382 [AOM, vol. 3, 11]: the Latin reads: "Duae veneres, duo amores, quarum vna antiquior & sine matre coelo nata quam caelestem venerem nuncupamus, altera vero Iunior a Jove & Dione prognata, quam vulgarem venerem vocamus."

highly abbreviated collection of Platonic *sententiae* compiled by the Italian humanist Niccolò Liburnio (1474–1557), the *Divini Platonis Gemmae* published in sextodecimo format in 1556.[51] Liburnio published several collections of *sententiae* in the early decades of the sixteenth century, some for the vernacular market and some for a Latinate audience, and also translated Virgil into Italian.[52] The "Gems of Divine Plato" consists of extremely brief summaries of each of Plato's works (compared to which the Platonic epitomes of Ficino seem positively lengthy), each no more than 6–8 sextodecimo pages. This is followed by a list of "remarkable places, or gems, even more succinct comments" (*locos insigniores, sive Gemmas, Commenta adhuc succinctiora*), which are basically single sentence tags extracted from the summaries.[53] These are followed by a set of thematic commonplaces listed under headings such as "Natura", "Bene Vivere" or "Deus".[54] Many of his contemporaries, Liburnio says, "are accustomed in their daily speeches to intermingle the thoughts of the divine Plato confusedly, in a piecemeal fashion, and in a corrupted form". His book will remedy the situation by publishing the best of his philosophical precepts collected from almost all of his works.[55] Through his work, he says, those "who have had little experience of literary studies, will be able to arrive at a clearer and fuller understanding of Plato."[56] That is to say, Liburnio was writing an early sixteenth-century *Plato for Dummies*. His printer, Benoit Prévost (who writes his own letter to the reader), puts it slightly differently. "I considered it a worthwhile thing to do", he says, "to present [Plato] to you, reduced into a compendium [*redactum in compendium*] in this way, so that you could have in your hands all those things which were written by him more diffusely [*fusiùs*] under the persona of Socrates and others."[57] Liburnio's book is a perfect example of the kind of reference works recently surveyed by Ann Blair, "designed to aid in reading and composing Latin

[51] Liburnio (1556), 24 verso: "Quoniam verò duae sunt Veneres geminum quoque amore necesse est. Geminam autem deam hanc esse quis neget? Nónne vna quaedam antiquor est, & sine matre Venus caelo nata, quam caelestem Venerem nuncupamus? Altera verò iuniore Ioue, & Dione progenita, quam vulgarem communémque vocamus?". I have underlined the phrases which Burton quotes verbatim.

[52] See, for example, Liburnio (1537), and Liburnio (1551). His Virgil translation was printed in Venice in 1543. On Liburnio's literary output see Peirone (1968).

[53] Liburnio (1556), 54 recto-65 verso.

[54] Ibid., 68 recto-127 verso. "Aliae sententiae ex eodem Platone depromptae" ("Other *sententiae* set forth from the same Plato").

[55] Ibid., 3 recto: "nonnulli ex eiusmodi hominum grege suis ipsorum quotidianis sermonibus, diuinos Platonis sensus confusè, intercisè, atque corruptissimè immiscere consuescant: his ergo remedio praesentissimo occurre posse indicarim, si optima quaeque talia ac tanti philosophi praecepta ex toto ferè eius opere collecta sub tuo nomine auspicatissimè publicarentur."

[56] Ibid., 3 recto-verso: "Eiusdem verò, qui literarum studia leuiter attigissent, commodissimum fore visum est, vt nostro hoc labore ad apertiorem & pleniorem Platonis cognitionem peruenire valerent."

[57] Ibid., "Typographus Lectori", 66 recto: "existimaui operae pretium me facturum, si *cum* vobis exhibarem ita redactum in compendium, vt in manibus omnes habere possent ea, quae fusiùs, sub Socratis & aliorum persona, ab eo conscripta sunt."

texts [...] used by students, teachers, and preachers, and also by scholars."[58] Liburno, and others like him, broke down complex texts into "more succinct", summarized fragments, which could then be reassembled and incorporated "more diffusely" into copious texts of their own composition – as Burton does here.

But Burton is not only in the business of filleting sextodecimo manuals – he moves directly from the Liburnio summary of *Symposium* 180D-E to "*Ficinus* in his Comment vpon this place", who (he says), "following Plato, calls these two loues, two Divells, or good and bad Angells according to vs, which are still houering about our soules."[59] This observation shows that Burton was aware of Ficino's interpretation of the Greek daimon (δαίμων) as something distinct from the meaning of "demon" in Christian theology. In Oratio VI.3 of the *Commentarium* Ficino notes that:

> Some Platonists and the Christian theologians claim that there are certain bad demons. For the present we are not concerned with bad demons. The good demons Dionysius the Areopagite is accustomed to call by their proper name, Angels, the governors of the lower world, and this differs little from the interpretation of Plato.[60]

In his *argumentum* to the *Apologia Socratis*, Ficino also writes, "if it displeases you to call a man's familiar guide a daemon, then at least – as it pleases people nowadays to do – call it a good angel."[61] If the retention of "Divell" shows that Burton is not happy to side unequivocally with Ficino, he does at least gesture towards his interpretation. He certainly follows this statement with a long quotation which fuses together elements of Oratio VI. 8, where Ficino says that one of the two demons of love "*reares to heauen, the other depresseth us to hell*" (*alter ad superna erigat, alter deprimat ad inferna*).[62] The better demon, as Burton puts it "*stirres vs vp to the contemplation of that divine beauty, for whose sake we performe Iustice, and all godly offices, study Phylosophy*, &c."[63] "So farre *Ficinus*", Burton adds, before supplementing Ficino's evocation of divine love with passages from several works by Augustine: book 15 of *De civitate dei*, Chap. 15 of *De moribus ecclesiae catholicae*, and his commentary on psalm 64 (*Enarratio in Psalmum 64*), where the twofold nature of love is similarly handled, and where it is asserted that the "foure cardinall vertues [... are] naught else but loue".[64]

---

[58] Blair (2010), 6.

[59] Burton, *Anatomy*, 382 [AOM, vol. 3, 11–12].

[60] Ficino, *Opera*, 1342: "Esse uerò alios quosdam malos daemones Platonici nonnulli, & Christiani Theologi voluerunt. Sed de malis daemonibus nulla ad praesens nobis est disputatio. Bonos autem nostri custodes, proprio nomine angelos inferiores mundi gubernatores. Dionysius Areopagitica, quod in Platonis mente minimè discrepat uocare solet." (Jayne, 80).

[61] Ficino (1532), 467: "At si minus tibi placet & familiarem hominis ducem daemonem appellare, saltem, ut placet nostris, bonum angelum appellato."

[62] Burton, *Anatomy*, 382 [AOM, vol. 3, 12]. Ficino, *Opera*, 1345.

[63] Burton, *Anatomy*, 382 [AOM, vol. 3, 12]. Cf. Ficino, *Opera*, 1345: "Profectò in hominis mente aeternus est amor ad diuinam pulchritudinem peruidendeum: cuius gratia, philosophiae studia, & iusticiae, pietatisque officia sequimur."

[64] Burton, *Anatomy*, 382 [AOM, vol. 3, 12].

A great part of the section on Love melancholy is given over to considering the deleterious effects of irrational love, and this is perhaps what Burton meant by the "delightsome" nature of this part of the *Anatomy*. Burton quotes widely from the classical amorists and satirists, as well as more contemporary poets who write in either Neo-Latin or the vernacular. Amongst English authors Burton cites Samuel Daniel, Edmund Spenser and Christopher Marlowe, amongst European neo-Latinists he cites the Dutch Catullan poet Johannes Secundus (the *Basia*), and Italian poets such as the Greek-born Michele Marullo and Giovanni Pontano, and quotes several times from Kaspar Barth's *Pornoboscodidascalus Latinus* (1624), a Latin translation of the late fifteenth-century Spanish comedy by Fernando de Rojas, *La Celestina*.[65] The sections on love melancholy and religious melancholy seem to mirror each other – beginning with evocations of divine love as it was understood by Renaissance neoplatonists and theologians, and then moving to consider the irrational abuses of love and religion.

Leone Ebreo and Lover's Melancholy

The *Dialogi d'Amore* written by the Portuguese Jewish physician, poet, and philosopher Leone Ebreo (Judah Leon Abravanel) in the early 1500s plays an important role in Burton's *Anatomy*. Burton clearly had no Italian and read the work in the Latin translation of Johannes Carolus Saracenus, published in Venice in 1564 (and later anthologized in Johannes Pistorius's *Artis cabalisticae* published in 1587).[66] Saraceno dedicated his translation to Carolus Perrenotus Granvellanus, Abbot of Fauverney and counsellor of Philip II of Spain, and sought to situate Ebreo's work in the context of both Christian and Platonic love. "Plato in his symposium on love," says Saraceno, "spoke eloquently and copiously of this great God diffusing himself widely through all things, both human and divine", and then goes on to insist on the importance of Christ for understanding the true nature of love.[67] It is in this Christian-Platonic sense that Burton understands Ebreo's work.

In the section on love melancholy Burton uses Ebreo as "the most copious writer on this subiect" alongside Ficino's translation of Plotinus's *Enneads* III ("De amore"), and Ficino's *Commentarium in Convivium*, to insist upon the idea that love is a "desire of enioying that which is good and faire."[68] He also uses Ebreo's "accurate Division" of love into the "*Naturall*, *Sensible*, and *Rationall*", with natural

---

[65] See, for example, Burton *Anatomy*, 457 [AOM, vol. 3, 131], 470–471 [AOM, vol. 3, 151–2], 482, 502. On Barth's translation see Fernández (2006).

[66] Ebreo's work was reprinted in Pistorius (1587), 331–608. The original Italian work was published posthumously in Rome: Ebreo (1535).

[67] Ebreo (1564), sig. [a vii] recto: "Plato in conuiuio de Amore disertè copioseque pertractans maximum hunc Deum per omnia tam diuina, quam humana latissimè sese diffundere pronunciauerit."

[68] Burton, *Anatomy*, 380 [AOM, vol. 3, 9]. Cf. Burton: "Amor est voluntarius affectus & desiderium re bona fruendi", and Ebreo (1564), 9 recto: "amorem verò affectum uoluntarium maxima quadam copulatione fruendi re, quae bona iudicatur, communiter esse definerem".

love defined as the "sympathy or Antipathy, which is to be seene in animate & inanimate creatures", such as stones tending naturally to move downwards. Sensible love is "that of brute beasts". "The third kinde", Burton says, "is *Amor cognitionis*, as *Leon* calls it, Rationall loue, *Intellectiuus amor*, and is proper to man, on which I must insist. This appears in *God*, *Angels*, *Men*. God is loue itselfe, the fountain of loue."[69]

It is Ebreo's insistence on divine love, the "copulation" (*copulatio*) or union between the soul of man and God that Burton seems most drawn to in the section on religious melancholy. After citing the "mysticall song of *Solomon*" and Augustine on the eternal and unchanging beauty of God Burton adds that in the after-life we shall be "perfectly inamored [...] and loue him alone, as the most amiable and fairest obiect, our *summum bonum*, or chiefest good."[70] Citing Ebreo in the margin, he adds "This likewise should we haue now done, had not our will beene corrupted." The marginal comment reads, "Leone Ebreo. It is doubted whether human happiness ends in understanding or loving God." (*Leon Hebraeus. Dubitatur an humana foelicitas Deo cognoscendo an amando terminetur*). This could refer to various passages in the *Dialogi*, but probably the most relevant is this passage in Dialogue 1:

> The greatest happiness consists neither in the act of understanding God, which engenders our love of him, nor in the love which succeeds understanding, but only in the act of inner copulation, and the united and divine understanding of him, which understanding is to be considered the highest perfection of the created intellect: and this act is the last and blessed end of that intellect, and in that state deserves to be called divine rather than human.[71]

Burton adds to this his own translations from Ficino's *Commentarium*, including the opening exhortation of Oratio II.8, where Ficino beseeches his readers: "I exhort and beseech you, that you would embrace and follow this diuine loue with all your hearts and abilities [...]."[72] Burton however deflects Ficino's passage which moves on to discuss Plato's views on reciprocal love, and pushes it instead towards the "louing" Christian God, before slotting in a passage from Ficino's translation of Plotinus's *Enneads* I.6.7, which urges us to "forsake the kingdomes and Empires of the whole earth", in favour of divine beauty.[73]

What I hope this rather close examination of Robert Burton's working methods has shown is that the *habitus* of the general scholar in the seventeenth century trans-

---

[69] Burton, *Anatomy*, 383 [AOM, vol. 3, 13]. See Ebreo (1564), 56 verso-60 verso.

[70] Burton, *Anatomy*, 576–577 [AOM, vol. 3, 333–335]

[71] Ebreo (1564), 38 verso-39 recto: "In summa faelicitas nec in actu Dei cognoscendi, qui eius amorem nobis ingenerat, nec in amore, qui huic succedit cognitionis, sed solùm in actu copulationis intimae, & vnitae diuinaeque illius cognitionis consistit, quae quidem cognitio summa perfectio intellectus creati esse censetur: & actus ille est vltimus, atque beatus ipsius finis & in eo statu intellectu noster diuinus potius, quam humanus vocari meretur." Cf, also earlier in the same dialogue, 36 recto-verso, where Philone tells Sophia: "De his autem proprius actus ipsius faelicitatis in cognitione, an in amore Dei reponatur, maxima fuit inter sapientes controuersia."

[72] Burton, *Anatomy*, 578 [AOM, vol. 3, 336]. Cf. Ficino, *Opera*, 1327: "Vos autem amici hortor & obsecro ut amorem rem profecto diuinam totis uiribus complectamini [...]." (Jayne, 49–50: "I urge and beg you all, my friends, to imbrace immediately this love, a thing certainly divine, with all your strength.")

[73] Burton, *Anatomy*, 578 [AOM, vol. 3, 336]. Cf. Ficino (1580), 56.

forms and shapes the materials that it works upon. John Nelson saw 1600 as a watershed beyond which Platonic love theory made less and less sense, because "the problems of the succeeding century were very different from those of the century which ended with Bruno's demise." He saw the philosophy of Galileo, Bacon and Descartes as definitive here, whereas I would suggest that the afterlife of Renaissance love theory in the milieux of general scholarship in Europe would make a worthwhile study. Burton draws together Renaissance theorists such as Ficino and Ebreo, but also later scholars working in medicine, like Vallès, or law, like Godeffroy, who had absorbed this love theory and interwoven it into their works. Drawing on sources as diverse as Piccolomini's weighty philosophical tome on civil science, to Liburnio's diminutive collection of Platonic commonplaces, Burton constructs his own copious discourse. Like his fellow Christ Church scholar Casaubon, Burton saw general learning as something which was the duty of a divine. "As a christian, & a Divine, I write to yow, who are a Diuine", Casaubon wrote,

> I am very well content as Plato woulde haue it, that nothing should be accounted learninge but what doth tend to the maine end. As a christian therefore it doth concerne mee to be well satisfied my selfe, & as a Diuine it is part of my charge to be able to satisfie others [...].[74]

The *Anatomy of Melancholy* concerns itself a great deal with medical matters, and Burton clearly saw general scholars working in other professions as fellow travellers, but the "maine end" of the *Anatomy*, I think, has much to do with his sense of himself as a Doctor of Divinity, and he seeks to give his readers spiritual succour as well as medical cures.[75] In the sections on love melancholy and religious melancholy, conceptions of divine love are drawn from Renaissance neoplatonists and from Church Fathers, to produce a Platonized *amor Dei*, which would have been recognised by pre-Tridentine Italian scholars as a viable synthesis. As with other general scholars of the seventeenth century, the work is more than the sum of its juxtaposed parts: the *Anatomy of Melancholy* is not so much a "Rapsody of Rags", as the work of a divine who sought to console those suffering under what he saw as the crushing weight of mortality. A comprehensive history of the reception of Renaissance love theory after 1600 remains to be written, and when the many tributaries of general scholarship have been reassembled I have no doubt that a very different picture of Early Modern Platonism will emerge.

Bibliography

Blair, Ann M. 2010. *Too Much to Know: Managing Scholarly Information Before the Modern Age*. New Haven: Yale University Press.

Brann, Noel. 2002. *The Debate over the Origin of Genius During the Italian Renaissance: The Theories of Supernatural Frenzy and Natural Melancholy in Accord and in Conflict on the Threshold of the Scientific Revolution*. Leiden: Brill.

[74] Casaubon, *Generall Learning*, 92–93.

[75] On Burton's mixture of medical and religious imperatives see Lund (2010).

Bravo, Benedetto. 2006. *Critice* in the Sixteenth and Seventeenth Centuries and the Rise of the Notion of Historical Criticism. In *History of Scholarship: A Selection of Papers from the Seminar on the History of Scholarship Held Annually at the Warburg Institute*, ed. Christopher R. Ligota and Jean-Louis Quantin, 135–195. Oxford: Oxford University Press.

Burton, Robert. 1628. *The Anatomy of Melancholy. What It Is, with all the Kinds, Causes, Symptoms, Prognostickes & Severall Cures of It. In Three Partitions, with Their Severall Sections, Members and Subsections. Philosophically, Medicinally, Historically Opened and Cut Up by Democritus Junior. With a Satyricall Preface Conducing to the Following Discourse. The Third Edition, Corrected and Augmented by the Author*. Oxford: Iohn Lichfield, for Henry Cripps.

Burton, Robert. 1989–2000. In *The Anatomy of Melancholy*, ed. Thomas C. Faulkner, Nicolas K. Kiessling and Rhonda L. Blair, 6 vols. vols. 1 and 3, Oxford: Clarendon Press.

Casaubon, Meric. 1638. *A Treatise of Vse and Custome*. London: Printed by I.L.

Casaubon, Meric. 1668. *Generall Learning*, ed. Richard Serjeantson. 1999. *Generall Learning: A Seventeenth-Century Treatise on the Formation of the General Scholar by Meric Causaubon*. Cambridge: RTM Publications.

Ebreo, Leone. 1535. *Dialogi d'amore di Maestro Leone Medico Hebreo*. Rome: Antonio Blado d'Assola.

Ebreo, Leone. 1564. *Leonis Hebraei Doctissimi, atque Sapientissimi viri, De Amore Dialogi tres, Nuper a Ioanne Carolo Saraceno purissima, candidissimaque Latinitate Donati*. Venice: Apud Franciscum Senensem.

Erasmus, Desiderius. 1511. *De ratione studii ac legendi interpretandique auctores libellus aureus*. Paris: G. Biermant.

Fernández, Enrique. 2006. *Pornoboscodidascalus Latinus (1624): Kaspar Barth's Neo-Latin Translation of Celestina*. Chapel Hill: University of North Carolina Press.

Ficino, Marsilio. 1532. *Omni Divini Platonis Opera translatione Marsilii Ficini*. Basel: Officina Frobeniana.

Ficino, Marsilio. 1580. *Plotini Platonicorum facile coryphaei operum philosophicorum omnium libri LIV. in sex enneades distributi. Ex antiquiss. codicum fide nunc primum Græce editi, cum Latina M. Ficini interpretatione et commentatione*. Basel: Ad Perneam Lecythum.

Ficino, Marsilio. 1576. *Commentarium in Convivium Platonis de Amore*. In *Marsillii Ficini Florentini Opera* […] *omnia*. Basel: Henricus Petrus.

Ficino, Marsilio. 1944. *Marsilio Ficino's Commentary on Plato's Symposium. Commentarium* […] *in convivium Platonis de amore*. (trans. Sears Reynolds Jayne). Columbia: University of Missouri.

Godeffroy, Pierre. 1552. *Petri Godofredi Carcasoniensis, Iureconsulti, Procuratoris Regij in fide, Dialogus de amoribus, tribus libris distinctus*. Lyons: Thibaud Payen.

Kraye, Jill. 2002. Eclectic Aristotelianism in the Moral Philosophy of Francesco Piccolomini. In *La presenza dell'aristotelismo padovano nella filosofia della prima modernità. Atti del Colloquio internazionale in memoria di Charles B. Schmitt (Padova, 4–6 settembre 2000)*, ed. Gregorio Piaia, 33–56. Rome/Padua: Antenore.

Liburnio, Niccolò. 1537. *Le motte et diverse virtu delli savi antichi da Greci & latini auttori in volgar sermone […] tradotte*. Venice: M. B. Zanetti.

Liburnio, Niccolò. 1551. *Diuini Platonis Gnomologia antea duobus libris distincta, nunc per locos communes perquàm apposite digesta*. Lyons: Apud Ioan. Tornaesium et Guilelmum Gazeium.

Liburnio, Niccolò. 1556. *Divini Platonis Gemmae, sive illustriores sententiae, ad excolendos, mortalium mores, & vitas rectè instituendas, à Nicolao Liburnio Veneto collectae. Quibus recèns accesserunt aliae sententiae, ex eodem platone depromptae*. Paris: Guillaume Cavellat.

Lund, Mary Ann. 2010. *Melancholy, Medicine and Religion in Early Modern England: Reading the Anatomy of Melancholy*. Cambridge: Cambridge University Press.

Nelson, John Charles. 1958. *Renaissance Theory of Love: The Context of Giordano Bruno's Eroici Furori*. New York: Columbia University Press.

Peirone, Luigi. 1968. *Tradizione ed irrequietezza in Nicolò Liburnio*. Genoa: San Giorgio.

Piccolomini, Francesco. 1583. *Vniversa Philosophia De Moribus a Francisco Piccolomineo Senense, Philosophiam in Academia Patavina e prima sede interpretante, Nunc primùm in decem Gradus redacta, & explicata*. Venice: Franciscus de Franciscis.

Piccolomini, Francesco. 1595. *Vniversa Philosophia De Moribus a Francisco Piccolomineo Senense, in Academia Patavina Philosopho primum in decem gradus redacta*. Geneva: Eustathius Vignon.

Pistorius, Johannes Nidanus. 1587. *Artis Cabalisticae: hoc est, Recondiate Theologiae et philosophiae scriptorum: Tomus I*. Basel: Sebastianus Henricpetrus.

Poppi, Antonio. 1997. *L'etica del Rinascimento tra Platone e Aristotele*. Naples: La Città del Sole.

Rohling, Detlef. 2012. *Omne scibile est discibile: eine Untersuchung zur Struktur und Genese des Lehrens und Lernens bei Thomas von Aquin*. Münster: Aschendorff Verlag.

Vairus, Leonardus. 1583. *De fascino libri tres. In quibus omnes fascini species et causae optima methodo describuntur, et ex philosophorum ac theologorum sententiis scitè et eleganter explicantur: nec non contra praestigias, imposturas, illusionesque daemonum, cautiones et amuleta praescribuntur: ac denique nugae, quae de iisdem narrari solent, dilucidè confutantur*. Paris: Nicolas Chesneau.

Vallès, Francisco. 1591. *Controuersiarum medicarum et Philosophicarum libri decem, Francisci Vallesij Couarruuiani edito tertia, ab auctore denuo recognita & aucta*. Lyon: Apud Haeredes Gvilielmi Rovillii.

Wecker, Johann Jacob. 1562. *Medicae Syntaxes Medicinam Vniversam ordine pulcherrimo complectentes, ex selectioribus medicis, tam Graecis quam Latinis & Arabibus collecta et concinnata*. Basel: Jacobus Parcus for Nicolaus Eiscopus.

Zara, Antonio. 1615. *Anatomia ingeniorum et scientiarum sectionibus quatuor comprehensa*. Venice: Ambrosius Dei et Fratrum.

Zonta, Giuseppe. 1975 [1910]. *Trattati d'amore del Cinquecento*, reprint ed. Mario Pozzi. Rome: Laterza.

Chapter 4
The Critique of Scholastic Language in Renaissance Humanism and Early Modern Philosophy

Lodi Nauta

Abstract This article studies some key moments in the long tradition of the critique of scholastic language, voiced by humanists and early-modern philosophers alike. It aims at showing how the humanist idiom of "linguistic usage," "convention," "custom," "common" and "natural" language, and "everyday speech" was repeated and put to new use by early-modern philosophers in their own critique of scholastic language. Focusing on Valla, Vives, Sanches, Gassendi, Hobbes, and Leibniz, the article shows that all these thinkers shared a conviction that scholastic language, at least in its more baroque forms, was artificial, unnatural, uninformative, ungrammatical, and quasi-precise. The scholastics were accused of having introduced a terminology that was a far cry from the common language people spoke, wrote, and read. But what was meant by "common language" and such notions? They were not so easy to define. For the humanists, it meant the Latin of the great classical authors, but this position, as the article suggests, had its tensions. In the later period it became even more difficult to give positive substance to these notions, as the world became, linguistically speaking, increasingly more pluralistic. Yet the attack on scholastic language continued to be conducted in these terms. The article concludes that the long road of what we may call the democratization of philosophical language, so dear to early-modern philosophers, had its roots – ironically perhaps – in the humanist return to classical Latin as the common language.

Introduction

Throughout the ages philosophers have questioned our common sense view of the world, claiming that the world is not as it appears to be. This claim is almost the philosopher's *raison d'être*. Philosophy thrives on the idea that there is a deep structure behind the phenomena we perceive and claim to know – matter, substance,

L. Nauta (✉)
Faculty of Philosophy, Rijksuniversiteit Groningen, Groningen, The Netherlands
e-mail: l.w.nauta@rug.nl

C. Muratori, G. Paganini (eds.), *Early Modern Philosophers and the Renaissance Legacy*, International Archives of the History of Ideas Archives internationales d'histoire des idées 220, DOI 10.1007/978-3-319-32604-7_4

powers, forms, Ideas, and so on; it would amount to naive empiricism to think that what we see is all there is to know, or would be enough to justify our claims to knowledge. As Robert Pasnau has rightly observed: "Over the centuries, it has been practically definitive of the philosopher's job to subject naive empiricism to a withering critique. Indeed, stages in the development of philosophy can be measured in terms of how far they depart, and in which direction, from our natural but naive pre-theoretical orientation toward empiricism."[1] This withdrawal from naive empiricism has often (though not necessarily so) gone hand in hand with the development of a language that likewise departs from the common way in which people speak about the world. Like scientists, grammarians, lawyers, theologians, and practitioners of other professions, philosophers too developed their own technical language, sometimes staying fairly close to the common parlance of the time but often introducing more technical, abstract, and formal terminology, needed, so it has always been thought, to refer to deeper levels of reality.

Scholastic Aristotelianism is a philosophical trend that scores high on both counts. The divergences between scholastic thinkers are immense but what these thinkers have in common is the conviction that an analysis of the world and of ourselves as knowing subjects and moral beings, while perhaps starting with what we daily perceive and think, will soon lead away from this common world, introducing all kinds of entities and corresponding vocabulary: form and matter, act and potency, universals, transcendentals, predicables, substantial and accidental forms, formal distinctions, intentions, species, active and potential intellect, categories, all kinds of distinctions in the analysis of language and argumentation, and so on – it makes reading scholastic authors philosophically immensely rewarding but often also very difficult and puzzling. What is true for almost any kind of theorizing is certainly true for the scholastic way of philosophizing: concepts require new concepts, and to clarify these new concepts still other concepts have to be introduced, and so on. The higher we come in this conceptual building the less we feel that we are still in the process of analyzing our initial object of study. It has become a game on its own, and even a highly sympathetic interpreter of scholastic thought such as Robert Pasnau must admit that "one risk this kind of analysis runs is that we will end up not just up to our necks in metaphysical parts, but positively drowning – that once we begin to postulate such entities, we will be forced to postulate infinitely many more." We might think "that nothing of any explanatory value has been achieved by all this philosophizing."[2] It is indeed "the timeless complaint made of all philosophy."

It was certainly a complaint voiced passionately by Renaissance humanists and early modern philosophers alike. They indeed thought that the scholastics had erected a conceptual building that was out of tune with its function and purpose. In this article I will study some moments in this long tradition of language critique. There are several reasons why this is an interesting theme worthwhile to explore. First, the critique of philosophical language is a clear example of continuity between Renaissance and early modern thinkers: not only were early modern thinkers

[1] Pasnau (2011), 115.

[2] Pasnau (2011), 211 and 210 (on Scotus's analysis of the inherence of accidents in a substance).

indebted to scholastic traditions – a historical fact widely acknowledged – but also to Renaissance humanism, a debt that is far less often recognized and appreciated by modern scholars. This critique of scholastic language is one of the factors that contributed to the demise of scholastic Aristotelianism, hence it is interesting to study how it developed in the period between, let's say, Petrarch and Leibniz. When Leibniz, for instance, claimed that in language "Der Gebrauch ist der Meister" ([linguistic]usage is the master) he repeated in the same words a common humanist point that philosophical language should follow the common language of the people though, as we will see, what was understood by "common language" shifted over the years.

This also suggests another reason why this is an interesting theme. Historians of philosophy have often dismissed the humanist critique of scholastic language as merely polemical and rhetorical, and as philosophically superficial and ill-informed.[3] Such a dismissal is understandable given the sometimes highly polemical nature of the humanists' invectives – a genre that was of course not meant to engage in deep philosophical argument. But behind the polemics a serious and age-old philosophical question looms large: What kind of language should be used in philosophy (and indeed in any kind of intellectual pursuit, including science)? Should we use the common language of the people or is this far too imprecise and should we develop our own technical vocabulary? If we plea for the first, then we will have to make clear what we mean by "common" (or the "ordinary"), "the people," "common usage" and so on, and also why this usage should be normative in our philosophizing. If we accept a technical language on the other hand, we must make clear, ideally, why common language does not suffice, and what the relationship is between this technical terminology and our common, non-technical language.[4] Seen from this perspective, the criticisms leveled against the scholastics raise philosophically pertinent and wide-ranging questions.

In what follows I can discuss only a small number of thinkers, from Lorenzo Valla to Leibniz, and many interesting and even major figures have to be left out of the picture (e.g. Petrarch, Agricola, Pico, Ramus, Cardano, Campanella, and Descartes). Nor can I pay attention, within the scope of this article to related debates, for instance, on Ciceronianism, or on the *questione della lingua* (on Latin versus the vernacular), or on later seventeenth-century attempts to construct a universal language (e.g. George Dalgarno and John Wilkins), let alone to debates and controversies that informed their positions. But the authors I will discuss sometimes refer to each other, and the earlier ones were among the (admittedly) many authors read, consulted, or at least known to early modern philosophers such as Gassendi, Hobbes, and Leibniz. Thus Valla is mentioned by Vives, and Vives by Sanches, and these Renaissance thinkers helped Gassendi, as he tells us, to break away from the Aristotelian-scholastic tradition. Between Valla and Hobbes some affinities have been detected. And omnivorous reader as he was, Leibniz had read many Renaissance authors, including Valla, and he had edited the work of the humanist Mario Nizolio,

[3] For discussion see Nauta (2009), 211–212.

[4] For a modern discussion see Hanfling (2000).

who in his turn was indebted to, among others, Valla, Agricola, and Vives. There are no straight lines of influence in history – the concept of influence is of course notoriously difficult to define – but a recognizable track can be discerned (or, at least, can be cleared) in the forest of Renaissance and early modern texts.

Lorenzo Valla (1406–1457)

Before Lorenzo Valla came on the scene in the 1430s, humanists such as Petrarch, Salutati, and Leonardo Bruni had already complained about the Latin of the scholastics, a language that they found ugly, ungrammatical and a far cry from the beautiful Latin that they wanted to revive and reinstall as language for communication, and literary and scholarly pursuits.[5] Their critique was primarily of a rhetorical kind. When Leonardo Bruni, for instance, criticized the scholastic translator of Aristotle's *Ethics* (whom we know was Robert Grosseteste, an identification unknown to Bruni), he focused on the lack of beauty of the translation: Greek words had been left untranslated in the translation, Latin words were used with different meanings than they had in classical Latin, and the clumsy style did not match Aristotle's copious and eloquent style.[6] A good translator must know both the source and the target language very well, a requirement the medieval translator clearly did not meet. Bruni's harsh words provoked a response from bishop Alfonso of Cartagena who defended style and terminology of the medieval translation. Though being primarily a debate on translation, it addressed the issue of philosophical language: should we give priority to rhetorical eloquence or to technical precision? All these categories were matters of contention, and Bruni would not recognize the validity of the opposition, claiming precision and exactness for his rhetorical approach, concerned with using words in the right context with their right meaning.

This debate on the language of philosophy was given powerful though controversial expression by Lorenzo Valla, who in his *Dialectical Disputations* subjected some core notions of Aristotelianism to a withering critique. While he presented himself as an orator, Valla was concerned with semantic precision rather than with the beauty of style: the term *elegantia* meaning first and foremost semantic precision.[7] Valla's profound studies of the Latin language and his vast reading in classical and post-classical works convinced him that the meaning of words and the use of grammatical constructions can be learnt only by careful observation of linguistic practice, that is, how classical authors had actually used language. Meaning is to be determined by linguistic usage (*consuetudo*), and for Valla this meant the usage of the great authors, the *auctoritates*, roughly from Cicero to Quintilian. It was during

[5] The literature is vast; for some excellent general works see Seigel (1968); Witt (2000); Rummel (2000); Wels (2000).

[6] Bruni in Griffiths et al. (1987), 213–229; Botley (2004), 41–62.

[7] Marsh (1979), 101–103.

these two centuries that Latin reached its peak, and while it continued to be used far and wide and for many centuries to come, Valla saw a gradual decline in the knowledge of good Latin, going downhill with Boethius's philosophical Latin, not to speak of what Valla regarded as the barbarous gibberish of the medieval scholastics.[8] This had disastrous effects on the arts and sciences, and especially for philosophy and theology where everything depends on words.[9] This is a prominent theme of the *Dialectical Disputations* in which Valla used linguistic usage as one of his main principles to attack Aristotelian-scholastic philosophy and logic. Thus, from a predominantly linguistic point of view, he criticized the ten Aristotelian categories, the six transcendental terms, important Aristotelian distinctions such as matter/form and act/potency, as well as what he thought the useless and abstract logic of the scholastics that had nothing to do with how people actually spoke, argued, and reasoned.

We may group Valla's criticisms in the following categories[10]:

(i) *Ungrammatical terms*. As is well known, Valla rejects terms such as "entitas," "haecceitas," "identitas," "quiditas," "iditas," "reitas" and "perseitas," since they are incorrectly formed. They cannot be formed from substantives such as "ens" and from pronouns such as "quid," nor from adjectives (with some exceptions).

(ii) *Superfluous terms*. While perhaps grammatically correct, many scholastic terms, especially those standing for categories and transcendentals, are superfluous. Transcendental terms such as "something," "one," "true," "good" and "being" are superfluous; "something" is nothing but "a certain thing" (*aliqua res*), "one" can be reduced to "one thing," "true" to "a true thing," "good" to "a good thing" and "being" to "that thing which is (*ea res quae est*)." Likewise, many of the nine accidental categories of Aristotle are superfluous and can be reduced to quality and action. Such qualifications as size, relationship, position and time do not differ from qualifications that refer to qualities such as white or smart: "big," "brother," "armed,", "in the house," and so forth all qualify a person or thing; from a grammatical point of view they are essentially qualitative terms. Valla's basic assumption seems to be that the categories should reflect or point to things in the world, and he therefore has no need for other categories than substance, quality and action, referring to a thing, how it is qualified and what it does or undergoes. Clearly, the grammatical categories of noun, adjective and verb lie behind these ontological categories.

(iii) *Words taken out of context*. Valla is in particular sensitive to this point. It is, for instance, an abuse of words to say that the senses are "being acted upon (*pati*) by an object," or that the soul is moved or is self-moving, or that inanimate

[8] Moss (2003), 36–37; Camporeale (1972), 181–182; Nauta (2007), 195–198.

[9] "Omnis enim huiusmodi questio, qua se philosophi theologique disputando torquent, de vocabulo est;" Valla (1982), 405.

[10] My examples come from Valla (2012), vol. 1. 54–62 (*haecceitas* etc.), 18–36 and 62–70 (transcendental terms), 276 (*pati*), 88 (*materia*), 270 (*prior*), 32 (*one*), 240 (*empty*); vol. 2, 18–142 (markers), 126–142 (modality); more examples are discussed in Nauta (2009), *passim*.

things have a final cause. He also thinks one cannot apply the matter/form distinction to God nor to the sun, with which God is compared: both God and the sun should be said to have an essence plus the qualities vibration, light and heat. (Valla almost takes the analogy between the Sun and the Trinity literally, speaking of "persons" of the Sun's essence.) He criticizes the application of "prior" and "prius" to a number of expressions found in scholastic literature such as "prior and posterior in nature," "the whole is prior to the part," "genus is prior to species." Many more examples could be given.

(iv) *Arbitrary restriction on the meaning or application of a term.* Valla's criticized, e.g., Aristotle's statement that one is not a number but the principle of number. Similarly, to say that a vessel can never be "empty" since there is always air in it, is an absurd restriction on the terms "full/empty" and "place."

(v) *Oversimplification and arbitrary restriction on a range of words.* Closely related to the previous point is Valla's criticism of the rather arbitrary restriction to a limited set of words in scholastic philosophy. Scholastics reduce the markers of quantity and quality to only a few, namely "all," "some," "none," and "no one," while Latin has a far richer arsenal of such words. Similarly, scholastics usually treat only the following six terms as modals: "possible," "impossible," "true," "false," "necessary," and "contingent." But again Latin is much more resourceful in expressing modality.

Valla thus aims at showing how Latin words – nouns, verbs, pronouns, and so on – were used and hence should properly be used, not only in literary studies and our own writings, but also in philosophy and other intellectual and literary pursuits. What he suggests then is that, once we create our own language or use words out of their "normal" context, we get a distorted picture of reality, that is, we will be searching for referents of these terms, or we will raise questions only because we have taken a metaphor literally or because we have applied a term outside its common domain. Valla's humanist assumption is of course that classical Latin should be our yardstick, because this is, according to him, the common, natural language in comparison to which post-classical forms of Latin and *a fortiori* the jargon of the scholastics can only appear as corrupt, depraved, distorted, unnatural, and artificial – words that abound in Valla's writings.[11] For us it is difficult to regard classical Latin as a common, normal, let alone ordinary language but, as I have pointed out elsewhere, this is how Valla indeed regarded it, especially when he compared it to the Latin of the scholastics. It was certainly the Latin of great authorities such as Cicero and Quintilian that constituted the norm of linguistic usage, but Latin had spread far and wide and had been used over a long period of time, so that for a humanist it could easily be regarded as the normal language, in opposition to scholastic Latin that was considered to be technical, artificial and "unnatural."

Valla's programme of ontological reduction was inspired and driven by his grammatical-rhetorical approach. It inspired – directly or indirectly – many later humanists as well as early modern philosophers such as Gassendi, Hobbes, and Leibniz.

[11] Nauta (2009), 274–279.

Juan Luis Vives (1492–1540)

One of the most influential humanists whose position who was much indebted to Valla is Juan Luis Vives. In his famous letter *Against the Pseudodialecticians* he takes up several themes we have already met in Valla: the importance of linguistic usage, the appeal to the common language of the people as a *sine qua non* for effective communication, the rejection of technical jargon, and so on. Vives had studied with the Parisian logicians, so more than Valla he knew what he was talking about when he polemically analysed late-scholastic sophisms. Just as for Valla, for Vives classical Latin was a storehouse of learning and erudition. To create one's own jargon, as the philosophers do, is to make communication impossible. Communication requires the use of one language, and access to one common source of learning[12]:

> if we all profess a Latin logic, words will have the meaning established by Latin practice and usage, not our own. It is unbecoming and foolish in Latin logic to use Getic or Sarmatian words, or not even those, but words belonging to no nation, which we have conjured up ourselves. Indeed, I should very much like to hear from these men: if they were to teach dialectic in Spanish or French, which is as feasible as in Latin or Greek, would they make up rules as they please rather than take them from the structure of the language itself?

The rules which the dialecticians derive from their own brand of Latin are not necessarily valid for other languages. But if we choose to use Latin in logic and communication, we should not make up the rules ourselves nor assign meanings arbitrarily, but rather stick to convention and linguistic usage.

> But if they would not be willing to accept rules from conventional discourse to teach logic in other languages, why do they want to exercise this tyranny over the language of the free Roman people, and force it to accept rules of speech from men as uncultivated and barbarian as themselves?[13]

To this question philosophers often give the answer that they speak "rigorously" (*de rigore*): making a distinction between "good," "common" or "everyday sense" and an "exact," "rigorous," or "philosophical" sense, they thus create room for themselves to uphold the truth of their claims (mainly in logic) that in everyday speech would be false, e.g. "You are not a man." But Vives thinks this self-acclaimed freedom is false. For him rigor can only mean "this very appropriateness, this distinct, innate, and genuine force, the right and true meaning of Latin discourse," which can be learnt only from good Latin authors.[14] We see here the same conflation of the learned language of the great authors with the common language of the people (the "free Roman people" in the quotation just given) as we saw in Valla.[15] But this conflation as well as the identification of "rigor" with "the right and true meaning of

---

[12] Vives (1979), 67. See Nauta (2015).

[13] Ibid., 67–69.

[14] Ibid., 69–71.

[15] On the distinction between exact and common manner of speaking, see Valla (2012), 266 (*populus an philosophus*); Nauta (2009), 108.

Latin discourse," is problematic. The uneducated masses, Vives says, sometimes use expressions that, strictly speaking, are not correct:

> Every language has its own appropriateness of speech, which the Greeks call *idioma*. Words have their own meaning, their own force, which the uneducated masses sometimes misuse.[16]

Vives then quotes from Cicero who had given some examples, also referred to by Valla, e.g. that the common people say "the vessel is empty," while strictly speaking this is not true; it still contains air. Valla was adamant in holding that everyday sense rather than the exact sense of philosophers should be the norm, and Vives wants to maintain the same, yet he also admits that "the better educated make some concessions to the common people in the use of language; among themselves they think and speak in a different manner."[17] (As, for instance, Bishop Berkeley was later to say: thinking with the learned, and speaking with the vulgar.) But to distinguish this situation – in which a normal word such as "empty" does not describe the facts correctly – from the jargon of the scholastics, he adds "though not to any great extent and mostly on abstruse and philosophical subjects which the people would not be in a position to know as precisely as the philosophers understand them." This seems to support the idea that philosophers may have their own "exact" way of speaking, but such freedom is apparently not allowed to scholastic logicians, who go far over the top in inventing rules of logic that allow them to say that "You are not a man" is, strictly speaking true (i.e. according to their rules).

Vives's position is somewhat ambiguous then. On the one hand, the common people sometimes misuse language, but the example of the vessel suggests that this misuse *is* actually not a misuse at all but rather the common way of talking (hence *consuetudo*), and moreover that this common way of talking *is* rigorous in Vives's sense of the word, namely proper and good Latin. Vives probably wants to make the point that "common language" does not always capture the facts right (the vessel is not really empty), though for the purpose of communication, in ordinary contexts, it is adequate because it captures our common sense feeling or perception of the matter (we see that the vessel is empty). Several issues seems to get mixed up here then.

Apart from the in-crowd character of scholastic language, it is also essentialistic, Vives thinks. It seems to lay bare the deep structure of reality that Vives, as a moderate sceptic, believes is impossible to know. We cannot know the essences of things: "what knowledge we have gained can only be reckoned as probable and not assumed as absolutely true."[18] All we can do is observe carefully the outer aspects of things (qualities, actions, their similarities with other things etc.), and from a careful comparison establish general patterns and laws, which however must remain provisional. Many scholastics would agree with this, as they also turned away from substance and essence, moving toward an examination – or at least a defense of such an

---

[16] Vives (1979), 69.

[17] Ibid.

[18] Vives (1971), 166–67. See Casini 2009.

approach – of sensible qualities.[19] But in Vives this move is closely connected to a reform of language. As he writes, in metaphysics the philosopher should have a good knowledge of language, for it is "the common meaning of words" rather than the technical terminology of the scholastics that should be followed (*communis verborum usus*; *sensus communis*; *verbis de vulgo sumptis*). Strange as it may seem, metaphysics is for Vives a discipline that must take its starting-point from common usage, laying bare (*enucleare*) the meaning of individual words, since "the rise and disappearance of nearly all problems in the disciplines are dependent on the way we phrase them in language."[20] Indeed, language is a shaping force: "the power of almost all knowing and understanding lies in words; for in words are perceptions (*sensa*) registered, and all that takes place in the mind and in thought is expressed in words."[21] When properly combined words will give us, "as far as possible," a description (*explicatio*) of the nature of whatever thing. And for him, as for Valla, it was classical Latin that serves us best in expressing our view of the world of observable phenomena.

Francisco Sanches (1550/51–1623)

The same intrinsic connection between a sceptic-empirical outlook and a critique of scholastic language can be found in the work of the Portuguese scholar, Francisco Sanches, who mentions Vives a few times. Sanches' *Quod nihil scitur*, first published in 1581, is well known as a skeptical treatise that aims at refuting what Sanches thinks are the pretentions of philosophers to arrive at the truth of things.[22] He is in particular very critical of the Aristotelian theory of scientific demonstration. It is a system based on definitions and demonstrations, not on observation of the facts, on *res*: "for other sciences are based on facts, whereas this one is a subtle invention, and quite useless – or rather most harmful inasmuch as it distracts me from the observation of facts and keeps me engaged in the study of itself."[23] Their system of logic prevents them seeing the facts (*res*): "They know nothing but a multitude of syllogistic inferences – no facts at all."[24] This is a running theme in Sanches, who was influenced by Galenic writings in which find we find a strong emphasis on the observation of facts and the importance of ordinary life experience[25]; hence, unsurprisingly, the word "*res*" abounds in Sanches's work. As he says in the address to the reader: "I would address myself to those who, 'not bound by an oath of fidelity to any master's words', assess the facts for themselves, under the guidance of

[19] Pasnau (2011), 115–134, and 634–635.

[20] *De prima philosophia* I, in Vives (1782–90; repr. London, 1964), vol. 3, 193.

[21] Ibid.

[22] On Sanches's debt to humanists such as Vives and Erasmus, see Limbrick in Sanches, Francisco (1988), 28–36, but see also Howald's cautious remarks in Sanches, Francisco (2007), ciii, and Lupoli (2009).

[23] Sanches, Francisco (1988), 103 (Latin)/186 (English).

[24] Ibid., 104/189.

[25] On ancient empiricism, see Frede (1987). On Sanches's affinity to it, see Caluori (2007).

sense-perception and reason."[26] This indeed could have been the words of Valla and Vives too. And like them, Sanches sees a close connection between the language of the philosophers and their claims to truth and certainty[27]:

> They distort words from their commonly accepted meanings (*a propria significatione*), and corrupt them in order to have another language of their own, quite different from their mother-tongue, yet the same. And when you go to them in order to learn something, they change the meanings of the words you had hitherto employed, in such a way that these no longer denote the same objects – that is, objects in the natural world – but instead the objects that they themselves have invented.

Greek terms such as "entelecheia" and Latin terms such as "essentia," "quidditas" and "corporeitas" have no meaning at all, he says, and "can be neither understood nor explained by anyone – much less rendered into everyday speech (*sermo vulgaris*), which is accustomed to assign only to *real* things (but not to invented things) names of their own."[28] The idea behind such an alleged exclusivity of Latin or Greek is that such a language has a particular efficacy to express the nature of things. Sanches rejects this idea, suggesting that languages, including Greek and Latin, have changed continuously: "Therefore there lies in words no power to explain the nature of things, except that which they derive from the arbitrary decision of him who applies them."[29] And it is popular usage, as he says elsewhere in his treatise, that determines meaning : "the meanings of words appear to depend, for the most part or wholly, on popular usage (*a vulgo*); and here, accordingly, is where we must look for them; for who but the populace (*vulgus*) taught us how to speak?"[30]

Sanches realizes however that there is no fixity and stability in the common language of the people either. Every question, every issue depends on words,[31] and words do not have fixed meanings: whatever meaning we give to words, these words will never be able to disclose the nature of things. In his commentary on Galen's *De differentiis morborum* he remarks that "Galen was right in striving to take the meaning of health and sickness from common linguistic usage (*a communi loquendi usu*); for linguistic usage (*consuetudo*) as well as the will (*voluntas*) of people gives speech its signification (*vim*)."[32] But this also represents a serious problem for the sciences (*scientiis*), Sanches continues

> since the populace (*populus*) does not use words appropriately nor does it understand the things referred to by those words, for while it speaks of health, it does not know at all what it is. Hence medical doctors, who understand or nearly understand the matter (*res*), are forced to use words in a different way than the populace does, or to use words with a different meaning (*ad alia significata transferre*) or, even, impose new meanings (*nova imponere*) after consultation of Galen, Cicero and other writers.[33]

---

[26] Sanches, Francisco (1988), 92/168.

[27] Ibid., 119/216–217.

[28] Ibid.

[29] Ibid., 121/219.

[30] Ibid., 97/177.

[31] Ibid., 101/183 ("almost every enquiry is about a name"); cf. 95–96/174; 97/177, and elsewhere.

[32] Ibid., 177 n. 34; my translation.

[33] Ibid.; my translation.

It seems then that what is forbidden to philosophers – inventing new words, or using words in a different way than common people do – is allowed to the medical doctors and practitioners of other crafts and professions, but of course the latter follow the correct authorities, Cicero and Galen.

The passage just cited complicates matters somewhat and shows how flexible (or ambiguous) phrases such as "common linguistic usage" and "the populace" are. Standing in the tradition of Galen and ancient empiricism, Sanches seems to defend ordinary life experience and the experience of professionals based on observation; hence common language should be the rule. But on the other hand, compared to medical professionals, the "common people" (*vulgus*) are not always a reliable guide to understanding the phenomena. But in spite of what Sanches suggests in this passage, namely that Cicero is an important authority for the use of words in a profession such as medicine, in *Quod nihil scitur* he dissociates himself from the rhetorical and polished style of Cicero and his followers: "You are not to look in me for an elegant, polished style (…) If that is what you want, seek it from Cicero, whose function it is; I shall speak prettily enough if I speak truly enough."[34] Elegant language, he says, "is seemingly for rhetoricians, poets, courtiers, lovers, harlots, pimps, flatterers, parasites, and people of that sort, for whom elegant speech is an end in itself."

Thus next to the similarities already mentioned, we also see clear differences between Sanches and humanists such as Valla and Vives: while the humanists still thought of a close intrinsic connection between elegance on the one hand and clear and truthful language on the other, Sanches no longer believes in such a bond.[35] Nor does he regard Latin (and hence the linguistic usage of the great Latin authorities) as a *sine qua non* for philosophizing and doing science. We need a sober, clear, unpolished language, avoiding both technical jargon and rhetorical elegance. This view was endorsed by many early modern philosophers: with the rise of the new science sober, unaffected language became an asset for most philosophers.

Pierre Gassendi (1592–1655)

Gassendi is a good example of an early modern philosopher whose views are indebted, not only to ancient sources such as Sextus Empiricus and Diogenes Laertius, but also to his humanist predecessors; modern scholars have even called him a "mitigated humanist."[36] He himself is quite explicit about his debts, stating that it was the reading of Renaissance authors such as Vives, Charron, Ramus, and Gianfrancesco Pico della Mirandola, that made him realize that "there was nothing

---

[34] Ibid., 171 for this and the next quotation.

[35] Lupoli (2009), 151 sees even "that (intrinsically anti-humanistic) resetting pattern (…) of philosophical reflection which was to characterize the Cartesian or 'modern' approach to philosophy in the seventeenth century."

[36] Osler (2003), 41. On a comparison between the humanism of Valla and Gassendi see Joy (1987).

wrong in supposing that this sect [of the Aristotelians] was not necessarily correct in all matters just because most men approved of it."[37] They showed him the importance of the *libertas philosophandi*, which often meant, as a first step, the liberty to attack Aristotelian-scholastic philosophy, and this is what Gassendi does in his first major work, *Exercitationes Paradoxicae adversus Aristoteleos*. Covering much the same ground as his Renaissance predecessors, Gassendi criticizes Aristotelian philosophy (logic, physics, natural philosophy, metaphysics, ethics), saying in words that remind us of Valla's, that he selected "just those opinions which were, so to speak, foundational doctrines of the Aristotelians."[38] Referring also to Valla's program of ontological reduction, Gassendi attacks the Aristotelian categories, the transcendental terms, and rejects Aristotelian logic and theory of demonstration as "artificial" and useless. In fact, in the words of a modern scholar he seems to attack the Aristotelians for "asking philosophical questions" at all, which illustrates "the difficulty, for an early seventeenth-century intellectual, in articulating just what was wrong with scholastic philosophy, and what ought to replace it."[39]

What Gassendi was to propose as a replacement, namely a Christianized version of the Epicurean system, need not detain us here. More relevant, though hardly surprising, is the fact that Gassendi appealed to "the common and accepted manner of speaking (*communis et protritus loquendi usus*)" that we have to employ in philosophy.[40] Of course, this no longer implies a Ciceronian or classical style, and concerning his own "style and manner of expression," Gassendi says that he is "neither Ciceronian nor the least bit scholastic," favoring "an unaffected (*illaboratum*) prose style which flows spontaneously."[41] Elsewhere he praises ancient authors who knew how to draw the attention from their audience, combining the useful with the agreeable in a pleasant prose style.[42] We should not strive for a grand style, but philosophers who have claimed that "solecisms are praiseworthy, and are the gems of philosophers" are to be despised: a decent style (*honestus cultus*) fits the philosopher, and "even the abstruse things can be presented in a decorous way (*cum ornatu*)."[43] The wording is vitally important, Gassendi writes, and debates on matters always turn out to be debates on words; if we do not use the proper meaning of words (*sermonis proprietatem*) we end up making our own idiosyncratic speech, that is, philosophical jargon. Like Valla and Vives, he thinks that the so-called "rigor" of the philosophers can only be defined in terms of the common and accepted

[37] Gassendi (1972), 18. Cf. his Letter to du Faur de Pibrac, in ibid. 5. Vives is mentioned also in Gassendi (1658), III, 119. Cf. Murr (1992) and Maclean (2006), esp. 264–267 on *libertas philosophandi*.

[38] Gassendi, Pierre (1972), 26. Cf. one of the titles of Valla's *Dialectical Disputations*, *Retractatio totius dialecticae cum fundamentis universae philosophiae*.

[39] Pasnau (2011), 93–94.

[40] *Exercitationes paradoxicae adversus Aristoteleos*, in Gassendi (1658), III, 151B.

[41] Gassendi (1658), III, 103; trans. Gassendi, Pierre (1972), 27.

[42] Gassendi (1658), III, 110A.

[43] Ibid.

manner of speaking, which he seems to equate here with the Latin of authorities such as Cicero and Livy.[44]

Much has been written on Gassendi's philosophy and his skepticism, called by Richard Popkin "constructive scepticism" or "mitigated scepticism," which he says "represents a new way, possibly the closest to contemporary empirical and pragmatic methods, of dealing with the abyss of doubt that the crisis of the Reformation and the scientific revolution had opened up."[45] Popkin describes this position as the "realization that the doubts propounded by the Pyrrhonists in no way affected *la verité des sciences*, provided that the sciences were interpreted as hypothetical systems about appearances and not true descriptions of reality, as practical guides to actions and not ultimate information about the true nature of things." Here too, however, we might point to Renaissance authors, and in particular Vives who had already formulated a philosophy that combines a sceptical attitude towards knowledge of the essences of things with a pragmatic empiricism.[46] As we have noted, Vives often emphasizes the importance of careful observation and the risks of hasty conclusions, though he would not have shared Gassendi's love for Epicureanism.[47]

Thomas Hobbes (1588–1679)

Friend of Gassendi, brilliant stylist and author of one of the most beautiful philosophical prose works written in English, the *Leviathan*, Hobbes too had no patience with the Aristotelian-scholastic philosophy of the schools and their language. If scholastic philosophers were forced to translate their barbarous, "insignificant speech" into the vernacular of the common people, we would immediately see, Hobbes suggests, how ridiculous and nonsensical it is.[48] They hide their confusion, ignorance or downright stupidity behind a fog of incomprehensible and obscure Latin and Greek words. They call the Lord, e.g., with a Latin or Latinized word "*verbum*," which sounds impressive, but when translated into ordinary French, "parole," gives something rather absurd.[49] Just try, Hobbes says, to translate a title of a chapter from the work of Suarez into "any of the modern tongues, so as to make

[44] Ibid.: "Quod vero interdum respondent loquendum esse *ad rigorem*, prorsus non diffiteor: quando sic apposite, vel nescientes, nominant suam illam insipidam marcidamque frigiditatem. Certe si cum tanto rigore isti Latine loquuntur, parum est M. Tullius, vel T. Livius loquutus Latine." (110B). Gassendi also gives here an etymology of "res" that he may have derived from Valla ("res" from "reor, reris," or from "ratus, rata, ratum"); Valla (2012), 124.

[45] Popkin (2003), 125 for this and the following quotation. Osler (2003), 32 argues that Gassendi's voluntarism led him to deny essences and necessary connections.

[46] Cf. a similar position in Campanella's *Metaphysica*, as discussed by Paganini (2009).

[47] See e.g. Vives (1971), 31 and 125 (expressing a negative view of Epicurus).

[48] Hobbes (1994), 21; I also refer to the new and definitive edition of the *Leviathan*: Hobbes (2012), 60. For a comprehensive treatment see Isermann (1991). See also Leijenhorst (2002).

[49] Ibid.

the same intelligible; or into any tolerable Latin, such as they were acquainted withal, that lived when the Latin tongue was vulgar: 'The first cause does not necessarily inflow any thing into the second, by force of the essential subordination of the second causes, by which it may help it to work.'"[50] This is nonsensical speech, certainly not the normal speech of man: "the common sort of men [...] seldom speak insignificantly."

Such criticisms was commonplace by now, but Hobbes transcends his humanist predecessors in giving a rather detailed analysis of insignificant speech: "insignificant sounds," he says, are of two sorts: "One when they are new, and yet their meaning not explained by definition; whereof there have been abundance coined by schoolmen, and puzzled philosophers. Another, when men make a name of two names, whose significations are contradictory and inconsistent."[51] Examples of this latter kind are "incorporeal body," or "incorporeal substance." As he explains:

> For whensoever any affirmation is false, the two names of which it is composed, put together and made one, signify nothing at all. For example, if it be a false affirmation to say *a quadrangle is round*, the word *round quadrangle* signifies nothing, but is a mere sound. So likewise, if it be false to say that virtue can be poured, or blown up and down, the words *in-poured virtue*, *in-blown virtue*, are as absurd and insignificant as *a round quadrangle*.

In the next chapter he develops this point by explaining how absurdities arise when we mix up words that belong to different categories, for example, when we give names of bodies to accidents, or vice versa ("faith is infused or inspired," "extension is body," "phantasms are spirits"), or when we give names of bodies to names or speeches ("there be things universal," "a living creature is genus, or a general thing"), or names of accidents to names and speeches ("a man's command is his will").[52] Another cause of "absurd conclusions" is "the use of metaphors, tropes, and other rhetorical figures, instead of words proper," which – though lawful in common speech – should not be admitted in the "reckoning and seeking of truth," that is in science and philosophy. A last source mentioned by Hobbes is the use of names "that signify nothing, but are taken up, and learned by rote from the schools, as *hypostatical*, *transubstantiate*, *consubstantiate*, *eternal-now*, and the like canting of schoolmen." By the standards of Hobbes's own materialistic philosophy, many things turn out to be absurd such as "incorporeal substance" or "incorporeal body," and his inclusion of the terms just quoted (and also "free will" since a person can be free, but not his will) clearly reveals Hobbes's ultimate aim, namely to criticize a number of philosophical and theological doctrines on transsubstantiation, the soul, separate essences, the Trinity – doctrines which he finds dangerous for the stability of the commonwealth. This critique forms the *basso continuo* of Part IV of the *Leviathan*.

---

[50] Ibid., 46 and Hobbes (2012), 122.

[51] Hobbes (1994), 21; Hobbes (2012), 60.

[52] Hobbes (1994), 24–25; Hobbes (2012), 68–70; probably adapted from a similar listing in Hobbes's *De corpore* I.v.3–9.

It has puzzled scholars that Hobbes recommends a proper, that is, non-figurative language in philosophy and science, because his own style abounds with metaphors and rhetorical tropes. But for Hobbes rhetoric was essentially a matter of style and presentation, but never part of the production of true knowledge and the construction of science. Science requires definitions, deductions and demonstrative reasoning, in which there is no place for metaphors and rhetorical embellishments.[53]

It is not difficult to point to similarities between Hobbes and Renaissance critics of scholastic Aristotelianism. Gianni Paganini has suggested that "the affinity is undeniable between the philological-linguistic argument defended in *Elegantiae* and *Disputationes*, on the one hand, and the many passages of the Appendix [to the Latin *Leviathan*] and the *Leviathan* itself," and more in particular that it is "very probable" that Valla was a source for Hobbes's daring views on the Trinity.[54] As author with a more than solid background in Renaissance humanist culture, Hobbes was surely indebted to a long tradition of Renaissance anti-Aristotelianism in which Valla was an important (early) voice, and Reformation authors referred to Valla's treatise against the authenticity of the Donation of Constantine. There is no evidence, however, that Hobbes had read Valla's *Disputationes*.[55] Important debates and controversies on religious and political matters had taken place in the two centuries that separate them – debates that enriched Hobbes's understanding and exegesis of these theological doctrines.[56] Moreover, there are important differences between the two thinkers, not only in their view of God, matter, body, soul and so on, but also in their linguistic approach. Hobbes's solution to the "insignificant speech" is of course not a return to classical Latin nor does he argue that the common language of the people is always correct. For Hobbes the solution lies in defining one's terms very carefully and avoiding the combination of words that belong to different categories. Moreover, Hobbes's own definitions of terms and concepts do not always reflect common usage but aim at reforming usage, often with the purpose of bringing them more in tune with the ultimate political aim he had in mind.[57] Valla does not distinguish between such categories. He did not share Hobbes's love for geometry, definitions, deductions and proofs. Valla's notion of linguistic usage (*consuetudo*) is defined in terms of the Latin of the great classical authors, an ideal that is certainly not central to Hobbes, who often criticizes these authors for having failed to understand the nature of morality and politics.

---

[53] Nauta (2002).

[54] Paganini (2003), 211.

[55] In private correspondence Noel Malcolm writes to me that in his transcriptions of the Hardwick Hall library catalogues he cannot find any reference to any work by Valla except his Latin translation of Thucydides. Of course, one would expect a well-educated man such as Hobbes to have encountered the *Elegantiae* at some stage, but still Malcolm sees no distinctive debt to it in Hobbes's writings.

[56] See Sommerville (1992), Chaps. 6 and 7; Malcolm in Hobbes (2012), vol. 1, 45–47, 106, 181–82.

[57] Curley in Hobbes (1994), x.

Gottfried Wilhelm Leibniz (1646–1716)

A philosopher who undoubtedly knew the work of Valla (viz. his dialogue on free will) was Leibniz, but it is in critical dialogue with a follower of Valla, the sixteenth-century humanist Mario Nizolio, that Leibniz developed his views on the requirements of clear speech and philosophical language. The Ciceronian Nizolio had published his *On the true principles and the true manner of philosophizing against the pseudophilosophers* in 1553. This work contains a radical critique of the ontology and conceptual armory of the scholastics, and in particular of universals, which Nizolio believed were accepted by virtually all philosophical schools except the nominalists.

In the preface to his edition, Leibniz discusses philosophical style. He shares a principle endorsed by Nizolio that "whatever cannot be explained in popular terms is nothing and should be exorcised from philosophy as if by an incantation, unless it can be known by immediate sense experience."[58] Like Nizolio, Leibniz thinks that the "passion for devising abstract words has almost obfuscated philosophy for us entirely." As we want to communicate our thoughts we must be very clear in our language: "the greatest clarity is found in commonplace terms with their popular usage retained." Since "usage is master" (*der Gebrauch ist der Meister*), Leibniz is sceptical about linguistic innovation, and he warns against assigning meaning to philosophical terms "which are not in conformity with usage from which one should not stray easily in writings intended for the common man."[59] Technical terms are therefore to be shunned "as worse than dog or snake, and one must abstain particularly from those words for categories which are far removed from Latin usage."[60] Yet Leibniz also realizes that technical terms cannot always be avoided "because of the prolixity which would result if popular terms were always used." In geometry e.g. "popular usage does not exactly fit the concepts of geometry." But technical terms, which can be convenient and handsome abbreviations of much longer descriptions in non-technical language, must ultimately be reducible to common terms. As Leibniz concludes[61]:

> There is certainly nothing which cannot be expressed in popular terms, at least by using many of them. Hence Nizolius rightly urges that anything be regarded as nonexistent, fictitious, and useless to which there cannot be assigned a word in the vernacular, however general; that is, as I interpret him, a word which joined together with other general words can express the matter.

Just like Hobbes, Leibniz argues that terms "ought to involve either no figures of speech or few and apt ones."[62] He criticizes the scholastics in particular, because "strange though this sounds, their speech abounds with figures. What else are such

---

[58] For this and the following quotation see Leibniz (1969), 124, 126, 123. For an analysis of Nizolio's work see Nauta (2012).

[59] Quoted by Laerke (2009), 942 n. 25.

[60] Leibniz (1969), 123 for this and the following quotation.

[61] Ibid., 124.

[62] Ibid., 126 for this and the following two quotations.

terms as *to depend, to inhere, to emanate* and *to inflow*?" And again like Hobbes, he gives as example Suarez' term "influx," which is metaphorical and more obscure than what it defines, viz. "cause." But it is not only the scholastics who use obscure language. In other letters and writings Leibniz frequently says that Descartes and especially Spinoza often used obscure terms and definitions.[63]

But while sharing Nizolio's plea for a common language in philosophizing, Leibniz omits elegance from the three praiseworthy marks of speech (clarity, truth, and elegance)[64]:

> since our discussion concerns philosophical discourse and the style that befits it, we shall omit elegance for the present, although we may admit that it can be of great service in securing attention, in moving minds, and in impressing things more deeply on the memory.

Not surprisingly, he considers Nizolio's principles of correct philosophizing, which included knowledge of classical languages and their literature as well as grammar and rhetoric, "principles of speech rather than of thought."[65] Thus, although he presents Nizolio as an excellent guide toward a "sober, proper, natural, and truly philosophical way of speaking," his omission of elegance from philosophical style suggests that the Ciceronian link between *verba* and *res*, style and content, elegance and clarity-truth, was no longer felt as intimate and intrinsic in the way in which humanists such as Nizolio had done.[65]

Conclusion

One of the aims of this article has been to show that the humanists, while most of them were not philosophers in any modern sense of the word, were important sources for later Renaissance thinkers and early modern philosophers for the way in which they formulated their critique of scholastic language. All shared a conviction – which we certainly do not need to share – that this language, at least in its more baroque forms, was artificial, unnatural, uninformative, ungrammatical, and quasi-precise. The scholastics were accused of having introduced a terminology that was a far cry from the common language people spoke, wrote, and read. They had erected a building in which no one but they wanted to live, full of invented notions and entities. But criticizing this language – and with that, a whole way of doing philosophy – is one thing, formulating an alternative is something else, and the notion of the common language turned out to be not so straightforward as it seemed.

For generations of humanists from Petrarch to Vives (and beyond) classical Latin was the common language, providing the norm for anyone who wanted to speak and write Latin. Their conviction was that it also provided a *sine qua non* for clear thinking. But as we have seen, common also meant what "everybody" would

[63] On this see Laerke (2009).

[64] Leibniz (1969), 121–122.

[65] Leibniz in Nizolio, Mario (1956), vol. 1, 30.

[66] These last sentences are taken from Nauta (2012), 62.

"normally" say in such and such a situation or how we "naturally" would argue or reason. Hence humanists emphasized that our grammars and handbooks of logic should be based on practice and usage rather than on theoretical rules of one's own making. What we see in the writings of these humanists is a smooth equation of these two senses of "common." Valla, for instance, frequently mentions "the speech that is common as well as learned" in one breath (*popularis sermo atque eruditorum*), speaking also of "speech that is natural, speech commonly used by educated people" (*ad naturalem et a doctis tritum sermonem*) or referring to "those who speak naturally (*naturaliter*), like orators."[67] In Vives we saw a similar merging of the category of the *auctoritates* with that of the common people. For two reasons this perhaps surprising interpretation of classical Latin as the common language seemed to these humanists a very natural one. First, they could believe that the day was near when an updated version of classical Latin would indeed be the language spoken and written again by a wide community of people. Second, since both classical Latin and our so-called common way of speaking were contrasted to the unnatural, "distorted" language of the scholastics, it was a short step for them to blur the distinction between the first two, particularly when such an elision aided in their fight against that scholastic language.

And yet they also realized that even this common language could be misleading in not always presenting the facts adequately (the vessel is empty, while it is, strictly speaking, not empty as it contains air). Cicero had already made this point, and, as we saw, Valla and Vives repeated the distinction between on the one hand an "exact" or "strict", also called "philosophical" description, and on the other hand an everyday or common sense one. They must have felt a bit uneasy about the distinction, for it seemed to leave the door ajar for the view that the common language was not always good enough for philosophical purposes, and that more exact uses of words had to be allowed for. Hence, we saw an attempt in Valla, Vives but also in Gassendi to deny the scholastic philosophers their self-proclaimed "rigor." For the humanists "rigor," if it meant anything at all, could refer only to the semantic precision of classical Latin. This equation was simply a repetition of their position and did not answer the issue raised by Cicero, for the rigor of classical Latin was apparently not always rigorous enough, as Cicero himself had in fact admitted.

Sixteenth-century Ciceronians bit the bullet and declared Cicero's Latin to be the only norm, a position deftly demolished by Erasmus, who pleaded – in the line of Valla and Poliziano – for a more flexible Latin, of course based (though not exclusively) on classical authors.[68] But the rise of the vernaculars made the equation of common language with classical Latin (of whatever stamp) increasingly difficult to defend. Common language could also mean the vernacular, though in practice Latin remained the lingua franca. But neither Latin nor the vernacular were always

[67] Valla (2012), vol. 1, 106; vol. 2, 208 and 228. For more references see Nauta (2009), 371 n. 36 where I explain that in Valla "natural" does not always refer to the "vernacular" as opposed to Latin, but to our common way of speaking and writing, irrespective of the particular language we use.

[68] There are several articles pertinent to this theme in Ford et al. (2014).

precise or reliable, as e.g. Vives, Sanches, and Leibniz pointed out. Hence, as before, "common" was an easy label used as weapon against what one thought was definitively *not* common, namely the language of the scholastics, but in a world that linguistically speaking became increasingly more pluralistic it became even more difficult to give positive substance to the notion. Sanches, for instance, no longer felt the intimate bond between classical Latin and the so-called common language as earlier humanists had done. Refusing Ciceronian Latin as much as scholastic terminology, he defended a language that he felt reflected the *res* (things) in a direct way, and Gassendi claimed the same for his manner of expression. But, as just noticed, the problem was that this language (Latin in their case) could not be equated with that of the common people, which was often imprecise for science and philosophy. From Valla to Leibniz we read that linguistic usage and convention should be followed – Leibniz said it in German (*der Gebrauch ist der Meister*) – but *whose* usage was thus not so easy to define.

The attack on what was considered the artificial and unnatural language of the scholastics continued in the seventeenth century: the alternative that was presented was the "common" language, either Latin as the lingua franca of the expanding Republic of Letters or the vernacular to address an even wider public. Philosophical and scientific language ought to be clear, plain and non-metaphorical – in the words of Leibniz "sober, proper, natural," or in the words of Gassendi "an unaffected prose style which flows spontaneously," an ideal that was endorsed later by the Royal Society.[69] This is not to say that philosophy was now always conducted in such a clear, "natural," and common language. But the point is that the long road of what we may call the democratization of philosophical language, so dear to early modern philosophers, had its roots – ironically perhaps – in the humanist return to classical Latin as the common language. It was the humanist idiom of "linguistic usage," "convention," "custom," "common" and "natural" language, and "everyday speech" that was repeated and put to new use by early modern philosophers in their critique of scholastic language. Here then is a clear line of continuity between the Renaissance and early modern philosophy.

Bibliography

Botley, Paul. 2004. *Latin Translation in the Renaissance: The Theory and Practice of Leonardo Bruni, Giannozzo Manetti and Desiderius Erasmus*. Cambridge: Cambridge University Press.

Bruni, Leonardo. 1987. In *The Humanism of Leonardo Bruni: Selected Texts*, ed. Gordon Griffiths et al. Binghamton, N.Y.: Medieval & Renaissance Texts & Studies in Conjunction with the Renaissance Society of America.

---

[69] The ideal of stylistic plainness also had strongly political and religious connotations; on these debates on rhetoric in seventeenth-century England, see Vickers (1985). Royalists and Conformists used it as a weapon to marginalize nonconformist sects. Such wider dimensions of the debates on language between humanists and scholastics are explored by Moss (2003) and in D'Amico (1983), dimensions which I have hardly touched on in this article.

Caluori, Damian. 2007. The Scepticism of Francisco Sanchez. *Archiv für Geschichte der Philosophie* 89: 30–46.
Camporeale, Salvatore. 1972. *Lorenzo Valla: Umanesimo e teologia.* Florence: Nella Sede dell'Istituto Palazzo Strozzi.
Casini, Lorenzo. 2009. Self-Knowledge, Scepticism and the Quest for a New Method: Juan Luis Vives on Cognition and the Impossibility of Perfect Knowledge. In *Renaissance Scepticisms*, ed. Gianni Paganini, José R. Maria, and José R. Maria Neto, 33–60. Dordrecht: Springer.
D'Amico, John F. 1983. *Renaissance Humanism in Papal Rome: Humanists and Churchmen on the Eve of the Reformation*. Baltimore: John Hopkins University Press.
Ford, Philip, et al. 2014. *Brill's Encyclopaedia of the Neo-Latin World*. Leiden: Brill.
Frede, Michael. 1987. The Ancient Empiricists. In *Essays on Ancient Philosophy*, ed. Frede Michael, 243–260. Oxford: Oxford University Press.
Gassendi, Pierre. 1972. *The Selected Works of Pierre Gassendi*, ed. and trans. Craig B. Brush. New York: Johnson Reprint.
Gassendi, Pierre. 1981. *Institutio logica* (1658), ed. and trans. Howard Jones. Assen: Van Gorcum.
Hanfling, Oswald. 2000. *Philosophy and Ordinary Language: The Bent and Genius of Our Tongue*. London: Routledge.
Hobbes, Thomas. 1994. *Leviathan*, ed. Edwin Curley. Indianapolis: Hackett.
Hobbes, Thomas. 2012. *Leviathan*, ed. Noel Malcolm. 3 vols. Oxford: Oxford University Press.
Isermann, Michael. 1991. *Die Sprachtheorie im Werk von Thomas Hobbes*. Münster: Nodus-Publikationen.
Joy, Lynn S. 1987. *Gassendi the Atomist: Advocate of History in an Age of Science*. Cambridge: Cambridge University Press.
Laerke, Mogens. 2009. The Problem of *Alloglossia.* Leibniz on Spinoza's Innovative Use of Philosophical Language. *British Journal for the History of Philosophy* 17: 939–953.
Leibniz, Gottfried Wilhelm 1969. *Philosophical Papers and Letters*. Trans. Leroy E. Loemker. Dordrecht: Reidel.
Leijenhorst, Cees. 2002. 'Insignificant Speech'. Thomas Hobbes and Late Aristotelianism on Words, Concepts and Things. In *Res et Verba in der Renaissance*, ed. Ian Maclean and Eckhard Kessler, 337–368. Wiesbaden: Harrassowitz.
Lupoli, Agostino. 2009. *Humanus animus nusquam consistit*: Doctor Sanchez's Diagnosis of the Incurable Human Unrest and Ignorance. In *Renaissance Scepticisms*, ed. Gianni Paganini and José R. Maria Neto, 149–181. Dordrecht: Springer.
Maclean, Ian. 2006. The 'Sceptical Crisis' Reconsidered: Galen, Rational Medicine and the *Libertas Philosophandi. Early Science and Medicine* 11: 247–274.
Marsh, David. 1979. Grammar, Method, and Polemic in Lorenzo Valla's *Elegantiae. Rinascimento* 19: 91–116.
Moss, Ann. 2003. *Renaissance Truth and the Latin language Turn*. Oxford: Oxford University Press.
Murr, Sylvia. 1992. Foi religieuse et 'libertas philosophandi' chez Gassendi. *Revue des Sciences Philosophiques et Théologiques* 76: 85–100.
Nauta, Lodi. 2002. Hobbes the Pessimistic? Continuity of Hobbes's Views on Reason and Eloquence Between *The Elements of Law* and *Leviathan. British Journal for the History of Philosophy* 10: 31–54.
Nauta, Lodi. 2007. Lorenzo Valla and the Rise of Humanist Dialectic. In *The Cambridge Companion to Renaissance Philosophy*, ed. James Hankins, 193–210. Cambridge: Cambridge University Press.
Nauta, Lodi. 2009. *In Defense of Common Sense: Lorenzo Valla's Humanist Critique of Scholastic Philosophy*. Cambridge, Mass.: Harvard University Press.
Nauta, Lodi. 2012. Anti-essentialism and the Rhetoricization of Knowledge: Mario Nizolio's Humanist Attack on Universals. *Renaissance Quarterly* 65: 31–66.
Nauta, Lodi. 2015. The Order of Knowing: Vives on Language, Thought, and the Topics. *The Journal of the History of Ideas* 76: 325–345.

Nizolio, Mario. 1956. *De veris principiis et vera ratione philosophandi contra pseudophilosophos*, ed. Quirinus Breen. 2 vols. Rome: Bocca.
Osler, Margaret J. 2003. Early Modern Uses of Hellenistic Philosophy: Gassendi's Epicurean Project. In *Hellenistic and Early Modern Philosophy*, ed. Jon Miller and Brad Inwood, 30–44. Cambridge: Cambridge University Press.
Paganini, Gianni. 2003. Hobbes, Valla and the Trinity. *British Journal for the History of Philosophy* 11: 183–218.
Paganini, Gianni. 2009. Tommaso Campanella: The Reappraisal and Refutations of Scepticism. In *Renaissance Scepticisms*, ed. Gianni Paganini, José R. Maria, and José R. Maria Neto, 275–303. Dordrecht: Springer.
Pasnau, Robert. 2011. *Metaphysical Themes 1274–1671*. Oxford: Oxford University Press.
Popkin, Richard. 2003. *The History of Scepticism, from Savonarola to Bayle*. Revised edition. Oxford: Oxford University Press.
Rummel, Erika. 2000. *The Humanist-Scholastic Debate in the Renaissance and Reformation*. Cambridge, Mass.: Harvard University Press.
Sanches, Francisco. 1988. *That Nothing is Known*, ed. and trans. Elaine Limbrick and Douglas F. S. Thomson. Cambridge: Cambridge University Press.
Sanches, Francisco. 2007. *Quod nihil scitur/Dass nichts gewusst wird*, ed. and trans. Kaspar Howald et al. Hamburg: Felix Meiner.
Seigel, Jerrold E. 1968. *Rhetoric and Philosophy in Renaissance Humanism: The Union of Eloquence and Wisdom, Petrarch to Valla*. Princeton: Princeton University Press.
Sommerville, Johann P. 1992. *Thomas Hobbes: Political Ideas in Historical Context*. Basingstoke: Macmillan.
Tavoni, Mirko. 1984. *Latino, grammatica, volgare: Storia di una questione umanistica*. Padua: Antenore.
Valla, Lorenzo. 2012. *Dialectical Disputations*. ed. and trans. Brian P. Copenhaver and Lodi Nauta. 2 vols. Cambridge, MA: Harvard University Press.
Vickers, Brian. 1985. The Royal Society and English Prose Style: A Reassessment. In *Rhetoric and the Pursuit of Truth: Language Change in the Seventeenth and Eighteenth Centuries*, ed. Brian Vickers and Nancy Struever, 1–76. Los Angeles: The William Andrews Clark Memorial Library.
Vives, Juan Luis. 1782–90. *Opera Omnia*, ed. Gregorio Myans y Siscar. 8 vols. Valencia: Monfort (reprint London, 1964).
Vives, Juan Luis. 1971. *De tradendis disciplinis*. In *Vives: On Education*. Trans. Foster Watson. Totowa: Rowman and Littlefield.
Vives, Juan Luis. 1979. *Against the Pseudodialecticians*. Trans. Rita Guerlac. Dordrecht: Reidel.
Wels, Volkhard. 2000. *Triviale Künste: Die humanistische Reform der grammatischen, dialektischen und rhetorischen Ausbildung an der Wende zum 16.* Jahrhundert. Berlin: Weidler.
Witt, Ronald G. 2000. *In the Footsteps of the Ancients. The Origins of Humanism from Lovato to Bruni*. Leiden: Brill.

Chapter 5
Henry More and Girolamo Cardano

Sarah Hutton

Abstract Henry More's view of Girolamo Cardano was ambivalent. On the one hand he regarded his philosophy as "false or uncertain" associating him with Vanini and Pomponazzi, yet he also regarded him as "that famous philosopher of his age" worthy of quoting on the title page of his *Immortality of the Soul* (1659). In my paper I discuss More's engagement with Cardano in that work, especially his comments on Fazio Cardano's dream. I argue that, for More, Cardano represents the obverse of the problem of materialism, for although he agreed on the existence of spirits, he filled his cosmos with all sorts of wrongly conceived spirits, making him guilty of heterodoxy and atheism. More sought to expose Cardano's errors by means of the same strategy used in the case of Hobbes: by using his adversary's own method of argument. He drew on modern philosophy (Cartesianism) to dispel the obscurantism and misbelief in Cardano, which he links to the atheism of Pomponazzi and Vanini.

The Cambridge Platonists and Renaissance Philosophy

The Cambridge Platonists were indebted to Renaissance Philosophy in two obvious respects. First, like all philosophers of their day, they owed to the labours of Renaissance Humanists the recovery of the corpus of ancient philosophy – especially the writings of Plato and Plotinus, but also Sextus Empiricus, Proclus and others. Secondly, among humanist philosophers, they owed a special debt to Marsilio Ficino for both his pioneering work in recovering the corpus of Plato's writings and the *Enneads* of Plotinus, and for the type of Christianised interpretation of Plato and Plotinus which he evolved.[1] But Ficino was not the only Renaissance philosopher with whom they were conversant: they also make reference to the

[1] The Cambridge Platonists were not in fact as deeply indebted to Ficino as is commonly assumed. They did not draw on his *Theologia platonica* and they used the Stephanus-Serranus edition of Plato, rather than Ficino's. See Hutton (2013b).

S. Hutton (✉)
Department of Philosophy, University of York, York, UK
e-mail: sarah.hutton@york.ac.uk

C. Muratori, G. Paganini (eds.), *Early Modern Philosophers and the Renaissance Legacy*, International Archives of the History of Ideas Archives internationales d'histoire des idées 220, DOI 10.1007/978-3-319-32604-7_5

writings of other Renaissance philosophers, among them Pietro Pomponazzi, Cesare Vanini, Girolamo Cardano, Iacopo Zabarella and Jean Fernel. Machiavelli too is mentioned. This list of names is not untypical of book collections in the period. Gabriel Naudé's 'wishlist' for the ideal library, his *Advis pour dresser une bibliothèque* (1627), which was translated by Evelyn in 1661, recommends that a library should contain a representative survey of philosophers from ancient to contemporary, including among the principal "novatores", "*Telesius, Patricius, Campanella, Verulamius, Gilbert, Iordanus Brunus*". He adds that "it would be a fault unpardonable in one who professes to store a Library, not to place in it *Piccolomini, Zabarell, Achillinus, Niphus, Pomponacius, Licetus, Cremoninus,* [...] *Montagne, Charon, Verulam*",[2] as well as Fernel, Cardano and others. From among these names he describes Cardano, Pomponazzi and Bruno as "curious and not vulgar Authors". The same cast of thinkers is also found in British libraries. One of the largest, was the library of the second Viscount Conway which contains many Italian philosophical books, including, besides Cardano, books by Giordano Bruno, Bernadino Telesio, Andrea Cesalpino, Giovanni Pico della Mirandola, Francesco Patrizi and Agostino Steuco, as well as Alessandro Piccolomini, Andrea Cesalpino, and Iacopo Zabarella.[3]

But engagement with the philosophy of the past did not preclude interest in contemporary philosophy. It is especially true of the Cambridge Platonists, that they took a clear – even pioneering – interest in contemporary new philosophy while at the same time drawing on the Platonic tradition revived in the Renaissance. However, an obvious debt to Renaissance philosophy is not, on the whole, regarded positively by historians of philosophy, but is often viewed as a debt to obscurantism. In large measure this is explicable by the fact that the most widespread narrative about the rise of modern philosophy treats the seventeenth century as a watershed, in which modernisers broke with the past.

A classic statement of the view that, as heralds of the new who retain a *penchant* for things Renaissance, the Cambridge Platonists were by definition muddled thinkers, is Alexandre Koyré's assessment of Henry More. In his *From the Closed World to the Infinite Universe* Koyré presents him as someone who struggled with modernity:

> Henry More enjoys a rather bad reputation among historians of philosophy, which is not surprising. In some sense he belongs much more to the history of the hermetic, or occultist, tradition than to that of philosophy proper; in some sense he is not of his time: he is a spiritual contemporary of Marsilio Ficino, lost in the disenchanted world of the 'new philosophy' and fighting a losing battle against it.

To be fair to Koyré he does actually credit Henry More with important insights, but he cannot reconcile these with the identifiably Renaissance elements of much of his

[2] i.e. Alessandro Piccolomini, Iacopo Zabarella, Achillinus, Agostino Nifo, Pietro Pomponazzi, Fortunio Liceti, Cesare Cremonini, Michel de Montaigne, Pierre Charon, Francis Bacon. Naudé (1661). For more information on these see Copenhaver (1992).

[3] The most well-represented Italian philosopher in Lord Conway's collection is Fortunio Liceti, with 18 books listed. Also listed is Julius Caesar Scaliger's attack on Cardano: Scaliger (1557).

thinking. Fundamentally he sees More as someone who achieved what he did, in spite of his Renaissance roots. Koyré goes on:

> And yet, in spite of his partially anachronistic standpoint, in spite of his invincible trend towards syncretism which makes him jumble together Plato and Aristotle, Democritus and the Cabala, the thrice great Hermes and the Stoa, it was Henry More who gave to the new science – and the new world view – some of the most important elements of the metaphysical framework which ensured its development: this because, in spite of his unbridled phantasy, which enabled him to describe at length God's paradise and the life and various occupations of the blessed souls and spirits in their post-terrestrial existence, in spite of his amazing credulity [...] which made him believe in magic, in witches, in apparitions, in ghosts, Henry More succeeded in grasping the fundamental principle of the new ontology, the infinitization of space, which he asserted with an unflinching and fearless energy.[4]

Although More's contribution to early modern philosophy is now receiving more acknowledgement, this is usually at the price of overlooking his relationship to Renaissance philosophy.[5] The same may be said of the other Cambridge Platonists. In this paper I want to challenge this negative view of the relationship of Renaissance philosophy to seventeenth-century philosophy by exploring Henry More's engagement with Girolamo Cardano, in the context of his lively interest in contemporary philosophy, especially Cartesianism. By examining more closely More's use of Cardano, we can perhaps view his engagement with Renaissance philosophy in a less negative light than Koyré did.

Henry More and Renaissance Philosophy

Henry More had an extensive knowledge of Renaissance philosophy from his undergraduate days. As he tells us in the preface to his *Opera Omnia*, he set himself to study not just Aristotle, but Cardano and Scaliger, in his eager search for "the Knowledge of natural and divine Things" which he "chiefly desired to be satisfied about", because they seemed to him "the highest Pleasure and Felicity Imaginable". However he recalls his disappointment after immersing himself

> Head and Ears in the Study of Philosophy; promising a most wonderful Happiness to my self in it. Aristotle, therefore Cardan, Julius Scaliger and other Philosophers of the greatest note I diligently perused. In which the Truth, though I met here and there with some things wittily and accurately, sometimes also solidly spoken; yet the most seemed to me either so false or uncertain or else so obvious that I looked upon myself as having plainly lost my time in the Reading of such Authors.[6]

After studying these philosophers "of greatest note" for four years, he writes, his studies "ended in nothing, in a manner, but mere Scepticism" (though he assures his

---

[4] Koyré (1957), 125–126.

[5] Important recent studies, especially in the history of moral philosophy are Gill (2006); Darwall (1995); Reid (2012); Schneewind (2003).

[6] Ward (2000), 17.

reader that he "never had the least Doubt" about the existence of God and moral obligations).[7] Nevertheless, he found more satisfaction when he discovered Ficino and the Neoplatonists – but it was only after graduating, that he came to read "Marsilius Ficinus, Plotinus himself, Mercurius Trismegistus; and the mystical Divines among whom there was frequent mention of the Purification of the Soul". The only mystical text he names is the *Theologia Germanica*, a firm favourite of his. There is a strong suggestion in this autobiographical sketch that Cardano, Scaliger and Aristotle had their limitations, and that Ficino and the Platonists took him to places that they didn't reach. I might add that he seems to have read Descartes not long after this, in the early 1640s.[8] We also know from his writings that he also read Machiavelli, Galileo, Pomponazzi and Vanini. However, neither here in his account of his studies, nor in his works, can one find direct references to Telesio and Campanella.

Apart from the special case of Galileo, the Italian philosopher to figure most significantly in More's writings is Girolamo Cardano (1501–1576), whom he cites or discusses in his three main English writings: *Antidote against Atheism* (1655), *Of the Immortality of the Soul* (1659) and *An Explanation of the Grand Mystery of Godliness* (1660). The Cardano works cited are *De subtilitate*, *De rerum varietate*, *De animorum immortalitate*. He also refers to *De vita propria* and he accords Cardano distinction of quoting from him on the title page of his *Of the Immortality of the Soul*.

More's decision to study Cardano when he was an undergraduate was not an idiosyncratic choice on his part. Although a controversial figure, Cardano was nevertheless considered an authoritative thinker and there appears to have been a revival of interest following Gabriel Naudé's publication of his life of him in 1643.[9] Cardano's writings were widely known in the seventeenth century. His *Opera Omnia* was reprinted as late as 1663.[10] *De subtilitate*, in particular, was an immensely popular work and remained in circulation well into the seventeenth century, when it was reprinted until 1664. Knowledge of *De subtilitate* would only have been intensified by Scaliger's 900-page vitriolic attack on it, *Exotericae Exercitationes*, another widely read work which was printed and reprinted well into the seventeenth century.[11] Opinion on Cardano was divided. Francis Bacon dismissed him as unreliable because his works contain "much fabulous matter, a great part only untried, but notoriously untrue, to the great derogation of the credit of natural philosophy with

[7] Quoted in Ward (2000), 18.

[8] More graduated BA in 1636 and MA in 1639. For the life of More, see Crocker (2003).

[9] See Maclean (1986). A physician by training, Cardano was a polymath, who published on a wide variety of topics, including astrology, mathematics, metaphysics and medicine. His intellectual formation is a complex interweaving of different strands – humanist, Aristotelian and Galenic, but he was a trenchant critic of Aristotle who sought a new methodology for the discovery of new knowledge, to be achieved by collaborative investigation based on experience. See Baldi and Canziani (1999); Giglioni (2013).

[10] Maclean (2007); Maclean (2009). Also Maclean (2005).

[11] Described by Anthony Grafton as "the most savage book review in the bitter annals of literary invective", this was a standard work in university reading lists. See Grafton (1999), 4.

the grave and sober kind of wits."[12] But others held him in high regard. In 1670 in his *The Voyage of Italy* Richard Lassels names Cardano along with Ficino and Pico della Mirandola as Italy's leading philosophers. In the same year, Robert Boyle cites *De subtilitate* as a source in *New Experiments Physico-mechanical*, while Locke's admirer, Richard Burthogge expressed the view that he was a "wonderfully knowing and learned man" as late as 1699.[13]

Given Cardano's fame, the fact that More read and cites him is not of itself surprising. We may surmise that More probably originally read Cardano *suo specie* (so to say) Julius Caesar Scaliger, since he cites Scaliger's *Exotericae Exercitationes* in the notes of his first-published work, his *Psychōdia Platonica*, which appeared in 1642.[14] Scaliger's text figures in university reading lists. More's scholarly interest in Cardano at such an early date was perhaps unusual, since Cardano's main readership was a general rather than an academic one, and More's interest in Cardano antedates Gabriel Naudé's publication of his life of him in 1643.

However, from the first, there was decided ambivalence in More's assessment of Cardano. As we saw his undergraduate "search for knowledge of natural and divine things" in Cardano, as well as Scaliger and Aristotle, ended in disappointment. He concluded that their grasp of the truth was patchy, and that their views were for the most part "false or uncertain". In his writings More refers to Cardano variously as "that famous philosopher of his age" and "that odd and crooked Writer".[15] On the one hand he uses him as an authoritative source for instances of natural and supernatural phenomena. For example he draws on him for examples of the orderliness of nature, including the natural architecture of martins' nests, honeycombs and spiderwebs.[16] And he also draws on Cardano for examples of supernatural events, such as the appearance of apparitions to warn of impending human disasters – for example, the prodigies reported to have been seen prior to the fall of Mexico.[17] Other Cardanesque prodigies are linked to more mundane circumstances, e.g. "a young man, [...] [on a] Cart all covered with fire", the apparition of "a man of a huge stature with his belly cut up and exenterated, and two children in his armes" seen by a couple of washerwomen; an account of man vomiting glass, iron, nails and hair. Cardano isn't his only source for natural phenomena and "prodigies", and he justifies mentioning such things on the grounds that it is difficult to explain them away

[12] Bacon, *Advancement of Learning*, 1.4.10.

[13] Lassels (1670), 8; Boyle (1682), 192; Burthogge (1699), 12.

[14] More (1642).

[15] More, *Immortality of the Soul*, 2.12.11, 114 (references to *Immortality of the Soul* are given by book, chapter and section number, followed by the page number from *A Collection of Several Philosophical Writings* (More (1662), abbreviated as *CSPW*, in which the constituent writings are individually paginated). More, *Immortality of the Soul*, 7.14.4, 336.

[16] *Immortality of the Soul*, 3.13.9. He also cites Scaliger, Fallopius and even Vanini for evidence of natural occurrences which can be explained in terms of the operation of some kind of spirit in the world, "if those Histories be true, of extemporary Salads sown and gathered not many hours before the meal they are eaten at: and of the sudden ingendring of Frogs upon the fall of rain [...]." Ibid., 2.15.6, p. 165, 267.

[17] More (1660), 219, 222.

as the effects of, e.g., melancholy or hallucination: "these effects extraordinary and supernatural being so palpable and permanent, they are not at all lyable to such Subterfuges as *Atheists* usually betakes themselves to as of *Melancholy*, and disturbance of *Phansie* [...]"[18]

By contrast, More was highly critical of Cardano's astrological theories of natural causality. The last four chapters of *An Explanation of the Grand Mystery of Godliness* (1660) are devoted to an attack on judicial astrology,[19] in the course of which he attacks Cardano's theories of sidereal influence. More judged that Cardano sometimes "writes more like a Priest of the Sun then a man of Reason or a sound Philosopher" for attributing understanding ("intellectus") to the sun, so making "Visible light and Intellect all one". He attacked Cardano's astrological determinism, and he subscribed to a long-standing general condemnation of Cardano for casting the horoscope of Christ ("Cardanus his high folly in calculating the Nativity of our Saviour").[20] More's condemnation of Cardano's heterodox astrology is understandable in theological terms, and nothing new.[21] His qualms about Cardano are fuelled by his view that Cardano gave encouragement to two atheists, Pomponazzi and Vanini, both equally far "laps'd into Atheism".[22]

The Immortality of the Soul

The complexity of More's attitude towards Cardano is most apparent in his *The Immortality of the Soul, so farre forth as it is demonstrable from the knowledge of nature and the light of reason* which was first published in 1659, and republished shortly afterwards in the first volume of his *A Collection of Several Philosophical Writings* (1662). The title page suggests approbation of Cardano – or, at the very least, invites comparison with Cardano:

> Quid jucundius quàm scire quid simus, quid fuerimus, quid erimus; atque cum his etiam divina atque suprema illa post obitum Mundique vicissitudines?

As translated by More's biographer, Richard Ward, this reads:

> What can be pleasanter than to know what we are, what we have been, what we shall be, and together with these those last and divine things which come after Death and the great Changes or Revolutions of the World?[23]

---

[18] More, *Antidote against Atheism*, 99, 121, in *CSPW*.

[19] These chapters were republished separately at a later date (1681) as tract against judicial astrology (*Tetractys Anti-Astrologica*). More was prompted to publish this in response to the attack on him by John Butler whose *Hagiastrologia. Or, The most Sacred and Divine Science of Astrology vindicated against the Reverend Dr. More's Calumnies* (1680), attacks More's treatment of judicial astrology in *An Explanation*, defending Cardano's use of it.

[20] More (1660), 60–61.

[21] See Ernst (2001). Also Grafton (1999), 151–154 and Grafton and Siraisi (2001).

[22] See Leech (2013).

[23] Ward (2000), 286.

This is preceded on the title page by a quotation in Greek attributed to Pythagoras, to the effect that the air is replete with souls, "That all the air is full of Spirits […] and that they are either demons or heroes".[24] Between them these epigraphs are the leitmotif of *The Immortality of the Soul*: the driving thesis of the book being that immaterial, or spiritual substances exist, and that souls have past, present and future life. The quotation from Cardano, so prominent on the title page, suggests that More felt an affinity with him. However, as I shall show, his subsequent references in the course of his discussion suggest otherwise.

Cardano's philosophy certainly had direct bearing on the subject matter of *Immortality of the Soul*. His *De animorum immortalitate* treats the same topic as More's *Immortality*. And there are many points of common ground between More and Cardano, not least their Platonising tendencies, and their readiness to venture into the supra physical realm of demons and spirits ("suprema illa post obitum Mundique vicissitudines"). More also finds Cardano shared with him opinions about the nature and pre-existence of the soul. Cardano, he claims "expressly concludes, that the rational Soul is both a distinct being from the Soul of the World and that it does praexist before it comes into the Body". Both Cardano and More conceived of the cosmos as filled with spirits of one kind of another. A large segment of Cardano's supra-lunary universe is reserved for spiritual beings which he calls *daemones* or *demons*. They both agreed that the natural abode for souls after death was the air. Furthermore, Cardano held that matter is inert and he explained natural phenomena in terms of spirit or soul pervading all things. Suffice to say that Cardano's is a far more complex theory, in which the functions which More attributes to his "Hylarchic Spirit" or "Spirit of Nature" are distributed between four different spirits.[25] The similarities with More are more apparent than real, and the differences between them not a simple matter of scale. In order to understand this we need to take cognizance of More's own theory of the soul.

The Afterlife of the Soul

More's *Immortality of the Soul* is a significant statement of his mature philosophy, in which he presents what might be called his trademark metaphysical doctrines. These include, first, his classic definition of the properties of incorporeal substance as the inverse of the definition of corporeal substance: both share the property of extension, but where body is impenetrable and discerpible (i.e. separable),

[24] Taken from Diogenes Laertius VIII.32, in Richard Ward's translation which renders *genii* as demons. Ward (2000), 286.

[25] Cardano proposes nine orders of celestial beings and "seven natures". These latter are "the infinite, or God", eternal in itself; "the soul of all things or lives", the soul of the world; first mover (the *primum movens*); the souls of the various planets; sentient minds (*mentes sensiles*); the common sentient that presides over all sentient lives (*communis sensilis*); the "common and vital soul" (*anima communis atque vitalis*); life or "the soul conceived in matter". See Giglioni (2013b)

immaterial substance, or spirit is penetrable and indiscerpible. Secondly, More's *Immortality* contains a first statement of his causal theory, his hypothesis of the "Spirit of Nature" ("the *Quartermaster General* of Divine Providence"). In addition to these, the book elaborates other theories of central importance in More's metaphysics – his revival and restatement of the doctrine of the vehicle of the soul, and his hypothesis of the "vital congruity" of soul and body in order to explain their cohesion. Many of the themes which he treats had been discussed in earlier works, including his philosophical poems. One such theme was mortalism.[26]

Immortality of the Soul is both a pneumatology and a work of rational religious apologetics, which seeks to defend the fundamental notion of the soul's immortality in non-theological terms, using only rational argument. In it More combines the modern mathematical methods (i.e. a set of axioms deduced from supposedly self-evident claims) with the arguments drawn from experience. These latter rely heavily on the testimony of witnesses to the activity of spiritual forces, and most of that testimony is derived from books. In the course of his discussion he invokes Descartes. It was in the Preface to *The Immortality of the Soul* that he famously recommended the teaching of Cartesian natural philosophy in the English Universities. Most attention accorded to this work has focused on More's refutation of the new 'mechanical' philosophy, particularly the materialist philosophy of Thomas Hobbes, but also Cartesianism which are discussed in the first two books. This has, therefore, become paradigmatic for More's confrontation with modern philosophy. Little or no attention has been accorded to Book 3 of *Immortality of the Soul* which deals with the life of the soul after its separation from its earthly body at death. It is this book which contains More's main discussion of Cardano.

For More, the afterlife of the soul is an essential aspect of his case for the soul's immortality. After all, if souls are immortal, it follows that they must continue in existence after death. And More's view of the life of the soul after death requires that this be continuous and that it involves more than bare existence. The afterlife is a full and active life where souls move, communicate with one another, feel pleasure or pain. For this reason he specifically rejects the hypothesis that souls sleep until judgement day ("psychathanasia") or that they die only to be resurrected at the last trumpet (mortalism). The activity of the soul, its capacity for sensation, imagination, and memory would not, however, be possible without a physical body.[27] It therefore follows that at no point, does the soul completely lose a material body: "And therefore if the Soul act at all after death, […] it is evident that she is not released from all *vital union* with all kind of *Matter* whatsoever."[28] In order to explain how souls can feel, enjoy sensations, and remember their past lives, More revived the Neoplatonic idea of the vehicle of the soul, according to which every soul, has a sort of "envelope", which contains it. After quitting its terrestrial body,

[26] *Antipsychopannychia, A Confutation of the Sleep of the Soul after Death* which was published in *Psychodia Platonica*. See Hutton (2013a). On More's metaphysics, see Reid (2012).

[27] *Immortality of the Soul*, 2. 11.1–6, p. 106–107.

[28] Ibid., 3.1. 2., 159–160.

the soul is attached to a "vehicle", a diaphanous body, first of air, then aetherial.[29] By means of the doctrine of the vehicle of the soul he could explain the continued life of the soul *post obitum* without attaching it to another, different earthly body – i.e. by transmigration which he regarded as an aberration of the Pythagoreans.[30] Furthermore by showing how the soul may live and act separately from its earthly body, he believed he could account for such para-normal phenomena as out-of-body experiences or *extasis*, apparitions, daemons, and witches. Such phenomena, in More's view constitute empirical evidence for the after-life existence of deceased souls.[31] But most importantly for More, there are moral reasons for the soul to retain the capacity for sensation and memory after death: without memory there would be no conscience, no awareness of past wrongs, and no sensation of the pain of punishment for them, or of the joy which rewards the good. Rewards and punishments in the afterlife would be useless and arbitrary if the soul had no memory of past actions, and could not feel the punishment or taste the rewards in which those actions incur according to the rule of divine justice. This is especially important in view of the fact that More held that souls could continue to improve themselves after death. As they do so they are able to progress from an aerial state to an aetherial one, exchanging their aerial vehicles for an aetherial ones. And their sensitivity is heightened as they do so: "The purer the Vehicle is, the more quick and perfect are the Perceptive Faculties of the Soul."[32] To deny punishment in the afterlife is to deny the providence of God, only one step from atheism.[33] Furthermore, *post-mortem* memory was an essential component of More's argument against the Averroist doctrine of the single universal soul.[34] For these reasons, More's theory of the soul invests heavily in arguments which explain the union of soul with body, and attack theories which either entail the separability of soul from body, or which claim that the soul can exist in a disembodied state. These include those who deny *post-mortem* memory (Aristotle and Averroes) and those who make a radical distinction between soul and body (Descartes). More's qualms about Cardano are fuelled by his view that Cardano gave succour to two atheists, Pomponazzi and Vanini, both equally far "laps'd into Atheism".

[29] For an account of More's theory of the vehicle of the soul, see Reid (2012), Chap. 10.

[30] The Pythagoreans he says in *Conjectura Cabbalistica*, "have mingled their own fooleries with it, either out of the wantonnesse of their Fancy, or mistake of Judgement; Such as are the Transmigration of Humane Souls into Brutes", *Defence of the Threefold Cabbala*, 43, in *CSPW.*

[31] Cf. *An Explanation*, 6.5.1, 226: "the Souls as well of the Good as the Bad after Death have an Aereal Body, in which, if Stories be true, they have sometimes appeared after their decease. And that they may act, think and understand in these Aiery vehicles, as well as other Spirits doe, is not at all incredible nor improbable […]." The doctrine of the vehicle of the soul is also used by More to refute deniers of the resurrection of the body, who claim that souls can exist without bodies.

[32] *Immortality of the Soul*, 3.11.1, p. 188.

[33] Ibid., 3.11.6, p. 189.

[34] Hutton (2013a, b).

Cardano in the Immortality of the Soul

In Book 3 More first discusses Cardano in relation to the question of "the natural abode of the Soul after death", which both he and Cardano think is the air. Cardano's "peculiar conceit", advanced in *De rerum varietate*, is that

> the *supreme Region of the Aire* was the only habitation of all Demons or Spirits whatever, and that their descent to us is as rare as the diving of Men into the bottome of the Sea, and almost as difficult, this *thicke Aire* we breath in being unsutable to their tenuious consistencies as the Water is to us.[35]

These spirits "bear us no good will", but occupying a different environment from us, when these demons attack us they do so like anglers, sending down "Dreams and Apparitions" to entangle us, "as we do the Fishes, by baits, and Nets, and Eel-Spears, or such like Engines which we cast into the bottom of the Water".[36] Although he agrees with Cardano that after death the natural element of the soul is the air, More denies that souls are confined to the upper regions of the air ("Cardan's conceit of placing all Daemons in the upper Region"), and that they are enemies of humans and delight in their destruction. He takes up Cardano's analogy with fish (which he refers to as a "Parable"), to offer an alternative explanation of why spirits are rarely seen in terms of his theory of the vehicle of the soul. Spirits are as difficult to render visible, as it is for divers to hold their breath under water. To deny their existence would be like fish disputing about whether men exist.[37] Continuing the fishing analogy, he dismisses the idea that aerial spirits are "Haters of Mankind" who enjoy destroying them:

> For Men do not hate Fishes because they live in another Element different from theirs, but catch them merely in love to themselves for gain and food.[38]

He further explains that spirits do little harm, not because they are insubstantial or too distant, but because of the universal law of justice, "the *Law of the Universe*, whose force penetrates through all the Orders of Beings". The general gist of More's critique is that many of Cardano's claims are "too trivial and idiotick, and far below the pitch of a philosopher". By contrast, his own explanation of the behaviour of demons is "a more rational hypothesis".

The next reference to Cardano in *Immortality of the Soul* comes in book 3 Chap. 12 in which More opposes Aristotle and Cardano's denial that the soul retains memory after death. This is pertinent to the issue of mortalism and is a key element in his refutation of Averroistic monopsychism, which he first proposed in his poem *Antimonosphychia*. Reiterating a point made earlier in book 2, that "the immediate seat of Memory is the Soul her self",[39] More argues that memory is essential for

---

[35] *Immortality of the Soul*, 3.3.4, p. 156.

[36] Ibid.

[37] Ibid., 3.3.7, p. 157.

[38] Ibid.

[39] Ibid., 3.11.1, p. 188.

conscience and the "inward sense of good and evil". Most of the discussion consists of an elaboration of the theory of how memory functions in relation to the operation of divine justice. This discussion is pertinent to More's conception of personal identity outlined in *Immortality of the Soul* Chap. 17, in which he opposes the Averroistic notion of a single universal soul (monopsychism) by arguing that awareness or consciousness is a key unifier of soul and body, which individuates particular souls, constituting "every man's personal Ipseity".[40]

The third, and most extensive discussion of Cardano in the same book of *Immortality of the Soul* also confronts the issue of mortalism, this time by reference to Cardano's account of a dream of his father's, Fazio (Facius) Cardano, which he recounts in *De subtilitate*. This is one of three dreams cited by More as the main evidence to be adduced in favour of mortalism ("the most notable Testimonies for the Mortality of Daemons" – the others being Hesiod and Plutarch). It consisted of a 3-hour conversation with seven "aerial Inhabitants" or demons on "The 13. day of August 1491 […], at the 20. houre of the day", when, according to Fazio Cardano:

> […] there appeared to me, after their usual manner, seven men cloathed in silk garments, with cloaks after the Greek mode, with purple stockings and crimson Cassocks, red and shining on their breasts; nor were they all thus clad, but onely two of them who were the chief. On the ruddier and taller of these two other two waited, but the less and paler had three attendants; so that they made up seven in all. They were about 40 years of age, but lookt as if they had not reacht 30. When they were asked who they were, they answered that they were *Homines aerii*, Aerial Men, who are born and die as we; but that their life is much longer then ours, as reaching to 300. years.[41]

The ensuing conversation with these "Aerial Men" in the dream yields a farrago of mutually inconsistent opinions. Among the topics discussed is "the Immortality of our Souls" and the eternity of the world. One airy man denied the eternity of the world and another advanced an occasionalist hypothesis, according to which God sustains the world continuously. They claimed to be a higher order of being than men, and that their "happiness or misery as much transcended ours, as ours does the brute Beasts". They can do no harm to men, because of their physical tenuousness, but they can transmit apparitions and knowledge to men. Two were university professors, each with a couple of hundred students. Some acted as the "genii' of good and noble men (rather as men who are trainers of dogs). They knew where treasure was hidden, but were forbidden by a special law from revealing its whereabouts.

In his response to Fazio Cardano's dream More does not dismiss it out of hand. Instead he tackles the various claims and interpretations made by the spirits in to show how they are "inveloped in obscurity", and concludes that to decide their veracity would be a waste of time, except in the case of one claim:

[40] Ibid., 212–212). See Hutton (2013a).

[41] *Immortality of the Soul*, 3.17.5, p. 218.

> how true they all are, it would be too much trouble to determine. But one clause [...] I cannot let pass, it so nearly concerning the present Subject, and seeming to intercept all hopes of the Soul's Immortality.[42]

The clause in question is that "our Souls are so farre mortal, as that there is nothing proper to them remaining after death". Thus the main concern of More is that these "Aerial Philosophers" (and by extension Cardano too) were Averroists.[43]

At first sight it seems astonishing that a philosopher who mustered rational arguments to refute Hobbes and to demonstrate the properties of corporeal and incorporeal substance should appeal to such unreliable evidence as dreams, or quibble about the nature and role of demons. However, More is in fact sceptical about whether these dreams can demonstrate what they claim to do – namely that spirits are mortal. In other words, he challenges the authority and value of such dreams as evidence. The point surely is that if this is the best evidence which the mortalists can muster ("These are the most notable Testimonies for the Mortality of Daemons that I have met withall", writes More),[44] it is very poor evidence. This is not to say that the reported dream does not have any significance. But the significance is not its capacity to demonstrate the truth of what is reported. Rather, its significance is what it reveals about the author who has recourse to dreams. More's comments indicate that he thinks the Fazio Cardano's dream is a guide ("cynosure") to Girolamo Cardano's opinions ("I am sure he most-what steers his course in his Metaphysical adventures according to this Cynosura, which is no obscure indication of his assent and belief.") More regarded Cardano's attitude as "shuffling" or equivocating, and the "dream" as full of "paradoxes". The danger of Cardano is not simply that he is wrong about many things, but that in his work, error is mixed with truth in order to deceive:

> if they had a design to winde us into some dangerous errour, it is very likely that they would shuffle it in amongst many Truths, that those Truths being examined, and found solid at the bottome, we might not suspect any one of their dictates to be false. Wherefore this vision being ill meant, the poison intended was, that of the Soul's Mortality the dangerous falseness of which opinion was to be covered by the mixture of the others that are true.[45]

There are two observations about the role of other philosophers in More's critique of Cardano in *Immortality of the Soul* which should be noted. The first is that he attributes Cardano's errors to his reading of Aristotle. This puts him in the same league as two "witty Fools in Philosophy", Pomponazzi and Vanini. More identified Aristotle as the origin of their naturalizing astrology, attributing "to that first Errour in the Aristotelian Philosophy that makes God and the Intelligences act from the heavenly sphears and so to produce all those Effects of Nature below". Aristotle failed to realize that these effects can only be achieved "by a present *Numen* and

[42] Ibid., 3.17.10, p. 221. More did apparently accept Cardano's dream as authoritative about some things: in *An Explanation*, he cites this passage as evidence that spirits "may be divided in their judgements" (ibid., 6.5.1, 226).

[43] On Cardano's Averroism, see Valverde (2013).

[44] *Immortality of the Soul*, 3.17.10, p. 221.

[45] Ibid., 223.

Spirit of Life that pervades all things' – i.e. the Spirit of Nature.[46] Another error of Aristotle's was his denial that souls have memory after death, a view with which Cardano concurred.[47]

A second feature of More's discussion is his invocation of modern philosophy, that is to say Cartesian physics, to refute Cardano, and to support his own theory of the soul. So, for example, one objection against souls being joined to bodies, aerial or other, is that they would be prone to separation from their bodies when buffeted by winds, "blown to pieces" by the "Windes and Tempests" which wrack the lower region of the air. This is an objection that originates with Lucretius. More counters it by invoking Cartesian physics: since according to the new cosmology of Descartes, "Windes are nothing else but Watery particles", the windes "do not so much drive the Aire before them, as pass through it, as a flight of arrows and showers of haile or rain", so "the Aire is not torn apieces thereby".[48] Another example of his use of Cartesianism, concerns the vehicle of the soul, which according to More contains as much solid matter as the bodies of men. He argues this from Descartes' conception of matter as extension.

> According therefore to his [Descartes'] Philosophy and the Truth, there is as much matter or Body in a cup of Aire as in the same cup filled with water, and as much in a cup of water as if it were filled with Lead or Quicksilver. Which I take notice of here, that I may free the imagination of men from that ordinary and idiotick misapprehension which they entertain of Spirits that appear, as if they were as evanid and devoid of Substance as the very Shadows of our Bodies cast against a Wall, or our Images reflected from a River or Looking-glass [...].[49]

Thus, in More's view Cardano is not just an obscurantist, but his errors can be linked to unreliable traditional philosophy, namely, Aristotle, who, as More observes "has the luck to be believed more than most Authors".[50] By contrast, More shows to his own satisfaction that the new natural philosophy of Descartes, which, as More notes in his *Epistola H. Mori ad V.C.* (1662) "saves" the phenomena of nature, is consonant with his own theories. We might also note that in the course of More's debates about the soul, he broaches ideas which would become significant in subsequent philosophical debates, notably his formulation of a theory of personal identity, based on consciousness.

---

[46] Ibid., 3.3.9, p.158.

[47] Ibid., 3.11.1, 187.

[48] Ibid., 3.3.11, p. 159.

[49] Ibid., 3.2.7, p. 153. Later in the argument, More invokes Descartes' vortical theory and his view that the sun is a star the light of which will eventually be smothered by maculae, in order to explain why the souls of the good have nothing to fear from the "extinction of the sun". Ibid., 3.19.1, p. 231.

[50] Ibid., 2.12.12, p. 114.

More Verses Cardano: Renaissance or Modern?

More as we know devoted his energies to working out a metaphysics which supported the new natural philosophy. His first step in that direction had been in his correspondence with Descartes in 1644, when he proposed the notion of incorporeal extension. *Immortality of the Soul*, published 15 years later, targeted the deniers of incorporeal substances. In it More was determined to oppose atheism by destroying a central plank in the atheist's case – that there are no spiritual substances. But to do so also required the formulation of a viable pneumatology. And that required correcting mistaken theories of the soul which might undermine the grand design to counter atheism, whether those erroneous or imperfect theories were proposed by Renaissance thinkers like Cardano or modern ones like Descartes.

More was, like Cardano, trying to reach the parts that other philosophies could not reach – the spiritual realm, with which neither Aristotelianism (according to Cardano) nor the mechanical philosophy (according to More) could deal satisfactorily. But for More, Cardano represents the obverse problem from Hobbes: the latter being a materialist who denied the existence of spirit, the former fully signed up to the existence of spirits, but so wildly that he tipped over into heterodoxy and atheism. Likewise, where More's quarrel with Descartes was that he could not find any place for the soul in his "nullibist" universe, his quarrel with Cardano was that he filled his cosmos with all sorts of wrongly conceived spirits, and packaged his arguments deceptively.

Against Cardano he uses the same strategy as he used when tackling Hobbesian atheists, by doing so in his adversary's own terms, using his adversary's weapon. In style Cardano's works are amorphous (Anthony Grafton aptly describes them as "omnium gatherums").[51] *De subtilitate* in particular seems disorganized and random – those who try to introduce it to modern readers apologise for this fact. By comparison with Cardano, More's English writings are models of clear organisation. To those readers more familiar with the written style of Descartes, Locke and even Hobbes, More's style seems baroque to excess. As modern readers of *Immortality of the Soul* we may indeed get lost in the arguments. But More is not lost, and the mazes are not of More's own making – or not entirely. Is it too much to suggest that baroque abundance of More's *Immortality* reflects the style of his Renaissance interlocutor, just as his adoption of arguments formulated *more mathematico* were tailored to his refutation of Hobbes?

More did not confront contemporary philosophy out of a dogmatic commitment to or, possibly, a nostalgic *penchant* for the philosophy of the past, either Renaissance or Ancient. What we have in More's *Immortality of the Soul* is not so much credulity but critique, not spiritual contemporaneity with Renaissance Florence but creative engagement with contemporary philosophy. More was trying to steer a course in between the excesses of different philosophies, correcting them by advancing metaphysical hypotheses to integrate body and soul (the vehicle of soul, vital congruity),

[51] Grafton (1999), 162.

and to retain a principle of life in nature (spirit of nature). He was not 'caught' between ancient and modern, grasping at new ideas against his better judgment. He was not trying to hold on to some things while rejecting others: his aim was to expose the dangerousness of Cardano's belief, and he uses modern ideas to dispel the obscurantism and misbelief in Cardano. He was out to show that his theory of the soul stands up to the scrutiny of "naturalists" by using arguments "onely within the bounds of Natural Light".[52] To that end he offered an updated natural theology, updated by bringing in the new philosophy to support it, to erect an "exoterick *Fence or exterior fortification about divinity*", as he put it in the introduction to *A Collection of Several Philosophical Writings*.[53] This was not a simple case of modern versus Renaissance, but of finding convincing arguments to combat equivocal or specious reasoning. It was a defence of sound theology, against the heterodox views of the likes of Cardano, Pomponazzi and Vanini. More did so by showing that his own views made sense in terms of modern theory, even if, in doing so he was defending rather unusual theories – the idea of the vehicle of soul was not widely accepted, and the doctrine of the pre-existence of the soul was regarded as heterodox. His own pneumatology has long since been discarded, but in the arguments which supported it, new themes emerged which would become a staple part of the philosophical debates of the later seventeenth century.

Bibliography

Baldi, Marialuisa, and Guido Canziani. 1999. *Girolamo Cardano. Le opera, le fonti, la vita*. Milan: Franco Angeli.

Boyle, Robert. 1682. *New Experiments Physico-mechanical Touching the Air*. Oxford: Miles Flesher for Richard Davis.

Burthogge, Richard. 1699. *Of the Soul of the World and of Particular Souls*. London: Daniel Brown.

Copenhaver, Brian P. 1992. *Renaissance philosophy*. Oxford: Oxford University Press.

Crocker, Robert. 2003. *Henry More, 1614–1687: A Biography of the Cambridge Platonist*. Dordrecht: Springer.

Darwall, Stephen. 1995. *The British Moralists and the Internal 'Ought', 1640–1740*. Cambridge: Cambridge University Press.

Ernst, Germana. 2001. "Veritatis amor dulcissimus": Aspects of Cardano's Astrology. In *Secrets of Nature: Astrology and Alchemy in Early Modern Europe*, ed. William R. Newman and Anthony Grafton, 39–68. Cambridge, MA: MIT Press.

Giglioni, Guido (2013a). Girolamo Cardano: University Student and Professor. *Renaissance Studies* 27:517–532.

Giglioni, Guido (2013b). Girolamo Cardano. In *Stanford Encyclopedia of Philosophy*. http://plato.stanford.edu/archives/sum2013/entries/cardano/. Accessed 4 Nov 2015.

Gill, Michael B. 2006. *The British Moralists on Human Nature and the Birth of Secular Ethics*. Cambridge: Cambridge University Press.

Grafton, Anthony. 1999. *Cardano's Cosmos: The Worlds and Works of a Renaissance Astrologer*. Cambridge, MA: Harvard University Press.

[52] *Immortality of the Soul*, Preface, section 2, p. 2.

[53] Ibid., vi.

Grafton, Anthony, and Nancy Siraisi. 2001. Between the Election and My Hopes: Girolamo Cardano and Medical Astrology. In *Secrets of Nature: Astrology and Alchemy in Early Modern Europe*, ed. William R. Newman and Anthony Grafton, 69–131. Cambridge, MA: MIT Press.

Hutton, Sarah. 2013a. The Cambridge Platonists and Averroes. In *Renaissance Averroism and Its Aftermath: Arabic Philosophy in Early Modern Europe*, ed. Anna Akasoy and Guido Giglioni, 197–211. Dordrecht: Springer.

Hutton, Sarah. 2013b. The Platonic Theology of the Cambridge Platonists. In *The Rebirth of Platonic Theology*, ed. James Hankins and Fabrizio Meroi, 295–310. Florence: Olschki.

Koyré, Alexandre. 1957. *From the Closed World to the Infinite Universe*. Baltimore: Johns Hopkins Press.

Lassels, Richard. 1670. *The Voyage of Italy*. Paris: Moutier.

Leech, David. 2013. *The Hammer of the Cartesians: Henry More's Philosophy of Spirit and the Origins of Modern Atheism*. Leeuven: Peeters.

Maclean, Ian. 1986. The Interpretation of Natural Signs. Cardano's *De subtilitate* and Scaliger's *Exercitationes*. In *Occult and Scientific Mentalities*, ed. Vickers Brian, 231–251. Cambridge: Cambridge University Press.

Maclean, Ian. 2005. Heterodoxy in Natural Philosophy and Medicine: Pietro Pomponazzi, Guglielmo Gratarolo, Girolamo Cardano. In *Heterodoxy in Early Modern Science and Religion*, ed. Brooke John and Maclean Ian, 1–30. Oxford: Oxford University Press.

Maclean, Ian. 2007. Girolamo Cardano: The Last Years of a Polymath. *Renaissance Studies* 21: 587–607.

Maclean, Ian. 2009. *Learning and the Market Place: Essays in the History of the Early Modern Book*. Leiden: Brill.

More, Henry. 1642. *Psychōdia platonica: Or a Platonicall Song of the Soul*. Cambridge: Roger Daniel.

More, Henry. 1659. *The Immortality of the Soul [...]*. London: James Flesher.

More, Henry. 1660. *An Explanation of the Grand Mystery of Godliness [...]*. London: James Flesher.

More, Henry. 1662. *A Collection of Several Philosophical Writings*. London: James Flesher.

Naudé, Gabriel. 1661. *Instructions Concerning Erecting of a Library*, (Engl. trans. John Evelyn). London: G. Bedle and T. Collins.

Reid, Jasper. 2012. *The Metaphysics of Henry More*. Dordrecht: Springer.

Schneewind, J.B. (ed.). 2003. *Moral Philosophy from Montaigne to Kant*. Cambridge: Cambridge University Press.

Valverde, José Manuel G. 2013. Averroistic Themes in Girolamo Cardano's *De immortalitate animorum*. In *Renaissance Averroism and Its Aftermath: Arabic Philosophy in Early Modern Europe*, ed. Akasoy Anna and Giglioni Guido, 145–171. Dordrecht: Springer.

Ward, Richard. 2000. *Life of Henry More*, eds. Sarah Hutton et al. Dordrecht: Kluwer Academic Publishers.

Part II
Natural Philosophy

Chapter 6
From *Attractio* and *Impulsus* to Motion of Liberty: Rarefaction and Condensation, Nature and Violence, in Cardano, Francis Bacon, Glisson and Hale

Silvia Manzo

Abstract There was a particular way of understanding and explaining changes in matter's quantity whose first exposition can be traced back to the Renaissance in Girolamo Cardano's classification of the natural motions of the universe, particularly in the motions of *impulsus* (impenetrability) and *attractio* (abhorrence of a vacuum). Cardano's exposition was read attentively by Francis Bacon, whose idea of "motion of liberty" both modified and retained elements of the Cardanian view. The Baconian treatment of the motion of liberty made its way well into the seventeenth century in the works of Francis Glisson and Matthew Hale, who draw heavily on it to provide their own account of rarefaction and condensation. The aim of this essay is to reconstruct the history of the accounts of the processes of rarefaction and condensation held by these authors in order to examine the ramifications of the Cardanian approach in the seventeenth century. This history will not only provide us with new instruments for understanding the intellectual relationship between the Renaissance and the early modern period but also improve our understanding of the transformation of the world picture across the emergence of early modern science.

The accounts of the changes of matter's quantity in terms of rarefaction and condensation were part and parcel of Renaissance and early modern natural philosophy. Being closely related to the conceptions of matter and motion, they express different views on nature as a whole. Antecedents of such accounts can be found in the Aristotelian corpus and in its re-elaborations by medieval commentators, which laid particular emphasis on distinguishing natural from violent motion when talking about the limits of quantitative material changes. In the wide range of alternative views on these

I want to thank José Manuel García Valverde for his helpful comments on an early draft of this essay. All translations are my own.

S. Manzo (✉)
IDHICS (Unversidad Nacional de La Plata – CONICET) – FAHCE, Ensenada, Argentina
e-mail: manzosa@yahoo.com.ar

C. Muratori, G. Paganini (eds.), *Early Modern Philosophers and the Renaissance Legacy*, International Archives of the History of Ideas Archives internationales d'histoire des idées 220, DOI 10.1007/978-3-319-32604-7_6

questions developed throughout the centuries, there was one particular way of understanding and explaining the changes of matter's quantity which extended from the Renaissance to the seventeenth century. In Girolamo Cardano's (1501–1576) classification of the natural motions of the universe we find its initial formulation through the motions that he called "attractio" and "impulsus". Shortly afterwards Francis Bacon (1561–1626) built his own classification of motions and introduced a kind of motion, namely the "motion of liberty", in which elements of the Cardanian account of rarefaction and condensation are clearly visible. In turn, the Baconian motion of liberty proved to be highly influential in Francis Glisson's (1597–1677) and Matthew Hale's (1609–1676) account of rarefaction and condensation.

In explaining rarefaction and condensation, Cardano, Bacon, Glisson and Hale assumed certain views on matter, motion, nature and violence, showcasing an interesting intellectual itinerary which I will try to reconstruct in this essay. I will argue that despite some relevant conceptual changes undergone by the explanations of rarefaction and condensation throughout the works of these authors, noteworthy continuities can be observed in the intellectual tools through which both processes were thought. These continuities show permanent concerns along with some slight changes in thought and context, but not abrupt breaks from the Renaissance to the early modern period.

Cardano and the Motions of *Attractio* and *Impulsus*

Although Cardano was a physician and a prolific author who published on several subjects, he was mostly known for his major work, the *De Subtilitate*, first published in 1550 and harshly attacked by Julius Caesar Scaliger in the equally famous *Exotericae exercitationes de subtilitate* (1557).[1] His work is not an exception to Renaissance eclecticism: it combines Late Aristotelianism – mainly in its Averroistic version – with Platonism, Epicureanism, and Stoicism.[2]

In the first book of *De subtilitate*, Cardano, provides the fundamental components of his natural philosophy that, despite remaining inside a mostly Aristotelian framework, intend to work out an alternative account.[3] It proposes five principles of nature (matter, form, soul, place, and motion)[4] clearly in contrast to the Aristotelian triad matter, form and privation.[5] Cardano's idea of matter is centred around a quantitative Averroistic approach, according to which matter has an undetermined

[1] On the strategies, targets and conflicting ontologies involved in this confrontation see Giglioni (2015).

[2] Giglioni (2013).

[3] Nenci (2004), 22–26 notes this questioning of the traditional Aristotelian view in regards to the concept of void.

[4] Cardano (1663), III, 368a.

[5] The first edition of *De subtilitate* (1550) includes the world soul among the principles. See Schütze (2000), 53–54.

quantity and receives a determinant form by which it turns to be an specific being.[6] From concrete examples that show that an amount of one substance cannot be transmuted into an equal amount of another substance, he concludes that prime matter must be endowed with some quantity.[7] He, therefore, maintains that prime matter lacks form but is not completely deprived of essence. The essence of prime matter *per se* consists of an undetermined or undefined quantity:

> But prime matter is not deprived of all things, for, as mentioned before, it is not possible to make a handful of iron from a handful of straw, because it is little matter; neither can a handful of straw be made from a handful of iron, as there is a surplus of matter. From this it follows that prime matter has a quantity that we call undetermined, since it does not determine for itself exact limits, because according to its form it takes up more or less space.[8]

On the basis of this quantitative character of prime matter Cardano draws important conclusions concerning the transmutation of bodies: a given substance cannot be transmuted into any other substance in any amount. Therefore, transmutations are regulated by definite proportions which establish the conditions under which bodies can change and receive new forms. For Cardano, the undetermined quantity of prime matter entails that once matter is determined by secondary forms, it is subjected to quantitative terms, as it is attested by empirically observed rarefactions and condensations. This quantity no longer belongs to the prime matter but to the matter specified by the diverse forms in the compounds. Cardano referred to the limits and changes of matter by means of the image of the god Proteus, who continually changes his figure. Thus, he states that matter "has circumscribed limits of abundance and scarcity, within which Proteus is subjected to infinite terms in its magnitude."[9] Soon after introducing Proteus's image in relation to the quantitative limits of matter, Cardano adds some lines suggesting somehow a correspondence between knowing nature and being able to alter it "by art or by chance". In this approach, the preservation of the form plays a central role and, as we shall see, lies behind Cardano's theory of natural motions:

> Therefore the essence of matter follows from the knowledge of matter, like when we hinder the coming forms to preserve the previous form […] Since given that the prime matter necessarily is always endowed with a form, if the imposition of a new form is hindered by art or by chance, necessarily the precedent form remains. And from there, every method of preserving the form arises.[10]

---

[6] The importance of matter's quantity in Cardano's *De subtilitate* is noted by Schütze (2000), 66–67 and Ch. 3.2.

[7] Cardano (1663), III, 359 a. Vd.. Schütze (2000), 59–67; Bianchi (1994), 118–119.

[8] Cardano (1663), III, 359a: "Sed nec materia prima omnibus est spoliata rebus: cum enim (ut dixi) neque ex pugillo paleae, ferri pugillus ob materiae paucitatem, nec rursus ex ferri pugillo, paleae pugillus ob illius redundantiam, sequitur ut materia prima quantitatem quandam retineat, quam indefinitam vocamus. Namque non sibi certos describit limites, cum modo sub forma maius complendo spatium, modo minus latitet."

[9] Ibid.

[10] Ibid., 359b: "Porro ad materiae scientiam, imo essentiam sequitur, vt cum formas ipsas aduenientes impedimus, prior remaneat […] Etenim cum necesse sit materiam primam semper sub aliqua

For this reason Cardano opposes the opinion of Alexander of Aphrodisias that matter has an indifferent power to acquire any form, regardless of the quantities required for such forms. Cardano claims that in Alexander's opinion a substantial change, v.g. from water to air, can take place through rarefaction regardless of quantitative constraints. By way of contrast, Cardano holds that a given amount of matter informed by a specific form can receive only certain forms on account of the proportions that exist in nature. Based on this assumption, he explains the changes of forms in rarefaction and condensation processes. Given that "bodies that are rarified or condensed, partially change the form" and that "the form of the body follows the form of the element", Cardano argues that rarefaction and condensation never allow the coexistence of two forms in the same matter. In this account, the penetration of the dimensions of one body by another body does not occur since "it is repugnant that two bodies be at the same place at the same time, not on account of matter – which is only potentially in one place and indeed is distinguished from quantity only potentially, not in the act – but rather on account of the diversity of forms."[11] Thus, the ultimate reason for the penetration of bodies' dimensions is not matter, but form, which determines the limits of matter's quantity required for actualizing that form.

Such a view of the proportions by which forms regulate body transmutations permeates Cardano's classification of motions, the main subject of book one. His classification distinguishes three kinds of natural motions in the universe: (1) *attractio*, (2) *impulsus*, and (3) downwards and upwards motion. Particularly, his concern with the proportions regulating transmutation lies behind his concepts of the motions that he calls *attractio* (attraction) and *impulsus* (repulsion). Thus, when defining the motion of *attractio*, Cardano assumes that there is a limit of rarity, which causes such a motion. According to this, the motion of *attractio* occurs when bodies are expanded to such an extent that it is nearly possible that vacuum takes place. The parts of the body do not tolerate being to be extended any more so that they keep united and a vacuum does not occur. In the same way, quantitative terms play a role in Cardano's definition of *impulsus*, in so far as he postulates that in nature there is a limit of density which incites this motion. Thus, the motion of *impulsus* is said to occur either when a body enters in the place of another body or when this body becomes dense to such an extent that a higher amount of density is not allowed.[12]

Attraction and *repulsion* are, therefore, considered as opposite motions, the first related to the limit of extensibility and the second related to the limit of compressibility tolerated by nature. Whereas *attraction* is associated with the rejection of vacuum (*fuga vacui*), *repulsion* is linked to the resistance to the penetration of the dimensions of the body (*repugnantia corporum*).

---

iacere forma, si subsequens impediatur arte, vel casu, priorem formam manere necesse est. Inde igitur tota praeseruandi methodus ortum habuit."

[11] Cardano (1663), III, 360b.

[12] Ibid., 360a, b.

Finally, along with these two motions, Cardano postulates the existence of the downwards and upwards motion discerned by the Aristotelian tradition as linked to gravity and levity respectively. In summary, he presents the three natural motions of the universe as follows:

> The first and certainly the most powerful motion stems from the rejection of a vacuum, but more exactly from the form of the element [à forma elementi], which does not admit further rarity, for the form does not allow higher rarity, and the parts of the matter can never be separated. […] The second […] is exactly the contrary to it, for the first one occurs due to the vacuum. This seems to happen so that bodies would not penetrate each other, but due to a reason contrary to that of the former, that is, so that the form would not obtain even a little more prime matter than it needs, as well as the first motion occurs so that it not obtains even a little less […]. The third is the motion of the heavy bodies downwards and the light ones upwards […].[13]

In presenting the three motions as *natural* motions, Cardano shows continuities as well as ruptures with Aristotelianism.[14] On the one hand, he agrees with the Aristotelian conception that natural motions are caused by an intrinsic principle of the substance – and this principle is the form of each substance. But, on the other hand, he distances himself from this tradition by adding that the form not only prescribes motions of gravity (downwards) and levity (upwards) but also involves a consensus with the universe tending to preserve the form. Thus, in Cardano's view, "natural places" established by the forms are not the only keys to explain natural motions. He adds that, besides being determined by natural places prescribed by forms, natural motions depend on a general consensus of individual bodies with the universe. Thus, there exists a universal "consensus" and "obedience" on behalf of which, under certain circumstances, bodies abandon their "proper motions" (*motus proprii*) in order to satisfy their consensus with the universe, so that heavy bodies move spontaneously upwards and light bodies move spontaneously downwards. If hermetically sealed bellows cannot be opened, unless air enters through some tiny opening or they are broken, it is because, on account of its consensus with the universe, the "form" of air is "not capable to tear itself apart or separate itself." The same explanation is true for the water contained in hydraulic machines which eventually moves upwards.[15] Those examples show, in Cardano's opinion, that the

---

[13] Ibid., 360b- 361a. I quote the critical reading of the text quoted by Schütze (2000), 73: "Primus quidem ac validissimus a vacui fuga, sed verius a forma elementi, cum maiorem raritatem non admittat, nec materiae partes separari unquam queant. […] Secundus […] e directo huic contrarius specie quidem ut primus a vacuo fit. Hic ne corpora se mutuo penetrent factus videtur, sed verius oppositam priori rationem, ne scilicet forma plus iusto, quam debeat materiae primae consequatur, sicut in priore ne iusto minus. […] Tertius autem motus gravium est ad inferna, levium ad superna […]."

[14] See Nenci (2004), 24–25 and Schütze (2000), 67–72.

[15] For examples of how this theory works in Cardano's explanations of hydraulic and mechanic machines see Nenci (2003), 68–71.

motion of *attractio* is a natural motion. Similarly, the "same reasons" prove that the motion of *impulsus* is natural too.[16]

By contrast, according to the Aristotelian orthodoxy *attractio* and *impulsus* should be considered violent motions, because the principle of the motion does not lie in the body which is moved. Only downwards and upwards motions – to and from the centre of the universe respectively – can count as natural motions[17] For Cardano, *repugnantia corporum* – linked to *impulsus* – and *fuga vacui* – linked to *attractio* – "are not principles of nature, but are dependent on the form".[18] Being dependent on the form – which has a consensus with the universe – they are spontaneous motions according to the nature of the substances.[19]

Thus, Cardano completes his explanation of body changes in rarefaction and condensation as well as his account of the motions intervening in such processes by combining forms – conceived in a particularly quantitative way[20]— with the consensus of particular substances with the universe. In doing this, he builds, inside the Averroistic pattern of his matter theory, an alternative theory of motion which redefines the ranges of nature and violence.

Francis Bacon and the Motion of Liberty

Bacon mentions Cardano on several occasions and judges his work both as a natural historian and as a natural philosopher, without citing any specific work by Cardano.[21] However, it can be stated with enough certainty that Bacon was acquainted with *De*

[16] Cardano (1663), III, 360b-361a. The argument based on the subordination of the "proper motions" of individual bodies to the universal consensus was quite widespread in the discussions on void from the Middle Ages to the seventeenth century. See Manzo (2013), esp. p. 25.

[17] Cardano (1663), III, 368b.

[18] Cardano (1663), III, 368a.

[19] For that reason Cardano overtly confronts Averroes's view that motions *ex necessitate vacui* are violent motions. See Cardano (1663), III, 368b and Nenci (2004), 24–25, note 23. It is probable that the terminology employed by Averroes (Aristotle [Aristotelis cum Averrois] (1550), vol. 5, 122r-123v) in commenting Aristotle's *De coelo* (IV, 5, 312b3-19) be the source which inspires Cardano to call *attractio* the motion related to *fuga vacui*.

[20] Bianchi (1994) 120—124 argues that although Cardano introduces form as one of the five principles of nature, his theories of fire, of elements, of soul and of animal generation as expounded in the second book of *De subtilitate* work towards the dissolution of the concept of form, by increasingly replacing qualitative by quantitative categories. Cardano's motion theory, I would suggest, as developed in book one provides another example of this tendency in Cardano's appropriation of the Aristotelian concept of form. Schütze (2000) 71—72 does not agree with Bianchi's interpretation by claiming that in Cardano's philosophy forms keep a substantial meaning in a traditional Aristotelian-Averroistic sense.

[21] Bacon's explicit references to Cardano are found in *Advancement of learning*, OFB, IV, 26; *De augmentis scientiarum*, SEH, I, 456; *Temporis partus masculus*, SEH, III; 530; *Redargutio philosophiarum*, SEH, III, 571 and *Cogitata et visa*, SEH, III, 603. On Bacon and Cardano see also Hutton in this volume.

subtilitate, for the work was immensely popular and was beyond doubt available to him. *De subtilitate* was Cardano's most published work in his lifetime.[22] After his death, it was reprinted four times in Latin before the *Opera omnia* of 1663 and thereafter once more in 1664. In contrast to his works related to medicine and mathematics, *De subtilitate* could not be easily suited to the 'map' of early modern universities.[23] While the statutes of the University of Cambridge of the 1540s and 1570s included Cardano's works for the arithmetic courses,[24] it was not until the 1630s that *De subtilitate* was officially read in metaphysics courses.[25] However, already in the 1570s, when Bacon attended Trinity College, *De subtilitate* was included in the catalogue of the university library, after being donated by Francis's father, Nicholas Bacon.[26]

A view that links Bacon's with Cardano's natural philosophy is his quantitative conception of matter. As a starting point, Bacon rejected completely the idea that, for instance, an amount of water can be transformed into the same amount of air. Assuming the transformation of substances without noticing the consequent changes in quantity would imply that something can be created from nothing or that something can be destroyed.

> Moreover, it is no less certain, even though not so clearly noted or asserted (whatever stories people make up about the impartial potential of matter towards forms) that more or less of this quantity of matter is contained in the same volumes of space according to the diversity of the bodies which occupy them, some of which we find to be very obviously more compact, others more extended or diffuse. For a vessel or cauldron filled with water and air does not hold an equal portion of matter, but more of the one and less of the other. Therefore if someone claimed that a given amount of water could be made from the same amount of air, it would be the same as saying that something can come from nothing.[27]

---

[22] On the reception of *De subtilitate* and Cardano's work in general see Jensen (1994); MacLean (1994); Blackwell (1994).

[23] MacLean (1994) 323–325.

[24] Dyer (1824), I, 161 and Oates (1986), I, 71. McKitterick (1992), 68–70.

[25] During the 1630s the B.A. students who attended the metaphysics courses taught by Joseph Mede read *De subtilitate* along with the critical reply from Julius Caesar Scaliger, *Exotericarum excertitationum libri XV De subtilitate ad Hieronymum Cardanum*. See Rogers (1988), 11 and Jardine (1974), 49.

[26] Cambridge University Library MS Oo. 752 (Donors Book), 19 (items printed in Parker (1729), xlii–xliii). Along with *De subtilitate*, the list of books donated by Nicholas Bacon in 1574 to the University Library includes two more titles by Cadano: the commentary to Ptolomeus, *De astrorum judiciis* and *De somniis* (a short title which probably refers to *Somniorum Synesiorum omnis generis insomnia explicantes*). Another item notes "Albert Dureus [sic] et Hieronimus Cardanus", most probably indicating a bounded volume containing Albert Durer's *Geometric* and Cardano's *De proportionibus*, as can be found in the inventory of the books of the University Library as in 1683 (see *A Table of the Books in the University Library* 1683, MS Cambridge University Library CUR 31.1 (10–12)). On the donation of books to Cambridge University Library by Nicholas Bacon see Durel (1998), 41–44.

[27] Bacon, *Phaenomena universi*, OFB, VI, 10: "Etiam illud non minus certum, tametsi non tam perspicue notatum, aut assertum sit (quidquid homines de potentia Materiae aequabili ad formas fabulentur), ex quanto illo Materiae sub iisdem spatiorum dimensionibus, plus & minus contineri, pro corporum diversitate a quibus occupantur, quorum alia magis compacta, alia magis extensa

Bacon became ever more insistent on the crucial view that the quantity of matter of every substance obeys definite proportions. By holding this view, he is opposing the doctrine, that he attributes to Plato and Aristotle, according to which matter is endowed with an appetite for receiving indifferently any kind of form (a view that, as we have seen, Cardano attributes to Alexander of Aphrodisias) and to the "Peripatetic decuple proportion" of the transmutation of the elements.[28] Bacon argues that postulating an indifferent matter entails ignoring its quantitative determinations. That would imply that any amount of matter is able to be informed by any form.

The association of Proteus with matter and nature was quite widespread across the Renaissance,[29] but the use of this classical figure to illustrate the quantitative limits is distinctive of Cardano and Bacon and, to my knowledge, is not found in any other contemporary. The coincidences between them are strong on this point. Not only did they attribute the same connotation to Proteus's image, but they also had a similar conception of the quantitative attributes of the matter symbolized by it. Like Cardano, Bacon refers to Proteus's transformations in relation to the changes in volume of bodies in experimental contexts in which the state of nature may be altered. One good example of this is what Bacon says in *Sylva sylvarum*: "But if Bodies may be altered by Heat, and yet no such Reciprocation of Rarefaction, and of Condensation, and of Separation, admitted; then it is like that this Proteus of Matter, being held by the Sleeues, will turne and change into many Metamorphoses".[30]

In Bacon's classification of motions, the exposition of the "motion of liberty" (*motus libertatis*) is reminiscent of Cardano's description of the motions of *attractio* and *impulsus*.[31] Bacon defines the motion of liberty as the motion by which "bodies

---

sive fusa evidentissime reperiuntur. Neque enim parem Materiae portionem recipit vas aut concavum aqua & aëre impletum; sed illud plus, istud minus. Itaque si quis asserat, ex pari aëris contento, par aquae contentum effici posse: idem est ac si dicat aliquid fieri posse ex nihilo". Cfr. *Cogitationes de natura rerum*, III, 23; *Historia densi et rari*, OFB, XIII, 38.

[28] Bacon, *Historia densi et rari*, OFB, XIII, 70. For the Aristotelian theory of transmutation see Needham (2006).

[29] Burns (2001).

[30] Bacon, *Sylva sylvarum*, SEH, II, 382. Cfr. Bacon, *Historia densi et rari*, OFB, XIII, 101: "that the separation and reciprocation of rarefaction and condensation be completely prohibited, [...] for perhaps this will keep the Proteus of matter in handcuffs and force it to act". On the other hand, Bacon associates Proteus with matter in general and with motion. See *De sapientia veterum*, SEH, VI, 652–3; *Filum Labyrinthi*, SEH, III, 625; *Cogitationes de natura rerum*, SEH, III, 20–21; *De augmentis scientiarum*, SEH, I, 632. For a study of this image in Bacon see Pesic (1999).

[31] Here I will deal only with the presentation of the motion of liberty in *Novum organum*, where it is developed in more detail. See also the shorter definition presented in Bacon, *Abecedarium novum naturae*, OFB, XIII, 192: "Corpora naturalia suam exporrectionern siue dimensum libenter tuentur, praeternaturalem siue pressuram siue tensuram fugiunt. Alia tamen alijs longe cedunt benignius aut obstinatius pro modo texturae suae; quinetiam postquam vim subierint, si detur copia, se in libertatern vindicant & restituunt. Hunc itaque motum, motum libertatis appellamus. Videtur enim libertatis quidam amor, qui se constringi aut trahi aegre patiatur. Duplex autem est motus iste; alius a pressura, alius a tensura; atque vterque eorum geminus, quatenus corpora cedunt, quatenus se restituunt. Quoniam autem iste motus constituit eum qui vulgo violentus vocatur".

exert themselves to be free from preternatural pressure or tension, and to restore themselves to a dimension convenient to their body".[32] This account might have been partly inspired in Hero's *Liber spiritualium*, which refers to the "preternatural" compression and separation of the parts of bodies, to which bodies answer by restoring their previous volume.[33] Bacon's exposition subdivides the motion of liberty in two further kinds: the motion of liberty from pressure (*a pressura*) and the motion of liberty from tension (*a tensura*). Bacon tells us that this second sub-group was labeled by "the School" (*Schola*) "motion produced by the form of the element" (*motus ex forma elementi*). He notes that, in doing so, the School wrongly associated the motion of liberty with the specific nature of air, water, and fire. By way of contrast, he claims that this motion belongs not only to those elements but also to all range of consistencies "in which each body has its own characteristic dimension, and is with difficulty forced from that to any noticeable degree".[34]

On the other side, Bacon adds that "some men" confuse the motion of liberty with the twin motions of *antitypia* and of *nexus* – both of them classified as diverse and fundamental kinds of motions by Bacon. Thus, the liberation from pressure is carelessly confused with the motion of *antitypia*, while the liberation from tension is wrongly confused with the motion of *nexus*, "as if compressed bodies would give way or dilate themselves to avoid *penetration of dimensions*" and "tensed bodies would spring back and contract to stop a *Vacuum* being formed."[35] Bacon distances himself from such an account by showing that neither the *fuga vacui* nor the rejection to the *penetratio dimensionum* really matter in the motion of liberty. Since, if compressed air wanted to acquire the density of water, there would be no need of penetration of its dimensions and yet there would be much greater compression of air than is actually sustained. Similarly, if water attempted to expand and achieve the rarity of air, there would be no need of a void and yet the extension of water would be much greater than it is in any way allowed. That entails, in Bacon's opinion, that the motion of liberty is not produced to impede the existence of a void or the penetration of dimensions, but it rather stops long before bodies reach such extreme degrees of rarefaction and condensation. Besides, unlike Cardano, Bacon admits the possibility of the penetration of bodies. What bodies "desire"

---

[32] Bacon, *Novum organum*, OFB, XI, 385.

[33] Hero (1583), 9: "quemobrem vi quadam accedente aerem densari contingit, et in vacuorum loca residere, corporibus praeter naturam inter sese compressis: remissione vero facta rursus in eundem ordinem restituitur, ob naturalem corporum contentionem, quemadmodum et in cornuum ramentis, et in spongijs siccis: quae si compressa remittantur, rursus in eundem locum redeunt, eandemque accipiunt molem. Similiter si aliqua vis aeris particulae a se invicem distractae fuerint, et maior praeter naturam locus vacuus fiat, rursus ad sese recurrunt, per vacuum enim celerem corporum lationem fieri contingit, nullo obstante, aut repellente, quo usque corpora ad sese applicentur." The *Liber spiritualium* was a major source for the debates on void, with which both Cardano and Bacon were very well acquainted. See, for instance, Cardano (1663), III, 369b, and Bacon, *Cogitationes de natura rerum*, SEH, III, 16–17.

[34] Bacon, *Novum organum*, OFB, XI, 386.

[35] Ibid., 387. Cf. *Phaenomena universi*, OFB, VI, 46.

in the motion of liberty is to keep their own "consistencies" (that is, the dimensions of bodies).

Finally, Bacon warns that the motion that is commonly called "violent motion" (*motus violentus*) and that "we call mechanical" is simply one instance of the motion of liberty, that is, from compression to relaxation.[36] Bacon gives new contents to the Aristotelian distinction between violent and natural motion, and understands violent motion in different ways which cannot be addressed in this essay.[37] Suffice it so say here that in the treatment of the motion of liberty the concept of violence is to be linked to the "preternatural" pressure and tension altering the "natural" dimensions of bodies. Violence denotes a kind of impediment to or alteration of the actual dimensions of a body that is freely enjoying its own nature. That is, a violence to the state of *natura libera* – to use the label coined by Bacon in his natural history – or nature in its ordinary course. The motion of liberty occurs when an external agent (be man or nature) forces this state of *natura libera* to a preternatural situation of tension or pressure, a situation that is outside the ordinary course of bodies. Thus, Bacon uses the word "violence" in talking about instances of the motion of liberty, despite giving this word a meaning that does not suit the Aristotelian and the Cardanian sense.[38]

The echoes of Cardano's presentation of *attractio* (or *motus a forma elementi*) or *fuga vacui* and *impulsus* or *repugnantia corporum* in Bacon's exposition of the motion of liberty are easily visible. However, the fact that Bacon ascribes the denomination *motus ex forma elementi* to the *Schola* is quite astonishing. I could not find neither direct nor indirect references to this kind of motion in the Late Scholastic texts with which Bacon had most probably been acquainted (namely, natural philosophical works of Agostino Nifo,[39] John Case,[40] Cornelius Valerius,[41] Benedictus Pererius, Jacopo Zabarella, the Coimbrans, Johannes Magirus, and Bartholomeus Keckerman).[42] In this literature there is no motion labeled *motus ex forma elementi*

---

36 Bacon, *Novum organum*, OFB, XI, 386.

37 *Ibid*., 386; *Abecedarium novum naturae*, OFB, XIII, 192. On Bacon's view of violent motion see Manzo (2004) and Pesic (2014).

38 See for instance, Bacon, *Sylva sylvarum*, SEH, II, 342–343.

39 Nifo's natural philosophical work was included among the books donated by Bacon's father and was a very widely read Renaissance Aristotelian source.

40 On the influence of John Case in Renaissance England see Schmitt (1983), 220–221 and *passim*.

41 Cornelius Valerius was author of an epitome of Aristotelian philosophy (*Physicae, seu de naturae philosophiae institutio*, 1567) who is named in a Letter to Fulke Greville (ca. 1589), attributed to Bacon (*Letter of advice to Fulke Greville*, OFB, I, 207).

42 I have checked those Late Scholastic books dealing with natural philosophy that circulated in English universities during Bacon's lifetime: Benedictus Pererius, *De communibus omnium rerum naturalium principiis et affectionibus* (Rome, 1562); Jacopo Zabarella, *De rebus naturalibus* (Cologne, 1590); Collegium Conimbricense, *Commentaria in octo libros physicorum* (Coimbra, 1592); Johannes Magirus, *Physica Peripatetica* (Frankfurt, 1597, later published as *Physiologia Peripatetica*); Bartholomäus Keckermann, *Systema physicum* (Danzig, 1610). See Schmitt (1975); Ashworth (1988); Rogers (1988).

(or similar); neither is a category of motion set forth which set limits to rarefaction and condensation processes. Besides, the account of the motion of the four elements by those authors was devoted to the central question of whether the cause of the motion of the elements is their intrinsic form or an external agent. Long and intricate discussions were devoted to this issue, but none of them refers to something similar to the *motus ex forma elementi*. For these authors, the form causes the motion of the elements in the sense that it is the intrinsic efficient cause by which heavy bodies tend to the center of the Earth and light ones tend to the heaven.[43] It seems beyond doubt that the source of Bacon's discussion must have been Cardano's *De subtilitate*.[44]

From Forms to Appetites

As we have seen, despite their similar views on the quantitative determinations of matter, Cardano and Bacon had different conceptions of motion. In order to clarify their theoretical differences in this regard, I would like to consider very briefly an experiment analyzed by both authors.[45] The first step of the experiment consists of extracting through a tiny hole the air contained in a glass egg. Immediately afterwards the hole is covered and then the glass egg is immersed in water leaving the hole opened again. As a result, it is observed that water enters the glass egg up to certain point, but without filling it completely.

According to Cardano, this experiment shows that a substance can be now rarer and now denser. By claiming that when bodies become rarer or denser, they "partly change their form", he apparently means that their form change not in its essential attributes but only in what it is not essential. Thus, the form of air establishes certain essential limits to rarity and density. When volume changes do not surpass those limits, bodies change its form only in part. Against this background Cardano's obscure and brief explanation of the glass egg experiment suggests that the quantitative constraints of the form of air cause a series of motions of attractio and

---

[43] A long disquisition on the subject can be found in Zabarella (1590), *De motu gravium et levium*, liber I, Chaps. 1–14.

[44] I have suggested the Cardanian inspiration of Bacon's motion of liberty for the first time in my PhD Dissertation defended at the University of Buenos Aires in 2000 (later on published in a shorter version in Manzo (2006), 191–196). The Oxford Francis Bacon editors of the Novum organum also note the coincidence between Cardano and Baconian motions in OFB, XI, 574–575

[45] One version of this experiment is exposed in Hero, *Liber spiritualium*, who drew on Phylo. See De Waard (1936), 67. Bacon mentions in passing the experiment as an example of the motion of liberty from tension in *Novum organum*, OFB XI 386. He deals with it in more detail in *Novum organum*, OFB XI 373; *Phaenomena universi*, OFB, VI, 42; and *Historia densi et rari*, OFB, XIII, 122.

impulsus.[46] On account of the suction, air becomes rarer and when it reaches a state of extreme rareness, it reacts by compressing itself in order to avoid a vacuum (motion of attraction). In this reaction, air compresses up to a certain point, because by repulsion it does not tolerate to be condensed beyond a certain limit. Since air occupies now less space than before, water is forced to enter the glass egg, again by attraction. In addition, Cardano makes the remark that the motion of repulsion, by keeping the due limits of air density, preserves the form of air, since if matter reached a higher level of density, then it would acquire the density of water, for when air condenses it passes over into the form of water. But that is impossible, because in such a case the same matter would receive two coexisting forms (the forms of air and of water).[47]

By way of contrast, in Bacon's opinion the cause of the motion observed in this experiment is the desire of air to restore its former dimension. Given that the suction provokes an extension of air beyond the limits of its own nature, the air, "racked and dilated beyond what was natural to it", reacts by "struggling to withdraw and contract itself" and as a consequence "drew in an amount of water enough to let the air regain its former sphere or dimension".[48] For Bacon, the metaphysical problem of the coexistence of two forms in the same matter, which worries Cardano, does not exist. The processes of rarefaction and condensation produce changes in bodies and even transmutations from a species to another but they do not imply the metaphysical incompatibility that Cardano pointed out. Besides, the penetration of bodies is possible, but in this particular instance the appetite of air to restore its dimension predominates.

Liberty and Beyond: Francis Glisson and Matthew Hale

Francis Glisson probably had knowledge of *De subtilitate* and certainly read Bacon's account of motion in *Novum organum*. In the *Tractatus de natura substantiae energetica* (1672) he works out a theory of nature according to which all matter, organic or inorganic, is endowed with life. Matter is an energetic substratum capable of perception, appetence and motivity on its own.[49] Like Cardano and Bacon, Glisson adopts an Averroistic view of matter (through the influence of Zabarella) by assuming that prime matter has an indeterminate quantity.[50]

---

[46] Cardano (1663), III, 360a-b: "Igitur aer ipse cogi potest, ac seipsum subingredi, eademque ratione rarior euadere: utque est terminus quidam in raritate, qui vacui rationem habet atque sic movet, ita densitatis alius, quem si quis praeterire nitatur, motum excitat, qui vocatur impulsus."

[47] Cardano (1663), III, 360a: "Cum vero corpora ipsa eiusdem non fuerint generis, eodem in loco esse non possunt: nam materia illa duas haberet simul formas." Cf. ibid., 360b: "At condensari et rarescere est formam ex parte mutare."

[48] Bacon, *Novum organum* OFB, XI. 373.

[49] Glisson (1672), 90–91.

[50] Ibid., 104–113. On the influence of Zabarella in Glisson's account of matter's quantity see Hartbecke (2006), 243–246.

Glisson argues that all the motions observed in nature are evidence of the fact that there is a material substratum endowed with life which constitutes the internal principle of every motion. Against this background, he proposes a complex classification of motions, very much influenced by Bacon's account. First of all, he distinguishes perpetual from non-perpetual motions. While the Baconian motions of *antytipia* and *nexus*[51] are said to be perpetual motions (along with the motion of celestial bodies), the motion of liberty is one of the many non-perpetual motions classified by Glisson.[52]

Non-perpetual motions depend on the four basic appetites that structure Bacon's classification of motions: appetite of self-preservation, of exaltation, of propagation and of self-enjoyment. Glisson fully agrees with Bacon on this point, and believes that the Lord Verulam's account "plainly insinuates" that "motions flow from [*dimanare*] or are at least regulated by an internal principle of life".[53] Non-perpetual motions are classified in genera according to two criteria: the origin of the motions (*ab intus*, *ab extra*, and *partim ab intus*, *partim ab extra*) and the subjects of the motions (inanimate, vegetal and animal beings).[54] The distinction of motions in internal (*ab intus*), external (*ab extra*) and mixed (*partim ab intus*, *partim ab extra*) parallels Glisson's distinction of *operationes* in *naturales*, *violentae*, and *mixtae*, which is conceived in Aristotelian terms: "operations originated from inside (*ab intus*) are called natural and according to nature; those which differ from this internal principle or inflict force on it, are called violent and preternatural". Besides, mixed operations are said to be those that partly come from inside and partly are pleasant.[55]

It is against this background that Glisson interprets Bacon's concept of the motion of liberty in an intricate way, in part due to the terminological labyrinth which, as we have seen, emerges throughout the exposition of *Novum organum*. Glisson's interpretation occurs when he tries to make sense of a number of traditional axioms attributed to Aristotle. In this context, he claims that the axiom "Whatever moves, is moved by other" applies only to that motion through which bodies' appetites are to some extent lessened or confined. This motion, Glisson holds, is called "violent motion" by Aristotle, whereas Bacon, in *Novum organum*, labels "motion of liberty" the motion which opposes it. Glisson adds that later on Bacon introduces the label "mechanical motion", in which he seems (*videtur*) to include at the same time "both the violent motion and the motion of liberty". Thus, in Glisson's opinion, Bacon's "mechanical motion" is a dual motion embracing at once the motion through which nature is acted upon (violent motion) and the motion through which it tries to vindicate itself against the inflicted violence (motion of

---

[51] Glisson (1672), 352–354 (*antytipia*), 354–355 (*nexus*).

[52] Henry (1987), 29–30; Hartbecke (2006), 253–254.

[53] Glisson (1672), 357. Cf. Bacon, *Novum organum*, XI, 412.

[54] Glisson (1672), 357–359.

[55] Glisson (1672), 256–257. Cf. ibid. 191, 357. In Glisson's metaphysics *operatio* is a category which includes *motus*, *actiones*, *passiones*, *cessationes*, and *quies* (ibid., 251). On Glisson's idea of natural and violent motion see Hartbecke (2006), 136–140.

liberty).[56] In doing so, Glisson's reading distorts the Baconian account, for, as we have seen, Bacon identifies the mechanical motion with the violent motion, as the relaxation from compression, and does not think of it as a dual motion.[57]

In Glisson's classification, the motion of liberty is a kind of non-perpetual motion of inanimate beings and concerns the "extension, or dimensions and positions of the portions and schematisms" of bodies. More specifically it has to do with density and rarity.[58] When dealing with rarity and density he also renames this motion as "vindication of liberty" and conserves his reading according to which it is a "complex" motion involving a violent motion "coming from outside" and a natural motion, that is, the vindication against the infliction of this enemy in order to recover liberty. Thus, external action conjoins with internal reaction.[59] Given that he describes the motion of liberty in this dual manner, Glisson recognizes the need to defend that the motion is commanded by an internal vital principle. Behind Bacon's affirmation that the motion of liberty has nothing to do with *fuga vacui* and *penetratio dimensionum* but rather with material desires, he sees once more a support for his conception of the energetic substratum of any motion: Bacon would attribute this motion to a vital principle endowed with desire. Although the motion of liberty starts with an external violence, the answer to this is guided by the internal vital principle which produces a "natural motion" of reaction against violence.

Glisson draws on Bacon's concept of *plicae materiae* (folds of matter) to argue for the essential elasticity of matter which allows the self-penetration of a substance and the interpenetration of bodies: "he who recognizes that the perpetual quantity of matter is indeterminate, and that only the determinate quantity is changeable, does not need to deny the self-penetration of the material substance."[60] In keeping with this, he adduces that both Bacon's motion of liberty and the experiments he gathers in *Historia densi et rari* testify in favor of the penetration of bodies. Like Bacon and unlike Cardano, Glisson thinks that the penetration of bodies neither entail the problem of the coexistence of two forms in the same matter nor the coexistence of two bodies in the same place at once.[61]

Further echoes of Bacon's conception of the motion of liberty can be found in Matthew Hale's *Observations touching the principles of natural motions* (1677). Hale praises Aristotle as the "Great Master in Natural Philosophy",[62] but in spite of

---

[56] Glisson (1672), 340: "Quid demum dicendum est de Axiomate Aristotelis *Quid movetur, movetur ab alio*?. Existimo restringi debere ad eum motum quo appetitus cujusvis corporis nonnihil infringitur seu cogitur. Aristoteles motum violentum nominat. [...] [Bacon] motum oppositum motum libertatis vocat, et paulo post, motum mechanicum. Verum hoc nomine motus mechanici videtur concrete simul includere tum motum violentum, tum libertatis".

[57] Bacon, *Novum organum*, OFB, XI, 386. In this context mechanical motion is associated with "Democritus' *motus plagae*."

[58] Glisson (1672), 360, 364.

[59] Ibid., 374–375.

[60] Ibid., 353.

[61] Ibid., Chap. 28.

[62] Hale (1677), 5. Aristotle is also called "the great priest of nature", ibid., 102. See Cromartie (1995), 196.

this, he considers the Aristotelian categories of matter and form more as intellectual tools than as real components of nature.[63] By way of contrast, he adopts fundamental components of Bacon's view of nature such as the distinction between pneumatic and tangible matter, the approach on rarefaction and condensation and the tables of specific weights.[64] He draws heavily on *Historia densi et rari* (1623) and recognizes that Lord Verulam was "a great inquisitor into Nature, and not very friendly to *Aristotle*."[65]

Hale is concerned with understanding the quantitative condition of matter across changes and giving accurate definitions of the concepts involved in the explanation of rarefactions and condensations. He provides an alternative view of *materia prima* by maintaining that matter is the "*substratum* of bodies", which is ever endowed with a particular extension, although it is "of it self indifferent to any particular extension or bodily *Concrement*." Matter is "ingenerable and incorruptible", a Proteus capable of any extension and form. The counterpart of matter is not form, but "body", which is defined as "Matter determined into a Body of that Nature, Figure, Texture, *Plexus*, Quality and Dimension." Finally, "quantity" is said to be an intrinsic accident of material substances, a "habitude [...] whereby a Material Substance under any determination is denominated more or less". Quantity must be distinguished from "extension", which refers to changes in volume, since changes of extension do not entail changes of body quantity. After this series of definitions, Hale can introduce "density and rarity" as qualities with which material substances are endowed in varying degrees, which can be altered by rarefactions and condensations mostly caused by external agents.[66]

Hale's analysis of experimental instances of rarefaction and condensation are very much in keeping with Bacon's account of motion, by referring to *motus nexus* and by reflecting the terminology characteristic of the motion of liberty:

> [5] that strong cohesion of the Filaments of the Air to the sides of the Vessel, when by the decay of Heat, it contracts it self with a *motus* or *conatus* of restitution, and with it the ambient Vessel by a *motus nexus*, yet if a small degree of Heat be moved to the Vessel, that contraction is relaxed, and the Vessel fall asunder, and the like is done by the smallest admission of the free or solute Air, though through a Pin-hole, for by the Heat the included Air is again dilated, left lax, as before it grew cold, and by the admission of foreign Air is relieved gradually to that expansion as is natural to it, [...] [8] the compressed Air in those Instances endeavours with great strength and force its relaxation from this compression, in so much as if it can get liberty.[67]

Like Cardano, Bacon and Glisson, Hale is concerned with defining whether or not processes of rarefaction and condensation are "violent" or "natural" in character. He claims that although rarefactions and condensations of bodies from their "natural size and dimension" are produced for the most part "by the agency or

[63] Cromartie (1995), 200.

[64] Hale (1677), 36–39, 44–47, 89, 91.

[65] Ibid., 105.

[66] Ibid., 113–118.

[67] Ibid., 52–54. Cf. ibid., 81–82.

efficiency of some external cause", rarity and density under such circumstances are "but natural affections" arising from the very texture of bodies.[68] From this point of view, he claims that the air contained in a vessel and expanded after receiving heat is "relaxed" when the heat diminishes and "endeavours its own contraction to its just and natural size and texture which it lost for the time, by the foreign violence of expansion by Heat or Tension, and this by a natural motion of restitution to its natural texture."[69]

In keeping with Glisson, whom he mentions, Hale maintains that the penetration of material bodies is perfectly possible.[70] Material penetration may occur by "contractions" or "compressions" caused by an "external efficient or force". Moreover, Hale affirms categorically that there are no limits to successive material penetrations: "in as much as the very same portion of Material Substance is successively capable of several textures, and consequently of several dimensions, there is nothing in Nature or Reason that prohibits a successive penetration of Material Substances under such a mutation of textures and actual dimensions, so that one and the very same numerical portion of Matter that this moment is under a texture accommodate to the nature of the most subtil Air."[71] In contrast to Glisson and Bacon, we don't find in Hale a strong vocabulary of material appetites and desires, since he tries to adopt a middle way (which he saw represented by the Helmontian philosophy) between what he judged to be "two extremes in the modern philosophy": the Cartesian mechanicism and the plastic vitalism represented by Henry More.[72]

Conclusion

What Cardano distinguishes as two kinds of motions aimed at preserving the form and at impeding the existence of void (*impulsus* and *attractio*), in Bacon's account becomes one single motion responding to specific appetites of matter (motion of liberty). Their accounts differ ultimately in the way in which each author conceives the reasons of motion in nature. By invoking the concept of form Cardano is much closer to the traditional Aristotelian view according to which the intrinsic principle or nature of substances determines their motions, although he does not reduce "natural motions" to the downwards and upwards motions as the Peripatetic school does. Bacon, instead, thinks that the ultimate reasons of changes in bodies are the appetites of matter, which diversify into several kinds. This last view had in Glisson one committed follower. Glisson expanded Bacon's analysis of rarefaction and

[68] Ibid., 88.

[69] Ibid., 101.

[70] Ibid., 123–124; 129–130.

[71] Ibid., 127. Hale's treatment of penetration is longer, but it cannot be discussed here in its full extent. He distinguishes three kinds of penetrations: of actual dimensions, of bodies and of material substances. Ibid., 128.

[72] Cromartie (1995), 206–208.

condensations in terms of material appetites, ever searching for evidence supporting the fundamental assumption of his metaphysics of nature: the vital energetic character of matter. In contrast, Hale is not committed to a metaphysics centred on material appetites, but is still very much concerned with giving an accurate account of the very nature of matter's quantity. That leads him to search for careful definitions which, for instance, distinguish "matter's quantity" from "matter's extension".

From this historical reconstruction it becomes apparent that along with the transformations above mentioned, one key element of Cardano's approach faded slowly away: the Aristotelian concept of form. Although the vocabulary of form certainly did not disappear at once (and is beyond doubt central to Bacon's natural philosophy), its Aristotelian nuances were increasingly diminished in subtle ways. Once the Aristotelian legacy started to lose the strong power that it still had in Cardano's natural philosophy, alternative views of motion, rarefaction and condensation could emerge. For instance, this loss allowed Bacon, Glisson and Hale to conclude that nature does not prohibit the penetration of bodies.

That notwithstanding, a number of issues vital to Cardano's approach persisted in the seventeenth-century discourse on condensation and rarefaction: the quantitative condition of matter, the existence of void, the penetration of bodies, and the distinction between natural and violent motion. Throughout this itinerary from the Renaissance to the early modern period, these were the intellectual tools which intervened in the discussion on rarefaction and condensation. It must be emphasised that in the seventeenth century, like in the Cardanian account, motion – be it conceived in terms of material appetites (Bacon and Glisson) or in terms of a combination of matter's appetites and mechanical properties (Hale) – still was thought inside the framework of the distinction between nature and violence, bearing the long-standing mark of the Aristotelian setting. As a result, the discussion on rarefaction and condensation in these seventeenth-century authors was still concerned with defining the boundaries of "nature" and "violence", and indirectly with defining the boundaries between the natural and the artificial, concerns shared with their Renaissance predecessor, which were fundamental issues in the emergence of early modern science and were to remain of central importance in philosophical and scientific thought until today.[73]

Bibliography

A Table of the Books in the University Library. 1683. MS Cambridge University Library CUR 31.1.

Aristotle [Aristotelis cum Averrois]. 1550. *Aristotelis opera omnia cum Averrois Cordubensis commentariis*. Venice: Iuntas.

Ashworth, Jennifer. 1988. Die philosophischen Lehrstätten. 1. Oxford. In *Grundriss der Geschichte der Philosophie. Die Philosophie des 17. Jahrhunderts, England*, ed. Jean-Pierre Schobinger, 6–9. Basel: Schwabe.

[73] On the historical relevance of the polarization "natural" and "artificial" see Bensaude-Vincent and Newman (2007).

Bacon, Francis. 1857–1874. *The Works of Francis Bacon*, vol. 14, ed. James Spedding, Robert Leslie Ellis, and Douglas Denon Heath. London: Longman. Facsimile reprint Stuttgart-Bad Cannstatt: Frommann – Holzboog (abbreviated in this essay as SEH).

Bacon, Francis. 1996. *The Oxford Francis Bacon*, vol. 15, ed. Graham Rees, and Lisa Jardine. Oxford: Clarendon Press (abbreviated in this essay as OFB).

Bensaude-Vincent, Bernadette, and William R. Newman (eds.). 2007. *The Natural and the Artificial: An Evolving Polarity*. Cambridge: MIT Press.

Bianchi, Massimo Luigi. 1994. Scholastische Motive im ersten und zweiten Buch des *De Subtilitate* Girolamo Cardanos. In *Girolamo Cardano. Philosoph, Naturforscher, Arzt*, ed. Eckhard Keßler, 91–114. Wiesbaden: Harrassowitz.

Blackwell, Constance. 1994. The Historiography of Renaissance Philosophy and the Creation of the Myth of the Renaissance Eccentric Genius-Naudé Through Brucker to Hegel. In *Girolamo Cardano. Philosoph, Naturforscher, Arzt*, ed. Eckhard Keßler, 339–369. Wiesbaden: Harrassowitz.

Burns, William E. 2001. "A Proverb of Versatile Mutability": Proteus and Natural Knowledge in Early Modern Britain. *The Sixteenth Century Journal* 32(4): 969–980.

Cardano, Girolamo. 1663. De Subtilitate Libri XXI. In *Opera omnia*. Lyon: Jean-Antoine Huguetan and Marc-Antoine Ravaud.

Collegium Conimbricense. 1592. *Commentarii Collegi Conimbricensis Societatis Jesu in octo libros Physicorum Aristotelis Stagyritae*. Coimbra: A Mariz.

De Waard, Cornelius. 1936. *L'expérience barométrique. Ses antécedents et ses explications*. Thouars: Imprimerie Nouvelle.

Cromartie, Alan. 1995. *Sir Matthew Hale, 1609–1676: Law, Religion, and Natural Philosophy*. Cambridge: Cambridge University Press.

Durel, Henri. 1998. Francis Bacon lecteur d'Aristote a Cambridge. *Nouvelles de la République des Lettres* 18: 29–60.

Dyer, George. 1824. *The Privileges of the University of Cambridge*, vol. 2. London: Longman.

Giglioni, Guido. 2013. Girolamo Cardano. In *Stanford Encyclopedia of Philosophy*. http://plato.stanford.edu/archives/sum2013/entries/cardano/. Accessed 4 Nov 2015.

Giglioni, Guido. 2015. Scaliger Versus Cardano Versus Scaliger. In *Forms of Conflict and Rivalries*, ed. David A. Lines, Marc Laureys, and Jill Kraye, 109–130. Goting: V & R unipress, Bonn University Press.

Glisson, Francis. 1672. *Tractatus de natura substantiae energeticae seu de vita naturae ejusque tribus primus facultatibus*. London: Flesher, Brome and Hooke.

Hale, Matthew [the Author of *Difficiles Nugae*]. 1677. *Observations Touching the Principles of Natural Motions; and Specially Touching Rarefaction and Condensation: Together with a Reply to Certain Remarks Touching the Gravitation of Fluids*. London: W. Godbid for W. Shrowsbury.

Hartbecke, Karin. 2006. *Metaphysik und Naturphilosophie im 17. Jahrhundert. Francis Glissons Substanztheorie in ihrem ideengeschichtlichen Kontext*. Tübingen: Niemeyer.

Henry, John. 1987. Medicine and Pneumatology: Henry More, Richard Baxter, and Francis Glisson's *Treatise on the Energetic Nature of Substance*. *Medical History* 31: 15–40.

Hero of Alexandria. 1583. *Spiritualium liber, a Federico Commandino urbinate ex graeco nuper in latinum conversus*. Paris: Gorbin.

Jardine, Lisa. 1974. The Place of Dialectic Teaching in Sixteenth-Century Cambridge. *Studies in the Renaissance* 21: 31–62.

Jensen, Kristian. 1994. Cardanus and His Readers in the Sixteenth Century. In *Girolamo Cardano. Philosoph, Naturforscher, Arzt*, ed. Eckhard Keßler, 265–308. Wiesbaden: Harrassowitz.

Keckermann, Bartholomäus. 1610. *Systema physicum*. Danzig: Hünefeld.

MacLean, Ian. 1994. Cardano and His Publishers 1534–1663. In *Girolamo Cardano. Philosoph, Naturforscher, Arzt*, ed. Eckhard Keßler, 309–338. Wiesbaden: Harrassowitz.

Magirus, Johannes. 1597. *Physica peripatetica ex Aristotele, eiusque interpretibus collecta, et in sex libros distincta*. Frankfurt: Zacharias Palthenius.

Manzo, Silvia. 2004. Francis Bacon y la concepción aristotélica del movimiento en los siglos XVI y XVII. *Revista de Filosofía (Universidad Complutense de Madrid)* 29(01): 77–97.

Manzo, Silvia. 2006. *Entre el atomismo y la alquimia. La teoría de la materia de Francis Bacon*. Buenos Aires: Biblos.

Manzo, Silvia. 2013. The Preservation of the Whole and the Teleology of Nature in Late Medieval, Renaissance and Early Modern Debates on the Void. *Journal of Early Modern Studies* 2(2): 9–34.

McKitterick, David. 1992. *A History of Cambridge University Press. Volume 1: Printing and the Book Trade in Cambridge*, 1534–1698. Cambridge: Cambridge University Press.

Needham, Paul. 2006. Aristotle's Theory of Chemical Reaction and Chemical Substances. In *Philosophy of Chemistry: Synthesis of a New Discipline*, ed. Davis Baird, Eric Scerri and Lee McIntyre, 43–67. Boston Studies in the Philosophy of Science. Dordrecht: Springer.

Nenci, Elio. 2003. 'Mechanica' et 'machinatio' nel *De subtilitate*. In *Cardano e la tradizione dei saperi*, ed. Maria Luisa Baldi and Guido Canziani, 67–82. Milan: Franco Angeli.

Nenci, Elio. 2004. Prefazione. In *Girolamo Cardano. De subtilitate.* Critical edition by Elio Nenci, Tomus I. Libri I–VII, 9–42. Milan: Franco Angeli.

Oates, John Claud Trewinard. 1986. *Cambridge University Library: A History from the Beginnings to the Copyright Act of Queen Anne*. Cambridge: Cambridge University Press.

Parker, Matthew. 1729. *De antiquitate Britannicae Ecclesiae*. London: Bowyer.

Pererius, Benedictus. 1562. *De communibus omnium rerum naturalium principiis et affectionibus*. Rome: F. Zanettum and B. Tosium.

Pesic, Peter. 1999. Wrestling with Proteus: Francis Bacon and the "Torture" of Nature. *Isis* 90(1): 81–94.

Pesic, Peter. 2014. Francis Bacon, Violence, and the Motion of Liberty: The Aristotelian Background. *Journal of the History of Ideas* 75(1): 69–90.

Rogers, G. A. John. 1988. *Die philosophischen Lehrstätten*. 2. Cambridge. In *Grundriss der Geschichte der Philosophie. Die Philosophie des 17. Jahrhunderts. England*, ed. Jean-Pierre Schobinger, 10–11. Basel: Schwabe.

Scaliger, Julius Caesar. 1557. *Exotericarum Exercitationum Liber XV De Subtilitate, ad Hieronymvm Cardanum*. Paris: Michel Vascosan.

Schmitt, Charles B. 1975. Philosophy and Science in Sixteenth-Century Universities: Some Preliminary Comments. In *The Cultural Context of Medieval Learning*, ed. John Murdoch and Edith Sylla, 485–530. Dordrecht-Boston: Reidel.

Schmitt, Charles B. 1983. *John Case and Aristotelianism in Renaissance England*. Kingston [Ont.]: McGill-Queen's University Press.

Schütze, Ingo. 2000. *Die Naturpilosophie in Girolamo Cardanos* De subtilitate. Munich: Fink.

Valerius, Cornelius. 1567. *Physicae, seu de naturae philosophia institutio, perspicue et breviter explicata*. Antwerp: Plantin.

Zabarella, Jacopo. 1590. *De rebus naturalibus libri XXX*. Cologne: Giovanni Battista Ciotti.

Chapter 7
Telesio Among the Novatores: Telesio's Reception in the Seventeenth Century

Daniel Garber

Abstract Bernardino Telesio was an important figure in Italian thought at the end of the sixteenth century, and his philosophy was thought to provide a genuine alternative to the Aristotelian natural philosophy then dominant. But by the middle of the seventeenth century, it was quite a different story. This essay examines two stages in the transformation of Telesio's later reputation. In Francis Bacon's *De principiis et originibus*, probably written in the early 1610s, Telesio is taken very seriously. While Bacon disagreed with Telesio in many respects, he was clearly an important interlocutor for Bacon. The essay then turns to an examination of the discussion of Telesio in Charles Sorel's 1655 essay, "Le sommaire des opinions les plus estranges des Novateurs en Philosophie." There Telesio appears as one of a long list of *novateurs*, an exhibition in a forgotten corner of a dusty *Wunderkammer*. By the middle of the seventeenth century, Telesio's philosophy is no longer a live option, part of the lively discussion about Aristotelian natural philosophy that dominated the intellectual world at that moment. He was remembered as a pioneer, the first to oppose the dominant Aristotelianism, but his doctrines were largely forgotten.

René Descartes is now usually considered the father of modern philosophy. (This is not just my opinion: it can now be substantiated scientifically. Google "father of modern philosophy" and up comes Descartes.) But Descartes' contemporaries didn't think so. For them the father of modern philosophy was Bernardino Telesio, a sixteenth-century figure now largely forgotten, except among scholars of Renaissance Italian philosophy. In this essay I would like to explore this curious figure, and how his thought was viewed in the seventeenth century.

All translations are my own unless otherwise stated.

D. Garber (✉)
Department of Philosophy, Princeton University, Princeton, NJ, USA
e-mail: dgarber@princeton.edu

C. Muratori, G. Paganini (eds.), *Early Modern Philosophers and the Renaissance Legacy*, International Archives of the History of Ideas Archives internationales d'histoire des idées 220, DOI 10.1007/978-3-319-32604-7_7

Telesio and His Project

Bernardino Telesio was born in Cosenza, in Calabria in 1509, and died there in 1588, though during his life he travelled widely in Italy and lived in a number of other places, including Milan, Rome, Padua and Naples. Coming from a privileged background, Telesio was able to devote much of his life to study.[1]

His major work is the massive *De rerum natura iuxta propria principia* [DRN]. The first version came out in 1565, in two books, with the title *De natura iuxta propria principia*. A second revised edition came out 5 years later, again in two books, but with a new title, echoing the title of Lucretius's famous poem: *De rerum natura iuxta propria principia*. Then in 1586, 2 years before Telesio's death, a much expanded edition in nine books appeared.[2] In addition to the *De rerum natura*, Telesio published a variety of smaller treatises on questions in natural philosophy, which were collected together after his death and published as *Varii de naturalibus rebus libelli* in 1590. This collection included essays on comets and the Milky Way ("lacteus circulus"), on meteors ("De his, quae in Aere fiunt"), on the rainbow, on the seas, on the soul, against Galen, on respiration, on color, on taste, and on sleep. The modern edition of the *Varii* (Telesio (1981)) adds some further medical writings, as well as Telesio's answers to criticisms by Patrizi and a poem dedicated to Giovanna Castriota.

Telesio's main focus was a complete system of natural philosophy, a systematic alternative to the Aristotelian natural philosophy that dominated the intellectual world of his day, and would dominate it for some time to come.[3] Telesio's orientation was resolutely empiricist. The *De rerum natura* begins with a call to investigate nature not through reason, but through the senses:

> The construction of the world and of the size and nature of the bodies contained in it should not be sought from reason, as the ancients did, but must be perceived by sense, and must be grasped from things themselves.[4]

Indeed, like Hobbes would later argue, Telesio argues for eliminating the intellect in favor of sense alone: "Aristotle shouldn't have attributed to man an intellect distinct from sense."[5] Books I–IV of the DRN is concerned with Telesio's basic physics; in book I Telesio gives an outline of his basic system, supplemented in books II–IV by elaborations and responses to others, mainly Aristotle. Standard Aristotelian text-

---

[1] For the biographical background to Telesio, see Mulsow (1998), 1–14 and the references cited there.

[2] The current standard modern edition of Telesio (1586) is Telesio (1965–1976), though volumes 1 and 2 of the set are sometimes difficult to find. In addition to the Latin text, it includes an Italian translation on facing pages. Telesio (2009) is a modern edition of Telesio (1570), which, again, contains both the original Latin text and an Italian translation on facing pages.

[3] For general accounts of Telesio's thought, see De Franco (1995) and Bondì (1997). For shorter accounts of his thought see Boenke (2013) and Leijenhorst (2010).

[4] DRN I proem, Telesio (1586), 1. In my brief account of the philosophy of the DRN, I will focus on the third edition of 1586.

[5] DRN VIII.11, Telesio (1586), 326. Cf. Hobbes, *Leviathan*, chs. 1–3.

books in physics begin with general principles of physics, definitions of space, time and motion, the principles of nature, and so on, and only then go on to discuss cosmology and sublunar physics.[6] In Telesio's exposition, though, the basic physics is deeply intertwined with his cosmology. On his view, there are three basic principles in the world. First there are heat and cold, which constitute the sun and the earth respectively (DRN I.1). These two contrary and competing principles are incorporeal, but they cannot exist apart from body or matter, his third principle (DRN I.4).[7] Heat is mobile, light, and associated with rarified matter, while cold is immobile, dark, and associated with condensed matter (DRN I.2). Furthermore, Telesio seems to attribute a kind of sensibility to all bodies, a kind of panpsychism (DRN I.6).[8] The surface of the earth, on which we live, is the zone where the two principles are both found, and where they exert their contrary influences. It is the battle between heat and cold that explains all the phenomena of the world, Telesio claims. As a result, there is no radical distinction in Telesio's cosmology between the heavens and the earth: it is the same elements in both (DRN I.1).

In book V Telesio turns to living things. There he argues that there are two kinds of souls, a soul "*e semine educta*," which is responsible for the vital functions, and in humans a special immaterial soul, given to us directly by God (DRN V.2-3; VIII.15). Other than that, the human and the animal are largely the same. Indeed, since Telesio eliminates intellect in favour of sense, as I noted earlier, he even argues that animals have a kind of capacity for reasoning not unlike ours (DRN VIII.14). In books V, VI and VII Telesio goes on to propose an account of sense perception that is based on the idea that sense is to be explained through the impingement of external bodies on the sense organs, and in book VIII, an account of reasoning. The *De rerum natura* ends with a discussion of the passions and ethics in book IX.

Throughout the *De rerum natura* are interspersed copious refutations of Aristotle and his followers: "perperam Aristoteles […]," "perperam Peripateticos […]" are repeated over and over. Telesio leaves no doubt that he is firmly opposed to Aristotle and his philosophy, and that he is proposing a thoroughgoing and systematic alternative to the accepted philosophy of the schools.[9]

During his lifetime, and in the years immediately following, Telesio was an important figure in the Italian context, with a number of prominent followers and opponents. In his hometown Cosenza, he transformed the local Accademia Parassiana into the Accademia Telesiana, an academy for the study of nature. (After his death, it became the Accademia Cosentina.)[10] Telesio's system was attacked by the Aristotelian Giacomo Antonio Marta, and by the anti-Aristotelian Francisco

[6] See, for example the *Physica* in Eustachius (1609), a popular textbook used in schools through much of the seventeenth century, both in Catholic and Protestant countries.

[7] The view here is actually rather complex. Telesio is unclear whether matter, heat, and cold are all equally well substances or whether matter is the only real substance. Furthermore, he isn't clear about the relation between *material* (*moles*), and *corpus*. On this question see Schuhmann (2004).

[8] On the significance of this position for Telesio's thought, see Giglioni (2010).

[9] Telesio's complex relation to the Aristotelian tradition is explored in Mulsow (1998).

[10] See Lupi (2011).

Patrizi, with whom he had an exchange in 1572.[11] But he counted among his supporters Antonio Persio, the editor of the posthumous 1590 collection of his *Opuscula*, Sertorio Quattromani, his successor as head of the Accademia, and perhaps his fellow citizen of Cosenza, the physician Agostino Doni. And, of course, Tomasso Campanella's *Philosophia sensibus demonstrata* (1591) offered a spirited defence of Telesio's philosophy.

Despite the initial stir that Telesio made, by 1612 or 1613, Francis Bacon could write in his *De principiis et originibus* that Telesio's philosophy is "not very famous or well accepted (*philosophia scilicet non admodum celebri aut recepta*)."[12] Indeed, Telesio never developed the kind of following that Bacon himself would, or Descartes, or Galileo, or any of the later stars of the so-called Scientific Revolution. But even so, he was by no means forgotten.

It has been argued that some of the better known figures of the century, such as Hobbes or Descartes were influenced by Telesio's views.[13] But outside of Campanella, a younger member of the broad circle of Italian thinkers that included Telesio, who lived well into the seventeenth century, I know of very few detailed discussions of his work in the seventeenth century. There is a paragraph of general overview in Tobias Adami's "Praefatio ad philosophos Germaniae" in his edition of Campanella's *Prodromus philosophiae instaurandae* (1617), hardly surprising given Campanella's relation to Telesio.[14] There is a longer overview of Telesio's physics in Gassendi's *Syntagma philosophicum*.[15] Given that Gassendi took it as his task to summarize every other important author on every main subject before presenting his own views, this isn't altogether surprising either. There are a certain number of more focused discussions as well. Jean-Cécile Frey was a Paris professor who, in 1628 offered his students a series of lectures refuting the views of a variety of *novatores* who had the temerity to oppose the philosophy of Aristotle, still then quite central to the philosophy curriculum at Paris and most other universities in Europe. In that series of lectures Frey offers a response to anyone who disagreed with Aristotle in any way. In his *Cribrum philosophorum*, the written version of those lectures later published by his students after his death, we find a long discussion and refutation of Telesio and Campanella on the claim that water is hot by its nature, and not cold, as Aristotle and his followers argue.[16] In Nathanael Carpenter's

[11] See Telesio (1981), 463ff for Patrizi's objections.

[12] OFB 6:258–259. The date of the essay is contested, but Graham Rees puts it in the early 1610s. The original Latin is given on facing pages with an English translation by Graham Rees and Michael Edwards. I quote the English translation, but the citation gives both the Latin and the English.

[13] For the claim about Hobbes and Telesio, see Schuhmann (1988); for the claim about Descartes and Telesio, see Hatfield (1992), 349.

[14] See Campanella (1617), B2v-B3r. For a brief discussion of Adami's remarks, see De Mas (1990), 176f.

[15] Gassendi (1658), 1:245–246.

[16] See Frey (1646), 46–49. For a discussion of the background to the *Cribrum*, see the introduction by Roger Ariew and Daniel Garber to Frey (2003). For a more general discussion of Frey, see Blair (1993) and Blair (1994).

Geography (1625), Telesio is one of the figures whose theories of the tides is mentioned in the course of his discussion of the question and survey of alternative points of view.[17] No doubt there are other mentions as well. But at this point I would like to focus on two of the most interesting of the discussions, one by Francis Bacon in the earlier part of the century, and the other by Charles Sorel in the middle of the century. The difference between the two is quite illuminating, and may illustrate a change in his status over the course of the century.

Bacon: Telesio as the First of the Moderns

Earlier I quoted Bacon's *De principiis et originibus* where he wrote that Telesio's philosophy was "not very famous or well accepted."[18] He continued:

> But I do not bother with such niceties. For I think well of *Telesio*, and recognize him as a lover of truth, a man useful to the sciences, a corrector of certain doctrines, and the first of the new philosophers [*novorum hominum primum*] [...].[19]

This is in the context of what is probably the longest and most serious explicit discussion of Telesio's philosophy in the seventeenth century. The full title of Bacon's essay reads (in English): "On the Principles and Origins according to the Fables of Cupid and Coelum, or, the Philosophy of Parmenides and Telesio and especially that of Democritus as it is treated of in the fable of Cupid."[20] It isn't entirely clear when or why Bacon wrote the essay, or at least the part of it that we have. (The essay was left incomplete, giving only part of the planned section on principles, and nothing at all on origins.) Graham Rees has suggested that the essay was written as part of a survey of current knowledge, which was intended to go in part I of his *Instauratio magna* project. He also suggests that it was probably written in 1613 or so. But none of this really matters for our purposes here. What is really important is the fact that significantly more than half of the essay is focused on the philosophy of Telesio.

The discussion of Telesio occurs about one third of the way into the text as we have it, after an analysis of the Aristotelian conception of matter and form, two of the three Aristotelian principles of nature. The treatment of Telesio's natural philosophy is intended to introduce an alternative to Aristotle's, though not one that Bacon himself would accept. Bacon begins by identifying Telesio as a follower of Parmenides,[21] an identification that a number of his and Bacon's contemporaries

---

[17] Carpenter (1625), bk. II, 89–90. (Note that the two books of Carpenter (1625) are paginated independently.)

[18] OFB 6:258–259.

[19] Ibid.

[20] OFB 6:196–197.

[21] OFB 6:224–225.

made, though modern commentators are not at all sure that that is fair.[22] In the course of his considerations, Bacon remarks:

> But our business is not with *Telesio* as such, but him as a restorer of the philosophy of *Parmenides*, to whom much respect is due. But my main reason for going into this so fully is that in dealing with the one that comes first, I speak of many things which can be carried over to the refutation of sects further down the list [...] and so I shall not have to say the same things time and time again.[23]

Even so, it is Telesio whom Bacon treats in detail, and who offers him the occasion to reflect on certain questions in natural philosophy.

The discussion of Telesio is in the context of a broader one about the principles of natural philosophy. After an introductory presentation of Cupid and Coelum, that is, principles and origins, and an analysis of Aristotle's principles of nature, form, matter, and privation, Bacon turns to a structured discussion of principles. He organizes this into four parts: two categories of thinkers who recognize only one kind of principle, a third category of thinkers who recognize "many principles of things," and a fourth category "of those who constitute infinite, or at least numerous principles of things [...]."[24] In the first two categories (which Bacon doesn't carefully distinguish) he discusses Thales and his view that all is water, Anaximenes, who took air as his principle, and Heraclitus, who held fire. Such one-principle natural philosophies are obviously inadequate, Bacon argues:

> But since such great armies of contraries appear throughout the universe, as of dense and rare, hot and cold, light and dark, animate and inanimate, and very many others, which attack, usurp, and slaughter one another in turn, to suppose that all these spring from some one source of material stuff, but still not disclose any of that stuff's mode of action, seems a kind of frantic speculation and a giving up of inquiry.[25]

And so Bacon turns to the next category of natural philosophy, those that recognize multiple, but not infinite principles.

What follows is an extended account of Telesio's views on principles. (Indeed, this is where the discussion ends, with a treatment of Telesio that is significantly longer than the rest of the essay; Bacon never gets to any other account of principles.) Bacon comments on Telesio's views in some detail. The focus, as one might expect, is the principles of heat and cold. Bacon gives a detailed account of these two contraries, and how it is that they are integrated into Telesio's cosmology. He gives special attention to the domain between the heavens and the earth, where Telesio "finds all tumult, conflict and internal disorder, as is the case in empires in which we find that the borders are troubled by incursions and violence, while the provinces inland enjoy a secure peace."[26] Bacon's exposition of Telesio exclusively concerns the doctrine of heat and cold and the related cosmology from the opening

---

[22] See, e.g., Patrizi in Telesio (1981), 463. On Telesio's Parmenideanism, see Lerner (1992).

[23] OFB 6:258–259. On Bacon's Parmenidean reading of Telesio, see Bondì (2001).

[24] OFB 6:210–211.

[25] OFB 6:222–223.

[26] OFB 6:230–231.

books of the *De rerum natura*; there is no discussion of living things or the passions, also important elements of Telesio's thought. But this may be only because in this particular essay, Bacon is focusing on principles, and not on the detailed account of the world that follows on the principles.

Bacon's long discussion of Telesio is followed by an even longer critique of Telesio's views. He begins as follows:

> Now what Telesio says would have been plausible if man, along with the mechanical arts which vex matter, were removed from nature, and the fabric of the World were regarded on its own [*Fabrica Mundi simpliciter spectetur*]. For his seems a kind of pastoral philosophy which contemplates the world calmly and as if in idleness.[27]

For Bacon this is a fundamental criticism: Telesio's is an arm-chair philosophy, one that teaches us how nature appears, but doesn't allow us to control nature. Bacon also criticizes Telesio for his empiricism, someone "who philosophizes according to the sense alone," a criticism that echoes his criticism of some of the ancients.[28] These two criticisms come together in one of Bacon's first mentions of Telesio in the *Advancement of Learning* (1605), where he refers to the philosophy of "*Thylesius*, and his Scholler *Donius*" as "a Pastorall Philosophy, full of sense, but of no great depth."[29] However, in the *De principiis et originibus* he also offers detailed criticism of Telesio's two principles, heat and cold. He sees four basic problems with the view: (1) there are certain phenomena of bodies that cannot be explained by heat and cold; (2) there are certain circumstances in which heat and cold are created anew, and so arise from something else; (3) some things that are correctly explained by heat and cold come from "their efficient and instrumental cause;" and (4) Telesio's association of heat with motion, light, and rarity and cold with immobility, darkness and density is confused.[30]

There are numerous other discussions of Telesio's philosophy in Bacon's writings, but none of them is longer than a few lines.[31] The discussion in the *De principiis et originibus* is by far the longest discussion of Telesio in Bacon's writings. Indeed, it may well be the longest sustained discussion of any other philosopher in Bacon's corpus. Bacon certainly took notice of Telesio's work. In the literature, there are various claims about the relevance of Telesio on Bacon's thought. Some commentators have focused on Bacon as a critic of Telesio's thought.[32] Others have seen Telesio as a positive influence on Bacon. Nicoletta Sciaccaluga sees Bacon's thought as reflecting Telesio's views on motion, for example.[33] Graham Rees, on the other hand, sees the influence of Telesio in Bacon's matter theory. Bacon's own

---

[27] OFB 6:250–251.

[28] OFB 6:250–251; cf. OFB 6:220–221.

[29] OFB 4:93.

[30] OFB 6:256–259.

[31] See Giachetti Assenza (1980) for a list and extensive discussion of all the references to Telesio in Bacon's corpus.

[32] See De Mas (1990), Margolin (1990), Posseur (1990).

[33] Sciaccaluga (1997).

matter theory posits two kinds of matter, spirit, which fills the heavens and dense, tangible matter, whose domain is earth, separated by an intermediate zone on the surface of the earth, where the two mix. Rees sees Telesio's mark on Bacon's account of spirit, and in his emphasis on the importance of the intermediate zone between the two where they mix, not unlike the intermediate zone in Telesio's philosophy where heat and cold mix.[34] I do not want to make any such assertions here, where my interest is less in Bacon than in Telesio and his later fate. Whatever influences there may have been in Bacon's thought, Bacon was not a simple follower of Telesio's philosophy. However, one cannot deny that Telesio was an important interlocutor for Bacon. Early in the century, then, in the generations of thinkers who followed the publication of the definitive edition of the *De rerum natura* in 1586, Telesio was a significant figure, someone who was taken seriously by other serious figures, like Bacon. He was for Bacon, in a way, the father of modern philosophy.

Sorel: Telesio Among the *Novateurs*

I would now like to turn to the treatment of Telesio's thought in Charles Sorel, the second figure I would like to discuss. Sorel is best known as a literary figure, the author of the daring romance *Francion* (1977 and 1633) and *Le Berger extravagant* (1627), among many other romances. But later in life, Sorel came to have more serious ambitions. In the 1630s, he began a project that he called the *Science universelle*, on which he worked for the rest of his career.[35] The first part, *La science des choses corporelles* came out in 1634. In the years that followed, Sorel published volume after volume, adding revisions and expansions. In the end it covered meteors, the vacuum, the immobility of the earth, cosmology, the stars, humans and animals, the immortality of the soul, inventions and arts that apply the universal science, and even ethics, in short, all the important topics covered in a natural philosophy and more. In 1655, at the culmination of the project, Sorel published a volume entitled *De la perfection de l'homme, où les vrays biens sont considérez et spécialement ceux de l'âme* [...] Included in that volume was a brand new and quite extended essay, "Le sommaire des opinions les plus estranges des novateurs modernes en la philosophie comme Telesius, de Patritius, de Cardan, de Ramus, de Campanelle, de Descartes, et autres; Et en quoy on les peut suivre."[36] Substantially the same essay was reprinted 13 years later, in 1668 in a volume Sorel entitled *La science universelle tome quatriesme*, with a slightly different title: "Des novateurs modernes en la philosophie [...] avec un examen sommaire de leurs principales opinions."[37]

[34] Rees (1977), 118; introduction in OFB 6:xxxvii–xxxviii. See also Weeks (2007), 55–61.

[35] See Picardi (2007).

[36] Sorel (1655).

[37] Sorel (1668). See Del Prete (2001), Picardi (2007), 255–297.

Sorel's use of the term *novateur* was quite deliberate and significant. By the early 1620s, there emerged in the philosophical literature of the day a group of thinkers that were together often identified as the "novatores," "novateurs," or in English, the "novelists," thinkers who were pioneers in rejecting Aristotle and Aristotelian natural philosophy in favor of something new. Unsurprisingly, Telesio was almost always associated with this group. In Mersenne's 1623 *Quaestiones* [...] *in Genesim*, Telesio was grouped together with Campanella, Bruno, Kepler, Galileo, "and other disciples of the moderns," objecting they are wrong in saying that all Catholics are dogmatic Aristotelians.[38] In his *Apologie pour tous les grand personnages qvi ont esté accusez de magie* (1625), Gabriel Naudé grouped Telesio with a variety of thinkers, including Patrizi, Campanella, Bacon, Bruno, and Basson, who have "no aim but to elbow out this philosophy [i.e. the philosophy of Aristotle] and to lay waste to this great building which Aristotle and the more than twelve thousand who interpreted him have been trying to build for a long time."[39] Despite these negative comments, in his *Advis pour dresser une bibliotheque* (1627), Naudé nevertheless recommends that this group of "novateurs" be included in a good library.[40] One can find very similar lists in numerous writers of the seventeenth century, including Bacon, Descartes, Adrien Heereboord, John Webster, and Robert Boyle.[41] There are many variations from one to another: Telesio, whom Bacon lists as "the first of the new philosophers" and "the best of the *Novellists*"[42] is on almost all the lists, as are Francesco Patrizi, Tommaso Campanella, Giordano Bruno and William Gilbert. Among older figures, Girolamo Fracastoro, Petrus Ramus and Girolamo Cardano occasionally appear, but, interestingly enough, rarely Paracelsus. Many later figures also appear with notable frequency. Among the better known figures there are Johannes Kepler, Galileo, and Bacon. As his reputation spreads, Descartes makes the list, and occasionally Pierre Gassendi. But there are many lesser-known figures who appear with great regularity, including Sebastien Basson, Nicholas Hill, Nathanael Carpenter, David Gorlaeus, and Godifredus Chassinus.

It is an interesting and diverse group. One might think of the *novatores* as a kind of alternative party to the Aristotelians. But there was an important difference. As different as the Aristotelians might have been from one another, they had texts in common: in natural philosophy the authoritative texts of the *De anima* and the *Physica* that they shared. Among the *novatores* the only thing that they had in common was that they rejected the authority of Aristotle and the Aristotelians; beyond that, there was little in the way of a common thread. Telesio explained everything in terms of hot and cold, Gilbert explained everything in terms of magnetism. Others, like Basson, Gorlaeus and Gassendi, were some variety or another of atomist. Galileo wasn't really a natural philosopher in the sense of offering a system of

[38] Mersenne (1623), "Praefatio et prolegomena ad lectorem," ér.

[39] Naudé (1625), 331–332.

[40] Naudé (1627), 135.

[41] See OFB 12:8–9; Descartes (1996) 1:158; Heereboord (1654), 28; Webster (1653), 106; Boyle (1674), 223.

[42] OFB 6:258–259; Bacon (1626), expt. 69.

explanation at all but did offer non-Aristotelian doctrines of cosmology and motion. Figures like Bacon, Galileo, and Descartes are usually grouped together as a kind of "progressive wing" of the new philosophers, what many twentieth- and twenty-first-century commentators have in mind when they talk about The New Philosophy. But when we examine them more carefully, we have to acknowledge that their programs were quite distinct and substantially different from one another. Though all the *novatores* from Telesio to Descartes and beyond agreed in rejecting Aristotle and Aristotelianism, they could hardly be said to form a uniform school of thought.

This group did not necessarily have a good reputation in the seventeenth century. In many camps, novelty was suspect. The term *novator* has its origin in the context of the debates between Catholics and Protestants in the sixteenth century.[43] For the Protestants, the Catholics were the innovators, bringing new doctrines and practices into the Church, and they – the Protestants – were simply returning the Church to its original state. For the Catholics, on the other hand, it was the Protestants who were innovating. When, in the early seventeenth century, those terms migrate into natural philosophy, they carry much of their negative connotation. Mersenne names the *novatores* in order to counter what he takes to be their slander against the Catholic church, that they are dogmatic Aristotelians. In a 1624 pamphlet written against Etienne De Clave and Antoine Villon, two *novatores*, Jean-Baptiste Morin wrote:

> There is nothing more seditious and pernicious than a new doctrine. I speak not only in theology, but even in philosophy. For if […] the true knowledge of visible and corporeal things, that is, the true natural philosophy raises and delights us toward the knowledge and love of invisible and incorporeal things, […] it is quite certain that the false philosophy or knowledge of the things in nature cannot lead the mind to the same end, and can only lead it to errors, heresies, and atheism.[44]

Morin goes on to observe that almost all heresies involve departures from the philosophy of Aristotle. This view was widely shared. It is no surprise that Telesio's philosophy had been put on the Index in 1596. Later, of course, innovation will gain a much better name. But in the beginning, at least, to be listed among the *novatores*, as Telesio characteristically was, was no compliment.

Sorel's 1655 essay "Le sommaire des opinions les plus estranges des novateurs modernes en la philosophie" was intended as a defense of the *novateurs*. In the essay, Sorel treats a large number of figures, almost all of whom appear regularly on various lists of *novateurs*: Telesio, Patrizi, Cardano, Ramus, Copernicus, Galileo "& autres Astonomes [Kepler is included in the chapter]," Bruno, Gorlaeus, Carpenter, "Enchyridion de la physique restituée [Jean d'Espagnet]", Basson, Campanella, Descartes, "les nouvaters chymistes, de Paracelse & autres, & particulairement d'Estienne de Claves, Henry de Rochaz [Rochas]." Sorel's essay contains both a criticism of Aristotle and Aristotelianism on the one hand, and a defense of the *novatores* whose writings he summarizes in some detail. (In the 1655 edition,

[43] For a brief account of the history, see Garber (2016).

[44] Morin (1624), "a Monseigneur Halligre […]," 3. (Note that the dedicatory letter is paginated separately from the rest of the pamphlet.) For an account of the larger context in which this pamphlet was written, see Garber (2002) and the references cited there.

the essay runs 66 pages.) Sorel has good things to say about Aristotle, and doesn't dismiss him by any means. Aristotle is to be taken seriously, he argues, though we must recognize that we have learned much since the time Aristotle wrote.[45] But Sorel does argue strongly against those who defend everything Aristotle said, and reject any kind of novelty: Aristotle shouldn't be considered infallible.[46] Sorel offers interesting analyses of why the Aristotelians are so resistant to allowing others to express their view. He suspects that the dogmatic Aristotelians think that they will lose students because of that. But, he thinks, the evidence of the success of the *novatores* in attracting students shows that they are wrong about that.[47] He suspects also that if the dogmatic Aristotelians discover imperfections in Aristotle, they will worry that they won't know whom to follow.[48] Sorel is not impressed by the fact that Aristotle is supported by the Roman Catholic Church: don't people know that in earlier times Aristotle had been rejected by the Church?[49] Furthermore don't they know that these days people aren't required to follow the opinions of the pagans?[50] Others are read and corrected, why not Aristotle? Sorel refers to those who refuse to correct Aristotle as "aveugles volontiares".[51]

Over and against the dogmatic Aristotelians, Sorel advocated for a tolerant attitude toward the *novateurs*. He admitted that while the views of some are "fantastiques et imaginaires," "the others address themselves to solid truths, and are to be praised the more for being hidden."[52] He continues:

> Although the very name of *novateur* might be odious to many people, we must be careful that even if it is to be feared in matters concerning theology, it isn't so in natural and human philosophy.[53]

He admits that there are some who are *novateurs* simply out of a spirit of contradiction.[54] But he praises others for their courage to point out the errors of Aristotle. Sorel ends with a plea for being open-minded. We shouldn't accept the ancients dogmatically, nor should we reject them all. "One should take the middle way in this matter," accept a view when it warrants being accepted, and suspend judgment in all things uncertain.[55]

It is in this context that we find Sorel's discussion of Telesio. Since, Sorel claims, "les premiers Novateurs ont paru en Italie,"[56] he begins his account with the Italians.

[45] Sorel (1655), 211–212; 273–274.

[46] Ibid., 211.

[47] Ibid., 270–271.

[48] Ibid., 271.

[49] Ibid.

[50] Ibid.

[51] Ibid., 271–272.

[52] Ibid., 210; cf. 267.

[53] Ibid., 210.

[54] Ibid., 267; cf. 210.

[55] Ibid., 273–274.

[56] Ibid., 215.

And among the Italians, Telesio is the first to be taken up.[57] The entry on Telesio is rather long; it goes for a full four pages.[58] The focus is on Telesio's *De rerum natura*, in the 1586 edition of nine books. Sorel goes through the text, book by book, offering brief summaries of its content. He then adds some criticisms of Telesio's philosophy. There is no single theme to his comments. He begins by claiming that heat and cold are not substances or principles, but qualities, and that cold, merely a privation of heat, in particular, cannot be considered a principle or an agent. He claims that the origin of the seas cannot be in question since Scripture tells us that they were created directly by God. He also disputes what Telesio says about heat and cold with respect to water. But he also adds some compliments, agreeing with Telesio in the claim that neither air nor fire enter into the composition of mixtures, about how the heat of stars diffuses below, and about the motion of celestial bodies, the generation of animals, and other matters.[59] Sorel ends with a summary of the criticisms that Jean-Cecile Frey presented in his *Cribrum philosophorum*, mentioned above.[60]

Though he disagrees with him on some points, Sorel is clearly sympathetic to Telesio. We are obliged to him for "having had the courage to collide with this ancient master of philosophy, to aid in freeing those who had subjected themselves entirely to his laws from his servitude."[61] It was Telesio's example that "excited many philosophers to search for knowledge different from that of the ancient."[62] There is no suggestion that Sorel thinks that we should revive Telesio's views: what was particularly important about Telesio was that he was the first: "We must praise the greatness of Telesio's courage to have dared to be the first to criticize the ancient errors."[63] His main virtue seems to be not in the specific doctrines that he held, but in the fact that he was the first to oppose the dominant Aristotelianism.

Sorel's treatment of Telesio suggests to me that he is directing his comments at an audience is no longer familiar with Telesio's views, and needs to be informed about what they are. Furthermore, Telesio is just one among many of the discussions of *novateurs*; even if he is the first, Telesio is part of a large group of other figures, special mainly for coming first. In Sorel's essay, Telesio is not an interlocutor in a serious conversation about the nature of things, as he was for Bacon. He is one rogue among many in a rogue's gallery of curious opinions. Or better still, he is an exhibit in a museum, hidden in a forgotten corner of a dusty *Wunderkammer*.

* *
*

[57] See Bianchi (1992); Picardi (2007), 259–264.

[58] Sorel (1655), 215–218.

[59] Ibid., 217–218.

[60] Ibid., 218.

[61] Ibid., 217.

[62] Ibid., 218.

[63] Ibid., 267.

It is striking the distance that Telesio fell since the days in the 1570s and 1580s when he had his own academy and was at the center of a lively debate, or since the 1610s when he was still a live option of sorts for a figure like Bacon. In his *The grand prerogative of humane nature namely, the souls naturall or native immortality* (1653) Guy Holland, an English Jesuit, included Telesio among a group of "soaring spirits" who "but newly sprung, yet are grown already into neglect."[64] By the 1650s, I think, Telesio had become a name on a list for some, a curiosity for others, but was no longer part of the active conversation among natural philosophers in Europe. Though his name was remembered, it isn't entirely clear how much his doctrines were.

Bibliography

Bacon, Francis. 1626. *Sylva Sylvarum, or a Natural History in Ten Centuries*. London: W. Lee.

Bacon, Francis. 1996. *The Oxford Francis Bacon*. Oxford: Oxford University Press (Abbreviated as OFB. References given by volume and page number).

Bianchi, Lorenzo. 1992. Des novateurs modernes en philosophie: Telesio tra eruditi e libertini nella Francia del Seicento. In *Bernardino Telesio e la cultura napoletana*, ed. Giuseppe Galasso, Raffaele Sirri, and Maurizio Torrini, 373–416. Naples: Guida editori.

Blair, Ann. 1993. The Teaching of Natural Philosophy in Early Seventeenth-Century Paris: The Case of Jean-Cécile Frey. *History of Universities* 12: 96–158.

Blair, Ann. 1994. Tradition and Innovation in Early Modern Natural Philosophy: Jean Bodin and Jean-Cécil Frey. *Perspectives on Science* 2: 428–454.

Boenke, Michaela. 2013. Bernardino Telesio. *The Stanford Encyclopedia of Philosophy (Fall 2013 Edition)*, ed. Edward N. Zalta. http://plato.stanford.edu/archives/fall2013/entries/telesio/. Accessed 4 Nov 2015.

Bondì, Roberto. 1997. *Introduzione a Telesio*. Rome and Bari: Laterza.

Bondì, Roberto. 2001. Bacon e la restaurazione di Parmenide. *Rivista di Filosofia* 92: 327–339.

Boyle, Robert. 1674. *The Excellency of Theology Compar'd with Natural Philosophy*. London: Henry Herringman.

Campanella, Tommaso. 1591. *Philosophia sensibus demonstrata*. Naples: Horatius Salvianus.

Campanella, Tommaso. 1617. *Prodromus philosophiae instaurandae, id est, Dissertationis de natura rerum compendium*. Frankfurt: Ioannes Bringerus sumptibus Godefridi Tampachii.

Carpenter, Nathanael. 1625. *Geography Delineated Forth in Two Bookes*. Oxford: John Lichfield and William Turner.

De Franco, Luigi. 1995. *Introduzione a Bernardino Telesio*. Soveria Manelli: Rubbettino.

De Mas, Enrico. 1990. Bernardino Telesio e la falsità di Aristotele: il giudizio di Bacone e di Tobia Adami. In *Convegno internazionale di studi su Bernardino Telesio (Cosenza, 12–13 maggio 1989)*, 167–179. Cosenza: Accademia Cosentina.

Del Prete, Antonella. 2001. Charles Sorel et l'Italie: une interprétation de la Renaissance. In *Sources antiques de l'irréligion moderne*, ed. Didier Foucault and Jean-Pier Cavaillé, 171–180. Toulouse: Université Toulouse-Le Mirail.

Descartes, René. 1996. In *Oeuvres de Descartes*, vol. 11, ed. Charles Adam and Paul Tannery. Paris: J. Vrin.

Eustachius à Sancto Paulo. 1609. *Summa philosophiae quadripartita*. Paris: Carolus Chastellain.

[64] Holland (1653), 89. Others included in this group are Patrizi, Petrus Ramus, Sebastien Basson, and Pierre Gassendi.

Frey, Ianus Caecilius. 1646. *Cribrum philosophorum qui Aristotelem superiore et hac aetate oppugnarunt*. In *Opuscula varia nusquam edita [...]*, 29–89. Paris: Petrus David.

Frey, Ianus Caecilius. 2003. *Cribrum Philosophorum*. Lecce: Conte Editore [This is a facsimile reprint of Frey (1646), with an introduction by Roger Ariew and Daniel Garber.].

Garber, Daniel. 2002. Defending Aristotle/Defending Society in Early Seventeenth-Century Paris. In *Wissensideale und Wissenskulturen in der frühen Neuzeit (Ideals and Culture of Knowledge in Early Modern Europe)*, ed. Claus Zittel and Wolfgang Detel, 135–160. Berlin: Akademie-Verlag.

Garber, Daniel 2016. Historicizing Novelty. In *What Reason Promises*, ed. Wendy Doniger, Peter Galison, and Susan Neiman, 186–94. Berlin: De Gruyter.

Gassendi, Pierre. 1658. *Opera Omnia*, vol. 6. Lyon: Laurentius Anisson and Ioan. Bapt. Devenet.

Giachetti Assenza, Valeria. 1980. Bernardino Telesio: Il Migliore dei Moderni. I Riferimenti a Telesio negli Scritti di Bacone. *Rivista Critica di Storia della Filosofia* 35: 41–78.

Giglioni, Guido. 2010. The First of the Moderns or the Last of the Ancients? Bernardino Telesio on Nature and Sentience. *Bruniana & Campanelliana* 16: 69–87.

Hatfield, Gary. 1992. Descartes' Physiology and Its Relation to His Psychology. In *Cambridge Companion to Descartes*, ed. John Cottingham, 335–370. Cambridge: Cambridge University Press.

Heerboord, Adriaan. 1654. *Meletemata philosophica, maximam partem, metaphysica*. Leiden: ex officina Francisci Moyardi.

Holland, Guy. 1653. *The Grand Prerogative of Humane Nature Namely, the Souls Naturall or Native Immortality, and Freedom from Corruption*. London: Roger Daniel.

Leijenhorst, Cees. 2010. Bernardino Telesio (1509–1588): New Fundamental Principles of Nature. In *Philosophers of the Renaissance*, ed. Paul Richard Blum, 168–180. Washington, DC: Catholic University of America Press.

Lerner, Michel-Pierre. 1992. Le 'parménidisme' de Telesio: Origine et limites d'un hypothèse. In *Bernardino Telesio e la cultura napoletana. Atti del Convegno internazionale, Napoli 15–17 dicembre 1989*, ed. Giuseppe Galasso, Raffaele Sirri, and Maurizio Torrini, 79–105. Naples: Guida.

Lupi, F. Walter. 2011. *Alle origini della Accademia telesiana*. Cosenza: Brenner.

Margolin, Jean-Claude. 1990. Bacon, lecteur critique d'Aristote et de Telesio. In *Convegno internazionale di studi su Bernardino Telesio (Cosenza, 12–13 maggio 1989)*, 135–166. Cosenza: Accademia Cosentina.

Mersenne, Marin. 1623. *Quaestiones celeberrimae in Genesim*. Paris: Sebastian Cramoisy.

Morin, Jean-Baptiste. 1624. *Refutation des theses erronees d'Anthoine Villon [...] et Estienne de Claves*. Paris: Chez l'Autheur.

Mulsow, Martin. 1998. *Frühneuzeitliche Selbsterhaltung: Telesio und die Naturphilosophie der Renaissance*. Tübingen: Max Niemeyer Verlag.

Naudé, Gabriel. 1625. *Apologie pour tous les grands personnages qui ont esté faussement soupçonnez de magie*. Paris: François Targa.

Naudé, Gabriel. 1627. *Advis pour dresser une bibliotheque*. Paris: François Targa.

Picardi, Mariassunta. 2007. *Le libertà del sapere: filosofia e "scienza universale" in Charles Sorel*. Naples: Liguori.

Pousseur, Jean-Marie. 1990. Bacon, a Critic of Telesio. In *Francis Bacon's Legacy of Texts: 'The Art of Discovery Grows with Discovery'*, ed. William A. Sessions, 105–117. New York: AMS Press.

Rees, Graham. 1977. Matter Theory: A Unifying Factor in Bacon's Natural Philosophy? *Ambix* 24: 110–125.

Schuhmann, Karl. 1988. Hobbes and Telesio. *Hobbes Studies* 1: 109–133.

Schuhmann, Karl. 2004. Telesio's Concept of Matter. In *Karl Schuhmann: Selected Papers on Renaissance Philosophy and on Thomas Hobbes*, ed. Piet Steenbakkers and Cornelis Hendrik Leijenhorst, 99–116. Dordrecht: Kluwer.

Sciaccaluga, Nicoletta. 1997. Movimento e materia in Bacone: uno sviluppo telesiano. *Annali della Scuola normale superiore di Pisa, classe di lettere e filosofia Ser.* 4(2): 329–355.

Sorel, Charles. 1623. *Histoire Comique de Francion*. Paris: Pierre Billaine.

Sorel, Charles. 1627. *Le Berger Extravagant*. Paris: Toussainet du Bray.

Sorel, Charles. 1633. *La Vraye Histoire Comique de Francion*. Paris: Pierre Billaine.

Sorel, Charles. 1655. Le sommaire des opinions les plus estranges des Novateurs en Philosophie. In *De la perfection de l'homme*, 209–275. Paris: Robert de Nain.

Sorel, Charles. 1668. Des Novateurs en Philosophie. In *La science universelle tome quatriesme*, 360–449. Paris: Theodore Girard.

Telesio, Bernardino. 1565. *De natura iuxta propria principia liber primus et secundus*. Rome: Antonius Bladus.

Telesio, Bernardino. 1570. *De rerum natura juxta propria principia, liber primus, & secundus, denuo editi*. Naples: Josephus Cacchius.

Telesio, Bernardino. 1586. *De rerum natura iuxta propria principia libri ix*. Naples: Horatius Salvianus.

Telesio, Bernardino. 1590. *Varii de naturalibus rebus libelli [...]*. Venice: F. Valgrisius.

Telesio, Bernardino. 1965–1976. *De rerum natura iuxta propria principia libri ix*, ed. and trans. (Italian) Luigi De Franco. 3 vols. vols. 1 and 2: Cosenza: Casa del Libro. vol. 3: Florence: La Nuova Italia.

Telesio, Bernardino. 1981. *Varii de naturalibus rebus libelli*, ed. Luigi De Franco. Florence: La Nuova Italia.

Telesio, Bernardino. 2009. *La natura secondo i suoi principi*, It. trans. and ed. Roberto Bondì. Milan: Bompiani.

Webster, John. 1653. *Academiarum Examen, or the Examination of Academies*. London: Giles Calvers.

Weeks, Sophie. 2007. *Francis Bacon's Science of Magic*. Unpublished Ph.D. Dissertation. Department of Philosophy, University of Leeds.

Chapter 8
Looking at an Earth-Like Moon and Living on a Moon-Like Earth in Renaissance and Early Modern Thought

Natacha Fabbri

Abstract The idea of an Earth-like Moon was the object of a lively debate throughout the Renaissance, a debate that was largely indebted to Plutarch's *De facie in orbe lunae* and *De Placitis philosophorum*, to Macrobius' *Commentary on the Somnium Scipionis*, as well as to Proclus' *Commentary on the Timaeus*.

The Earth-like Moon argument was formulated not only against the ontological difference established between the two realms of the Aristotelian-Ptolemaic cosmos, but also as evidence in favour of heliocentrism. Both Galileo and Kepler stressed this connection to such a degree that they "saw" the motion of the Earth reflecting on the face of the Moon. The mutual relationship between Earth and Moon implied indeed the existence of a Moon-like Earth, namely of an Earth displaying the same phases and luminosity of the Moon, capable of illuminating the other celestial bodies and, in the end, endowed with motion.

In considering some Renaissance and Early Modern philosophers (such as Francesco Patrizi, Giordano Bruno, Michael Maestlin, William Gilbert, John Wilkins), who dealt with the idea of an Earth-Moon identity or of a more generic kinship between them, this chapter shall elucidate to what extent Galileo and Kepler intertwined the lunar cosmology that emerged during the previous century with the *historia* discovered by the telescope.

All translations are my own unless otherwise stated.

N. Fabbri (✉)
Galileo Museum. Institute and Museum for the History of Science, Florence, Italy
e-mail: n.fabbri@museogalileo.it

C. Muratori, G. Paganini (eds.), *Early Modern Philosophers and the Renaissance Legacy*, International Archives of the History of Ideas Archives internationales d'histoire des idées 220, DOI 10.1007/978-3-319-32604-7_8

Lunar Landscape

The definition of the Moon as another Earth – and, conversely, of the Earth as another Moon – gave rise to a lively debate throughout the Renaissance. Even after the telescope confirmed the earthly status of the Moon, several authors supported that idea by drawing on ancient and Renaissance sources and by recasting them into the new frame provided by the *perspicillum*.

It might not be generally known that this definition of the Moon played a central role not merely in the rebuttal of the ontological difference existing in the Aristotelian-Ptolemaic cosmos, but also in the fulfilment of the heliocentric view. Although the *De Revolutionibus* had not analysed this matter in depth, the two most fervent advocates of Copernicanism, namely Galileo Galilei and Johannes Kepler, gazed at the motion of the Earth on the face of the Moon.

This chapter is divided into two parts. In the first one, it considers some Renaissance philosophers who developed the idea of an Earth-like Moon, or of a more generic kinship between those two celestial bodies. It will be argued that one of their main aims was to introduce several compelling arguments to reappraise the statute of the Earth. In the second part, the paper focuses on Galileo and Kepler's thought in order to evaluate whether they were affected by such readings, as well as to show how crucial the new ontological status of the Moon was in arguing the truthfulness of heliocentrism.

Before turning to the philosophical debate that arose over the earthly Moon, I would like to point out that in the Renaissance the rise of an Earth-like Moon occurred in different fields. In painting, the first known naturalistic representation of the Moon (with spots, reliefs, the terminator line) was depicted in the *Crucifixion of Christ* (1420–1425) by the Flemish artist Jan van Eyck.[1] One century later, Ludovico Ariosto's *Orlando furioso* (1532) assigned a rough surface to the Moon, which was furrowed by mountains, lakes and rivers, and interlaced with cities. This description of the Moon – which was deeply indebted to Leon Battista Alberti's *Intercenales*, and mainly to his *Somnium*[2] – might have affected Galileo, as also Tommaso Campanella noticed.[3] Indeed, Galileo had great familiarity with the *Orlando furioso*, as is attested by his *Postille all'Ariosto* and also by his comments to the excerpts describing Astolfo's voyage to the Moon (canto 34).[4]

It is worth mentioning that in the years between Alberti's *Intercenales* and Ariosto's *Orlando*, Leonardo da Vinci devoted a significant number of folios to the survey of the earthly Moon: these folios are now gathered in the so-called *Codex*

[1] It also appeared in other canvases by Jan van Eyck, such as in *Saint Barbara*, in *The Knives of Christ* (belonging to the Ghent Altarpiece) and in the *Madonna of Chancellor Rolin*. See Montgomery (1994).

[2] See Martelli (1964); Segre (1986).

[3] Galilei (1890–1909), XII, 287.

[4] Galilei (1890–1909), IX, 125.

Atlanticus, *Arundel* and *Hammer*, as well as in the *Manuscript F*.[5] On one folio belonging to the *Codex Arundel*,[6] Leonardo drew two very similar globes, namely the Earth and the Moon, by presenting both covered with water and earth: "Here you will conclude that what of the Moon is shining is water similar to that of our seas and so it is flooded; what of her does not shine are islands and mainland."[7] The lunar body was therefore made of earth, fire and water[8]: its surface was not smooth and clean, but rough and rugged, since the inequalities were formed by the movement of the waves that, at least partially, covered the Moon.

Leonardo's drawings and survey of the Moon are very significant for four reasons. First, he overcame the difference between celestial and terrestrial realms, although he did not work out a wide-ranging rebuttal of the Aristotelian physics. Secondly, he refuted the objection that if the Moon had been a heavy body like the Earth, it would have fallen down on the Earth, just as water would have done had it been on the Moon.[9] Leonardo pointed out that in the Moon the centre of magnitude did not coincide with the natural gravity: it was indeed put in the centre of the Moon, thus preventing the lunar water from falling on the Earth.[10] Thirdly, by stating that the Moon was not bright on its own as it reflected both the solar light and the light coming from the Earth, Leonardo described the "secondary light" for the first time: this phenomenon occurs when the sunlight hits the Moon after having been reflected by the Earth. Specifically, "So much light has the side of the moon which is turned to us as is that of the sun reflected on it from our seas. And this light is so much greater as our ocean receives greater light from the sun, that is, when the moon is new and sets just after sunset; and this light lessens as the moon grows older."[11] Finally, according to Leonardo's studies, the Moon did not have a crystalline surface, rather a rough one stirred by winds and surrounded by waves.[12] While the presence of water on the Moon might have played a role in the elaboration of Francesco Patrizi's cosmological model, it is equally possible that Leonardo's explanation of secondary light might have affected Galileo's view – yet no evidence has been found so far.[13]

[5] See Reaves and Pedretti (1987), Laurenza (2004).

[6] Leonardo da Vinci (1998), 104r.

[7] Ibid., fol. 28r.

[8] Leonardo da Vinci (1973–1980), 112v.

[9] In the *Codex Hammer*, Leonardo (1987), 2v, claimed that the Moon is a heavy body which is "dressed" – just like the Earth – with the four elements.

[10] Leonardo (1987), 2v.

[11] Ibid., 1r. On Leonardo's study of the secondary light, see Reeves (1997), 29–31.

[12] Leonardo (1987), 1r.

[13] On Leonardo's alleged presence in Galileo's writings, see Reeves (1997), 29–31, 113–118; Dupré (2003), 375–385. It is highly possible that several studies of Leonardo were spread among sixteenth-century philosophers and mathematicians thanks to Vincenzo Pinelli, an erudite man who owned one of the largest private libraries in Europe at the time and who was in touch with, among others, Antonio Persio, Paolo Sarpi, Justus Lipsius, Francesco Patrizi and Galileo. Pinelli's collection numbered ancient books, prohibited books, manuscripts on optics, botany, music, geography, as well as Leonardo's treatise on painting (in addition, realistically, to other folios by the same author). On Pinelli's library, see Nuovo (2007).

Besides Leonardo's sketches of the Earth-Moon likenesses and the similarity between the description of the Moon presented by Ariosto and the one provided by the telescope, it is important to bear in mind the significant role the recovery and translation of Plutarch's *Moralia* played throughout the sixteenth century. Plutarch was considered the most relevant Neoplatonic theologian and his writings covered a significant part of the *curriculum studiorum*.[14] His dialogue *De facie in orbe Lunae* – which was included in the *Moralia* – is centred on the topic of the terrestrial Moon, a topic also discussed in chapters 25-30 of the second book of the *On the Opinions of the Philosophers* (*De Placitis philosophorum*), which were often paraphrased by sixteenth- and seventeenth- century philosophers and mathematicians. The description of the Moon provided by Plutarch was rapidly employed as an argument against the Aristotelian vision of the cosmos, as we can read in Francesco Patrizi, Giordano Bruno, Kepler and Galileo's works.[15]

Directing our attention to Patrizi and Bruno's writings, it may be noticed that they were not limited to a description of the Moon based on a series of similarities with the Earth, just as it occurred in Plutarch; on the contrary, they capsized the terms of comparison, by choosing – first and foremost – to describe the Earth on the basis of several likenesses with the Moon. As we shall see afterwards, going beyond some stereotyped readings of Galileo's survey of the Moon, it is possible to claim that this purpose was also shared – yet with some significant differences – by Galileo and Kepler.

Two Terraqueous Globes

In the *Pancosmia* of the *Philosophia Nova* (1591), Francesco Patrizi provided a doxographical excursus on the "ethereal Earth", which was largely based on the *De facie* and on the Neoplatonic cosmology of Proclus' *Commentary on Timaeus*. Patrizi started describing the Moon's spots and its changing colour (as it can become ash-grey, green and black), in order to introduce the definition of the Moon as *altera* or *antichtona Terrae*[16] – that is, the definition given by the Pythagoreans and Philolaus.[17] Patrizi then used the similarity between Moon and Earth to clarify the status of the Earth, although he did not develop every part of this argument and failed to tackle some of its consequences. He went on to say that the Earth, just like the Moon, had dark and luminous parts: the dark parts were everything that surfaced from the water; the luminous ones were oceans, seas, lakes, rivers, ponds, and so on, since water can reflect light. The Moon did not shine with its own light, but rather it reflected the one coming from the Sun, just as our Earth returned the solar rays. The "luminous" water, which largely composed the Moon, was nevertheless onto-

[14] See especially De Pace (1996), (1998).

[15] See Fabbri (2012).

[16] Patrizi (1591), 113r: "Luna ergo, aetherea terra esto. Et terra nostra, elementalis esto Luna".

[17] Ibid., 112v.

logically different from the water present on the Earth: since the Moon was an ethereal globe, its water was ethereal, too. Patrizi also sought to explain why the water on the Moon did not fall on the Earth, as instead it should have done according to the Aristotelian theory: by relying on the idea of the multiplicity of centres of gravity – which went back to Martianus Capella and showed significant likenesses with Leonardo da Vinci's analysis – he claimed that not only did the Moon have an external centre of gravity – the Earth – but it also had an internal one, to which its water referred.

Patrizi's cosmos abandoned the Aristotelian structure of crystal spheres and the distinction between sublunary and celestial world, without however reaching a perfect homogeneity. The ensuing hierarchical structure in which the Moon was an "ethereal Earth" and the Earth was an "elemental Moon"[18] prevented him from claiming the perfect identity between Earth and Moon. From the point of view of optics and astronomy, the fact that this similarity was "mutual" might have led Patrizi to introduce the phenomena of reflection and of secondary light (*i.e.* from the Earth toward the Moon). Instead, he limited his survey and described only the brightness of the Earth and the fact that it was in some way a "star": specifically, he never regarded the idea of the Earth's mobility. Patrizi listed a series of actions-reactions, which pertained more to phenomena of *simpathia* than to quantitative phenomena: after all, as Patrizi noted, the bond of sympathy was also possible between bodies that did not have the same substance, as it happened to be with the Earth and the Moon. Generally speaking, the Earth and the Moon heated, influenced, and reinforced each other, as is shown by the ebb and flow of seas, plants, animals, as well as by the humours of the human body.[19]

Giordano Bruno went one step further: he based the Earth-Moon kinship not merely on similarity, but rather on their perfect ontological identity. He had already defined the Moon as "another Earth" in the *De umbris idearum* (1582),[20] later unfolding this topic in the Italian dialogues *De l'infinito* and *Cena delle Ceneri* (1584), then tackling the subject again in *Articuli adversus Peripateticos* (1586) and *Articuli adversus Mathematicos* (1588), and finally devoting to it a large number of pages in *De immenso* (1591). Bruno's astrological drawings have been already examined by many scholars. My aim is not so much to present a further reading of them as to show that his analysis of the Earth-Moon relationship was extremely close to those of Galileo and Kepler, and marked a significant point in the achievement of the heliocentric view.

"And just as of the moon from the earth, which is another moon, some different parts appear more bright and others less bright, like so different parts of the earth appear from the moon, which is another earth, because of the variety and difference of the spaces of its surface."[21] The survey of the Moon sketched out in the third book of the *De immenso* clearly aims at pointing the reader toward the thorny problem of

[18] Ibid., 113r.

[19] Ibid., 132r.

[20] Bruno (1961–1962), II/1, 8. On Bruno's earthly Moon, see Tessicini (2001).

[21] Bruno (2007), 100.

the Earth's status: if one went up to the Moon (or went down to the Moon, depending on the point of reference), he would notice that the Earth had light, spots, and phases that are very similar to those of the changeable Moon.

> From the high shores of the sky you can see the Earth with the same picture of the changing Moon, so that the Earth with respect to the Moon and the Moon with respect to the Earth are mutually sky, low, middle and high, and with the same alternations, with the same light and with the same image they also look with respect to the Sun.[22]

To an observer placed on the Moon, everything would seem to revolve around him, and he would see the Earth with the same changes he could observe on the Moon: "Such is the Earth when compared to the Moon, sky and star, just as the Moon is to the Earth."[23] Bruno therefore emphasized the fact that the Earth seemed to return light, just as the Moon did, since both surfaces reflected sunlight back: the Moon received the nocturnal light from the Earth and the diurnal one from the Sun. In the third chapter, Bruno clarified that the Moon "is a companion (*comes*) and one with the Earth", and together they complete the yearly revolution[24]: namely, the Moon describes an epicycle around the Earth moving on a deferent centred on the Sun.

Unlike Patrizi, Bruno linked this topic to the one of heliocentrism. In the *De Revolutionibus*, the *cognatio* (namely, the kinship) had been the justification for the lunar revolution around the Earth, yet this change had not led Copernicus either to define the Moon as a satellite, or to deepen the nature of its familiarity with the Earth. Bruno explicitly developed such connection in the well-known diagram of the *Ash Wednesday Supper*, which presents a significant variation with respect to the model displayed in *De immenso* and *Articuli*.

In the fourth day of the *Supper* the Copernican diagram is the object of two different readings: one is explained by Torquato, a character who is convinced he is interpreting Copernicus correctly; the other is upheld by Bruno, who claims that the spot at the centre of the orbit of the Moon does not represent the Earth, rather it is only "the tread of the compass" ("pedata del compasso"). The inferior drawing placed indeed the Earth on the same epicycle of the Moon, but on the opposite side, as if the Moon were an Anti-Earth.[25] Bruno arrives at such cosmological model in an attempt to explain the apparent variation of the Sun's diameter when it is observed from the Earth: the Sun looks bigger when the Earth is closer and it appears smaller when our planet is farther.[26] This would not have been possible if Bruno had postulated the existence of the Earth's circular orbit around the Sun, as Torquato instead

---

[22] Bruno (1961–1962), I/1, 324.

[23] Ibid., 328.

[24] Ibid., 330–331.

[25] Tessicini (2001) regarded this definition as Bruno's attempt to bring the Copernican model back to the Pythagorean cosmology, which had presented the relationship between Earth and Anti-Earth (*i.e.* Moon).

[26] With regard to the heliocentric drawing of the *Cena*, see, among others, Badaloni (1955), 85; McMullin (1987); Aquilecchia, (1997), 153–157; De Bernard (1987), 168–169; Gatti (1999), 62–68; Granada (2010), 33–35.

does. If it is true that Bruno, in this dialogue, seems to misunderstand Copernicus (because of the fact that in the *De Revolutionibus* the Earth and the Moon are not on the same epicycle), it is equally true that Torquato's model is wrong, too, since it refers to the diagram of the *De Revolutionibus*, which is an oversimplification of the heliocentric system. As Copernicus explained in the third book of the *De Revolutionibus*,[27] according to his system it is possible that the Earth does not move on a deferent around the Sun (as someone would expect when considering only the famous diagram), but rather it moves on a combination of two epicycles, the centre of which describes the deferent. In light of this, Bruno's first assertion that Moon and Earth share the same epicycle can find a justification: specifically, the centre of the smallest epicycle run by the Moon describes the first epicycle of the Earth, and they then move together on a second epicycle; or rather, to be exact, they do not move *on* the same epicycle, but *according to* the same epicycle. Some decades later, while challenging the Copernican system, Lodovico delle Colombe, too, resorted to a description that was very similar to Bruno's, and claimed that in the heliocentric system "the Earth and the Moon are in the thickness of a Heaven, like in an epicycle […]."[28]

The fact that the drawing described by Bruno in the *Cena* may be considered as an inaccurate reproduction of the astronomical system discussed therein is confirmed by the *De immenso*, where he admitted to have presented a scheme with many fewer epicycles than there actually were. The astronomical model that both the *De immenso* and the *Articuli* put forward features a Moon that revolves around the Earth and an Anti-Earth played by Mercury (which, in turn, is at the centre of the epicycle of Venus). Although the Earth no longer moves on an epicycle, the apparent variation of the Sun's diameter is warranted by the fact that – again according to one of the explanations Copernicus advanced in chapter 20 – the Sun is not at the centre of the astronomical system, but moves on a circle that is concentric with it. In order to oversimplify the diagram of the *De immenso*, Bruno leaves out the circular path along which the Sun runs – which is instead represented in the *Articuli adversus Mathematicos* (art. 160)[29] – and passes over many other epicycles: "[…] as we show in this scheme, although in a rather brief and confusing way, as it is not easy for us, given the multiplication of circles or spirals (as it is really more conform to nature), to be able to reach a happy composition."[30]

In spite of some not so negligible differences between *Supper*, *De immenso* and *Articuli*, all the astronomical models therein share Bruno's aim, that is to assert the movement of the Earth and the perfect identity of Earth and Moon – to which he then added Mercury and Venus as well. Taking advantage of the fact that the

[27] Copernicus (1543), 91v-92r.

[28] Delle Colombe (1611), 269.

[29] Even the *De immenso* (Bruno (1961–1962), 396–398) does not show the circle of the Sun, although it describes it in the text. For a detailed explanation of the diagrams of the *Articuli adversus Peripateticos* and *adversus Mathematicos*, see Granada (2010), 40–43.

[30] Bruno (1961–1962), 397–398. Therefore Bruno does not seem to definitively rule out the movement of the Earth on an epicycle, as the *Supper* alleged.

Moon had been considered as a planet (and not a satellite) until then, Bruno first assimilated the Earth to the Moon by relying on their numerous similarities, and later he argued for the sameness between them and other planets (*i.e.* Venus and Mercury) even from the viewpoint of the planetary theory.

Earthly Moon and Copernicanism

Notwithstanding Bruno's broad survey of the Earth-Moon system, hardly anyone quoted him explicitly when wondering about the existence of an Earth-like Moon: Galileo, Kepler, Paolo Sarpi, Gilbert, Maestlin, and other less known philosophers (such as Tommaso Gallaccini, Alimberto Mauri, Raffaello Gualterotti) preferred to draw on Plutarch and Proclus' writings rather than hint at Bruno's. As we will see later on, in this regard John Wilkins represented an exception: he mentioned Bruno by taking great care to separate the topic of the earthly Moon from the discussions on the infinite universe. For instance, one of Galileo's closest friends, Raffaello Gualterotti, built the description of the earthly Moon drawing on Patrizi and giving it a Copernican outcome: both Earth and Moon were dense and opaque, and our planet, as it revolved around the Sun, was illuminated by the Sun in the same way as the Moon was.[31] The premises of this similarity were the fluidity and the ontological homogeneity of the cosmos, even if with a different degree of purity.[32]

Before the *Sidereus Nuncius*, Kepler himself had debated at length about the lunar statute without ever mentioning Bruno – yet he knew most of his cosmological writings – and widening his theoretical horizon with Michael Maestlin and William Gilbert's work. Maestlin's *Epitome Astronomiae* (1597) provided the first – although not detailed – explanation of the secondary light phenomenon, or Earthshine. As Kepler stressed in his *Astronomia pars optica*, in the *Disputatio de eclipsibus solis et lunae* (Tübingen 1596) Maestlin had already devoted three theses (n. 21, 22 and 23) to this subject.[33] A comprehensive analysis was later carried out by Gilbert in the *De magnete* and in the studies published posthumously in 1651 under the title of *De mundo Nostro Sublunari Philosophia Nova*. The latter dedicated two parts to the Earth-like Moon, once again stressing the Earth's luminosity.[34] In what was published under chapter 13 of the second book – titled "De luna seu telluris socia" –, Gilbert presented a doxographical excursus on the lunar statute,[35] closely following Plutarch's *De Placitis* and defining the Moon as "companion", "friend", "follower" of the Earth, and, in the end, as "another Earth" ("tellus altra"): the spots represented the earthly parts and the luminous areas were related to the water on the surface. The Moon had a completely different nature from the Sun, as it was considered as a

[31] Gualterotti (1605), 12.

[32] Ibid., 19.

[33] See Kepler (1604), 266; Maestlin (1596), 7–8.

[34] Gilbert (1651), 173, 202. On Gilbert's Moon, see Pumfrey (2011).

[35] Gilbert (1651), 170–171.

smaller Earth, and the Earth another Moon – as it reflected the solar light.[36] Gilbert emphasized the harmony and symmetry existing between the motion of the Earth and the Moon (as is attested by the tides), that is, a strong bond that the *De magnete* based on the "magis rationi" between these two celestial bodies.[37]

Gilbert's theory did not go unobserved.[38] Specifically, Francis Bacon alluded to Gilbert's lunar map, which would be published many years later in the *De mundo*: "And that *Geography*, or *Map of the Moon*, which Gilbert conceived in his Mind, may now, by the Industry of *Galilaeo*, and others, seem to be actually making."[39] Bacon continued his analysis by connecting the question of the earthly Moon to the statute of the other planets, that is, to establish whether they too were solid and opaque as the Moon and the Earth, and whether their luminosity might be considered as "splendor", namely reflected light.

Although Bacon did not trace any connection between this topic and Copernicanism, the followers of the heliocentric system saw in it a crucial argument in favour of heliocentrism, as is attested by the title page of John Wilkins' best known work – *A New World in the Moon* – which was published for the first time in 1638 and again in 1640 with the addition of a second book. The frontispiece shows a heliocentric cosmos with the Moon rotating around the Earth, the phases of Mercury, Venus, Moon, Mars, Jupiter and Saturn, and the motto "mutuo se illuminant" featuring the Earth-Moon relationship. At the bottom of the page, there are Copernicus, Galileo and Kepler who are caught in the act of discussing the true structure of the universe.[40]

The belief that the lunar issue was decisive for the Copernican cause was indeed upheld by both Galileo and Kepler. Their interest in the Earth-Moon similarities started long before telescopic observations and was not a consequence of them. Kepler dealt with this matter in his 1596 theses and in the *Astronomiae pars optica* (1604), whereas in the 1610 letter to Belisario Vinta Galileo wrote: "that the moon is a body very similar to the Earth, I had already made sure of it, and in part I had shown it to our Lord, although imperfectly, because I still didn't have a glass as excellent as the one I have now".[41] Galileo regarded the telescope as the indispensable tool to mark the transition from *fabula* to *istoria*, since it enabled him to formulate apodictic arguments: "and if someone said to believe that it was harsh and mountainous, he was reputed to speak too soon, fabulously rather than philosophically. Now of this same lunar body, which we saw by means of the light of the sun, I state the first, no longer by reason of imagination, but by reason of sensible experience and necessary demonstration."[42]

---

[36] Ibid., 173.

[37] Gilbert (1600), 86, 232. Gilbert (1651), 184.

[38] Galileo was a reader of Gilbert and had a copy of his *De magnete*. See Favaro (1886), 200.

[39] Bacon (1858), 449.

[40] See Nonnoi (2001).

[41] Galilei (1890–1909), X, 262.

[42] Galilei (1890–1909), XI, 142. On the connection between the improvement of the telescope and the elaboration of Galileo's theory of the celestial light, see Dupré (2003).

According to Kepler, the telescope provided evidence that definitely strengthened a rather widespread belief, although – unlike Galileo – he always stressed the subordinate position of an inquiry that relied on senses (*i.e.* telescopic observations) with respect to a theoretical approach. Such statement sprang also from the fact that telescopic observations were not by themselves sufficient to 'prove' heliocentrism: as the Tychonian astronomers showed, Galileo's discoveries (such as Jupiter's satellites and Venus' phases) fit in with the geo-heliocentric system as well.

Notwithstanding the fact that Galileo and Kepler agreed about the general reading on the Earth-Moon identity, the reasons they highlighted were nevertheless significantly different. Kepler stated that the dragging of the earthly element through the sky – which had been reached by means of the Earth-like Moon – laid the groundwork for the acceptance of the heliocentric system, according to which the Earth occupied an intermediate position in the sky, just between Venus and Mars. Even for this reason, Plutarch should have been deemed the precursor of Copernicus.[43] Galileo instead believed that thanks to the secondary light it would have been possible to attribute the status of star to the Earth, thereby increasing its nobility and showing its affinity with the other planets: the Earth was one of the most dazzling stars as it was endowed with light, and it lit the Moon by reflecting sunrays.[44]

Kepler's resolution to claim the existence of inhabitants on the Moon might also be numbered among the so-called "heliocentric arguments". Indeed, ascribing this topic merely to the literary genre of utopia would be misleading, as it would disregard its connection to physical arguments and to Kepler's belief concerning the use of rhetoric in natural philosophy.[45] The statements contained in the *Somnium* and its *Notes*, in the *Appendix Selenographica* and in the *Notes to the Appendix*, and concerning the existence of cities on the Moon, actually required accepting the idea that the motion of the Moon did not impede the presence of living beings on it: despite the fast double motion of the Moon (on its axe and around the Earth), buildings, trees, stones, living beings were not scattered in the sky by centrifugal motion, as they could remain on the Moon's surface thanks to the action of a sort of gravity. If this was possible in the case of an Earth-like Moon, it was equally possible on a Moon-like Earth. By dealing with the earthly Moon, Kepler could also deepen some issues concerning the Earth, and precisely, the diurnal and annual movements of the Earth. With this "evidence", he could thus rebut several traditional arguments that followers of the Ptolemaic and Tychonian universe had put forward against Copernicus.

When reading the *Nuncius* more in depth, we notice that even in that writing, as well as in the *Dialogo*, the description of the earthly Moon was undertaken in order to elucidate the status of the Earth, namely its perfect ontological identity with the

[43] See Kepler (1604), 224.

[44] See Galilei (1890–1909), III/1, 75.

[45] On Kepler's awareness of the importance of rhetoric for natural philosophy, see, among others, Moss (1993), 65–96; Rothman (2009); Voelkel (2001), 211–253.

other celestial bodies. Galileo thus adopted an argumentative style that was absent from his main source, namely Plutarch's *De facie*.

In a letter to Gallanzoni (16 July 1611), Galileo claimed that the purity of celestial bodies and the ontological difference between them and terrestrial ones was a consequence of the circular motion Aristotle had attributed to them. When someone will demonstrate – just as Galileo had been trying to do – that the circular motion also belongs to the Earth, the ontological difference the Peripatetics have been upholding will be destroyed. And in the end, according to Galileo, the study of the secondary light achieved precisely that aim.

The "Evidence" of Secondary Light

In some notes to the letter addressed in 1640 to Leopoldo dei Medici and concerning the *Litheosphorus* by Fortunio Liceti, Galileo made it clear that the true bond between Earth and Moon was not so much the similarity of their morphology as the fact that both displayed the same phases[46]:

> among the phenomena that induced the very great philosophers and Aristotle himself, the greatest among all, to acknowledge great sympathy and correspondence between the Moon and the Earth, were not only the similarity of shape and the spotted face [...], but much more the correspondence of this threefold illumination.

The triple lighting Galileo mentioned here concerned the direct illumination of the Sun (the strongest light), the feeble luminosity caused by the reflection of the solar light on the atmospheric vapours of Earth or Moon (the weakest light), and the light coming from the reflection of sunrays on the lunar or terrestrial surface.

Referring to the secondary light, both Cesare Lagalla (in his *De phaenomenis in orbe lunae*) and Antonio Rocco (in the *Esercitazioni filosofiche*) placed Galileo within a "heliocentric tradition": the mutual illumination phenomenon was neither "invented" nor "discovered" by Galileo, rather it dated back to the Pythagoreans.[47] Plutarch, indeed, in *De Placitis philosophorum* (chap. XXIX), had ascribed it to the Pythagoreans. In his copy of Rocco's book, a rather annoyed Galileo jotted down: "But where did you read that you would see the Earth from the Moon first that it reflects the light of the Sun, just as the Moon does for us, and then that it has a period of light that is similar, in figures and in time, to that of the Moon?"[48] It was precisely with regard to this aspect that in the *Nuncius* Galileo announced that a forthcoming work, titled *Systema cosmicum*, would provide further demonstrations.

The explanation of the ashen light was the first piece of "evidence" Galileo put forward in favour of the earthly movement. In the *Nuncius*, Jupiter's satellites were merely reckoned as an argument supporting the Copernican system: they acquired

---

[46] Galilei (1890–1909), VIII, 554.

[47] See Lagalla (1612), 328. Rocco, (1633), 641.

[48] Galilei (1890–1909), VII, 640–641.

the epistemological status of "evidence" only around 1612, when Galileo was able to calculate their periods of revolution around Jupiter and in order to measure them he needed to take into account the motion of the Earth as well.

In order to reach a thorough understanding of the reasons underlying the importance of the earthly Moon for the Copernican cause, we need to briefly turn to the issue of the brightness of planets. According to the traditional view, planets were endowed with a light similar to the one of the stars, and they could be partly or wholly self-luminous[49]: Macrobius and Avicenna believed that planets and stars were self-luminous; Averroës and Albert the Great argued that planets received their *lumen* from the Sun (the unique source of celestial light) and, being translucent and not opaque, they absorbed (rather than reflect) the solar light; Witelo instead upheld that planets reflected solar light. In the *De Revolutionibus*, Copernicus mentioned – most likely, by drawing on Plutarch's *De Placitis* – that Pythagoreans had claimed the darkness and opaqueness of all planets, their ability to reflect the solar light, as well as to display the same phases as the Moon.[50]

Galileo embraced this last opinion when trying to explain how something dark could be bright and could brighten other bodies. By tracing a series of similarities between Earth and Moon, first he established that the body of the Earth was luminous, capable of illuminating and of displaying phases; secondly, he regarded as highly possible that the planets' brightness might be the same as the Moon's and therefore as the Earth's,[51] just as it occurred with Jupiter and the shining of its four stars. Finally, he confirmed the Pythagorean theory by discovering, at the end of 1610, the phases of Venus,[52] thus proving that this planet revolved around the Sun and had a body that was just as obscure and dense as the Earth's and Moon's.

Galileo's survey of lunar morphology, namely the examination of the secondary light and of the spots formed by the shadows of mountains, in the end represented a strong argument to prove that the Earth was endowed with circular motion. A manuscript dating back to 1604–1605[53] confirms that Galileo was already interested in the Moon's luminosity at that time and that such study was strongly connected to the issue concerning the motion of the Earth. The fol. 21*v* of Gal. 70 presents a structured discussion and complex calculations on the *Lucida Luna* in Latin; the recto of the same folio contains sketches of the lighting phenomena on the lunar surface, and, on the left margin, bears a note on the movement of our planet: "[…] the rest is not less opposed to the movement downward than to the circular motion, although there are given in the Earth two principles, one to move, and the other to

[49] See Grant (1994), 393–402.

[50] Copernicus (1543), 7v.

[51] Galilei (1890–1909), III/1, 76; Kepler (1610), 118.

[52] See Palmieri (2001), 109–117; Gingerich (1984).

[53] This dating is suggested by the presence, on this folio, of a short writing in Venetian dialect inspired to the characters of the *Orlando Furioso* and considered as going back to the Paduan years. See Tomasin (2008).

stay at rest. the right and circular motion sympathize with each other, but not the circular one and the rest."[54] From the beginning of 1600 until the 1640 letter to Leopoldo de Medici, Galileo carried out an in depth analysis of the variation of the secondary light on the lunar surface in order to establish that the different grades of luminosity of the Moon depended on the phases of the Earth, and precisely on a new, full, waxing or waning Earth.[55]

John Wilkins stressed the importance of this point and developed it in proposition 11 of the first book of *A New World in the Moon*: "That as their world is our Moone, so our world is their Moone."[56] According to Wilkins's reappraisal of the ancient and Renaissance sources that could have upheld Galileo and Kepler's arguments, Macrobius had been the first to claim the Earth's brightness, despite not admitting that the Earth was able to illuminate other celestial bodies. Lodovicus Caelius Rhodiginus (Ludovico Ricchieri) was to be numbered among the first to state that the Sun made the Earth bright. In his *Antiquae lectiones*, published in 16 books in 1516 in Basel, Ricchieri had written the following while quoting Plotinus: "If you did conceive your selfe to bee in some such high place, where you might discerne the whole Globe of the earth and water, when it was enlightned by the Sunnes rayes,'tis probable it would then appeare to you in the same shape as the moone doth now unto us."[57] It was again Ricchieri (*Progym. 1.*) who had ruled out the possibility that the secondary light hitting the surface of the Moon could come from other planets. Such light, Wilkins went on, did not come from Venus, because sometimes that planet was beyond the Moon and thus it was not able to illuminate the lunar face that wasn't turned towards it. Nor could that light come from the Fixed Stars either, because otherwise the phenomena of lunar eclipse would not have existed.

Besides Ricchieri, Wilkins quoted Cusano, Maestlin and Bruno. Also following the second book of Cusano's *De docta ignorantia*, the Earth appeared as bright as a star, not only because of the reflection of sunlight, but also because it reflected the light coming from the region of fire. Wilkins then summed up that, in the opinions of Bruno and Cusano, the Earth was indeed a "noble starre having a distinct light, heat and influence from all the rest"[58]: they regarded the Earth as a star not only because of its light, but also because it was able to influence – and thus to act upon – the other celestial bodies. With Bruno and Cusano, the Earth no longer had a passive role in the history of the universe.

---

[54] Ms. Gal. 70, folio 21*r*.

[55] Galilei (1890–1909), III/1, 74–75; ibid., VII, 88–90; ibid., VIII, 496–519.

[56] Wilkins (1638), 143–165.

[57] Ibid., 149.

[58] Ibid., 85.

Anti-Copernican Outcomes: Godwin and Kircher's Earthly Moon

I have examined so far only those philosophers who reappraised the new status of the Moon in Copernican terms. It is worth mentioning briefly some cases attesting a different outcome in order to show that, although that topic was regarded as decisive for the heliocentric cause, its cogency was nevertheless merely stated. Thomas Godwin's *Man in the Moone*, published posthumously in 1638 and one of Wilkins' main sources, incorporated most of the lunar descriptions that are within Kepler's *Paralipomena* and Galileo's *Nuncius*, without however adopting their heliocentric arguments. Godwin's voyage – also greatly indebted to Lucian's *True Story* – presented the Earth as being bright as the Moon and, partially following the Copernican theory, endowed with a rotation motion (yet devoid of any revolution), in this way explaining the difference between the spots of the Moon and those of the Earth:

> the Earth (which ever I held in mine eye) did as it were mask it selfe with a kind of brightnesse like another Moone; and even as in the Moone we discerned certaine spots or Clouds, as it were, so did I then in the earth. But whereas the forme of those spots in the Moone continue constantly one and the same; these little and little did change every hower. The reason thereof I conceive to be this, that whereas the Earth according to her naturall motion, (for that such a motion she hath, I am now constrained to joyne in opinion with Copernicus) turneth round upon her owne Axe every 24. howers from the West unto the East.[59]

Godwin also mentioned the topic of the Earth-Moon mutual illumination, however without hinting at the Earth's revolution, nor analysing the secondary light.[60]

A similar attempt to separate telescopic observations from astronomical models was made some decades later by Athanasius Kircher (and also by one of the fathers of selenography, Giovanni Battista Riccioli): Kircher rejected the Aristotelian view of the cosmos – and, therefore, also the ontological difference between Moon and Earth[61] – by putting forward a panspermic interpretation of the Creation. In the *Iter extaticum coeleste* (1660) Kircher supported the idea of an Earth-like Moon, despite laying emphasis on some significant differences, which made it impossible to claim that other human beings existed on the Moon. Specifically, he admitted that the Moon had the same geological features as the Earth,[62] but not identical meteorological phenomena; moreover, the elements that made up the Moon had the same substance – but not the same accidents – as the ones on the Earth.[63] After all, as Galileo had also acknowledged in the *Dialogue*, one of the most significant differences between the Earth and the Moon was the period of their phases: in the Moon they

---

[59] Godwin (1638), 61. The impact that modern science and Galileo's discoveries had on literature has been discussed in the classical – but still enlightening – studies of Nicolson (1948, 1950, 1956).

[60] See Godwin (1638), 92–93.

[61] Kircher (1660), 19–20, 151–152, 374–375, 445–446.

[62] Ibid., 86, 89.

[63] Ibid., 90.

occurred over the course of a month, whereas in the Earth they needed a 24-hour lapse of time, consequently having a different impact on meteorological phenomena.[64]

The cases of Godwin and Kircher epitomized how, although the phenomenon of dual illumination was almost always regarded as evidence, that topic – as well as the one concerning the Moon-Earth identity – was neither a prerequisite nor a consequence of the Copernican system. It required Galileo and Kepler's theory and work to transform it from being a strong argument into "evidence" (at least if we take the term "evidence" in line with the use these authors made of it). This also shed more light on the reason why Kepler felt the need and the urgency to complete Galileo's work: with the *Dissertatio cum Nuncius Sidereo*, Kepler aimed at going beyond what seemed to him as a bare description of the phenomena observed thanks to the telescope, by adding a sort of historical and philosophical counterpoint to the observational data. The *Dissertatio*'s stylistic composition shares many elements with the style of juridical writings. While drawing on Cicero's lesson, Kepler displayed two kinds of *testimonia*: the first were Galileo's observations of spots, phases, eclipses, and so on; the second were the theories of ancient and modern natural philosophers, such as Plutarch, Della Porta, Maestlin and Kepler himself, which served as evidence of the reliability of Galileo's statements and strengthened their trustworthiness.[65]

The reading of the Moon that Galileo, Kepler and other contemporaries adopted could not have left aside the observations provided by the telescope. Nevertheless, the composition of the small quarters of the lunar disk observed through the telescope called for an interpretation of the observational data according to a well-defined astronomical and cosmological model. For that reason, very different theories concerning the Moon arose following the introduction of the telescope, and some of them acknowledged the Earth-Moon likeness, but without accepting their ontological identity and the movement of the Earth at the same time. Unlike most contemporaries, Galileo and Kepler chose to see the evidence of heliocentrism in the Moon – and this happened not least because they looked at the Moon through the reading lenses of Renaissance philosophers.

Bibliography

Aquilecchia, Giovanni. 1997. Astri, plebe e principi nella "Cena" di Bruno. In *Sguardi sull'Italia*, ed. Gino Bedani et al., 146–157. Leeds: Society for Italian Studies.

Bacon, Francis. 1858. Description of the Intellectual Globe. In Bacon, Francis. 1857-74. *The Collected Works of Francis Bacon*, ed. James Spedding, Robert Leslie Ellis and Douglas Denon Heath, 14 vols., vol. X. London: Longmans.

Badaloni, Nicola. 1955. *La filosofia di Giordano Bruno*. Florence: Parenti.

Bruno, Giordano, 1961–1962. *Opera Latine Conscripta*, facsimile ed. Fiorentino and Tocco et al. (1879–1891), 8 vols. Stuttgart-Bad Cannstatt: F. Frommann.

[64] Galileo (1632), 125–126.

[65] See Kepler (1610), 107.

Bruno, Giordano. 2007. De l'Infinito. In *Opere italiane*, ed. Aquilecchia Giovanni. Turin: Utet.
Copernicus, Nicolaus. 1543. *De revolutionibus orbium coelestium libri VI*. Nuremberg: I. Petre.
De Bernard, Luciana. 1987. *Immaginazione e scienza in Giordano Bruno*. Pisa: ETS.
Delle Colombe, Lodovico. 1611. *Contro il moto della Terra*. Manuscript. In Galilei (1890–1909), III/1.
De Pace, Anna. 1996. Ficino e Plutarco: storia di un equivoco. *Rivista di storia della filosofia* 51: 113–135.
De Pace, Anna. 1998. Plutarco e la rivoluzione copernicana. In *L'eredità culturale di Plutarco dall'antichità al Rinascimento*, ed. Italo Gallo, 320–331. Naples: D'Auria.
Dupré, Sven. 2003. Galileo's Telescope and Celestial Light. *Journal for the History of Astronomy* 117: 369–399.
Fabbri, Natacha. 2012. The Moon as Another Earth: What Galileo Owes to Plutarch. *Galilæana* 9: 103–135.
Favaro, Antonio. 1886. La libreria di Galileo descritta e illustrata. *Bollettino di Bibliografia e di Storia delle Scienze Matematiche e Fisiche* 19: 219–293.
Galilei, Galileo. 1632. *Dialogo sopra i due massimi sistemi del mondo*. Florence: Gio. Batista Landini. In Galilei (1890–1909), VII.
Galilei, Galileo. 1890–1909. *Opere di Galileo*, ed. Antonio Favaro, 21 vols. Florence: Barbèra.
Gatti, Hilary. 1999. *Giordano Bruno and Renaissance Science*. Ithaca: Cornell University Press.
Gilbert, William. 1600. *De magnete […]*. London: Peter Short.
Gilbert, William. 1651. *De mundo Nostro Sublunari Philosophia Nova*. Amsterdam: Elzevir.
Gingerich, Owen. 1984. Galileo and the Phases of Venus. *Journal for the History of Astronomy* 15: 98–104.
Godwin, Thomas. 1638. *The Man in the Moone*. London: John Norton.
Granada, Miguel Angel. 2010. L'héliocentrisme de Giordano Bruno entre 1584 et 1591. *Bruniana & Campanelliana* 16: 31–50.
Grant, Edward. 1994. *Planets, Stars, and Orbs: The Medieval Cosmos, 1200-1687*. Cambridge: Cambridge University Press.
Gualterotti, Raffaello. 1605. *Discorso […] sopra l'apparizione de la nuova stella*. Florence: Cosimo Giunti.
Kepler, Johannes. 1604. Astronomiae pars optica. In Kepler, Johannes. 1939. *Gesammelte Werke*, ed. Franz Hammer, vol 2. Munich: C.H. Beck.
Kepler, Johannes. 1610. *Dissertatio cum Nuncio Sidereo*. Prague: Typis Danielis Sedesani. In Galilei (1890–1909), III/1.
Kircher, Athanasius. 1660. *Iter extaticum coeleste*. Würzburg: Sumptibus J. Andr. & Wolffg. Endterorum Haeredibus.
Lagalla, Giulio Cesare. 1612. *De phaenomenis in orbe lunae*. Venice: Apud Thomam Balionum. In Galilei (1890–1909), III/1.
Laurenza, Domenico. 2004. Il sole e la luna in Leonardo: frammenti di un discorso. *Micrologus* 12: 565–572.
Leonardo da Vinci. 1973–1980. *Codice Atlantico*, ed. Augusto Marinoni. Florence: Giunti Barbèra.
Leonardo da Vinci. 1987. *The Codex Hammer* (Eng. trans. Carlo Pedretti). Florence: Giunti Barbèra.
Leonardo da Vinci. 1998. *Codice Arundel*, ed. Carlo Vecce. Florence: Giunti.
Maestlin, Michael. 1596. *Disputatio de eclipsibus solis et lunae*. Tübingen: George Gruppenbach.
Martelli, Mario. 1964. Una delle *Intercenali* di Leon Battista Alberti fonte sconosciuta del Furioso. *La Bibliofilia* 66(2): 163–170.
McMullin, Ernan. 1987. Bruno and Copernicus. *Isis* 78: 55–74.
Montgomery, Scott L. 1994. The First Naturalistic Drawings of the Moon: Jan van Eyck and the Art of Observation. *Journal for the History of Astronomy* 25: 317–320.
Moss, Jean Dietz. 1993. *Novelties in the Heavens*. Chicago: University of Chicago Press.
Nicolson, Marjorie Hope. 1948. *Voyages to the Moon*. New York: Macmillian.

Nicolson, Marjorie Hope. 1950. *The Breaking of the Circle*. Evanston: Northwestern University press.

Nicolson, Marjorie Hope. 1956. *Science and Imagination*. Ithaca: Great Seal books.

Nonnoi, Giancarlo. 2001. La scienza e la filosofia galileiane nel *New World* di John Wilkins. *Nuncius* 16: 49–84.

Nuovo, Angela. 2007. The Creation and Dispersal of the Library of Gian Vincenzo Pinelli. In *Books on the Move: Tracking Copies Through Collections and the Book Trade*, ed. Giles Mandelbrote et al., 39–68. New Castle/Delaware/London: Knoll Press and The British Library.

Palmieri, Paolo. 2001. Galileo and the Discovery of the Phases of Venus. *Journal for the History of Astronomy* 32: 109–129.

Patrizi, Francesco. 1591. *Nova de universis philosophia*. Ferrara: Benedetto Mammarelli.

Pumfrey, Stephen. 2011. The *Selenographia* of William Gilbert: His Pre-telescopic Map of the Moon and His Discovery of Lunar Libration. *Journal for the History of Astronomy* 42: 193–203.

Reaves, Gibson, and Carlo Pedretti. 1987. Leonardo da Vinci's Drawings of the Surface Features of the Moon. *Journal for the History of Astronomy* 18: 55–58.

Reeves, Eileen. 1997. *Painting the Heavens*. Princeton: Princeton University Press.

Rocco, Antonio. 1633. *Esercitazioni filosofiche*. Venice: Francesco Baba. In Galilei (1890–1909), VII.

Rothman, Aviva. 2009. Forms of Persuasion: Kepler, Galileo and the Dissemination of Copernicanism. *Journal for the History of Astronomy* 40: 403–419.

Segre, Cesare. 1986. La favola della luna (Ariosto, *Sat.*, III, 208-31) e i suoi precedenti. In *Book Production and Letters in the Western European Renaissance*, ed. Anna L. Lepschy et al., 279–283. London: Modern Humanities Research Association.

Tessicini, Dario. 2001. Pianeti consorti. La Terra e la Luna nel diagramma eliocentrico di Giordano Bruno. In *Cosmología, teología y religión en la obra y en el proceso de Giordano Bruno*, ed. Miguel Angel Granada, 159–188. Barcelona: Universitat de Barcelona.

Tomasin, Lorenzo. 2008. Galileo e il dialetto veneziano. *Atti e memorie dell'Accademia Galileiana di Scienze, Lettere ed arti in Padova* 119: 3–16.

Voelkel, James Robert. 2001. *The Composition of Kepler's "Astronomia Nova"*. Princeton: Princeton University Press.

Wilkins, John. 1638. *The Discovery of a World in the Moone*. London: E. G. for Michael Sparl and Edward Forrest.

Part III
Changing Conceptions of the Human

Chapter 9
Descartes, the Humanists, and the Perfection of the Human Being

Emmanuel Faye

Abstract In this study, I reconsider the question of the continuities and discontinuities between the Renaissance and the Early Modern period, putting the accent on the idea of "the perfection of man," as René Descartes conceived it in his *Meditations* and his *Principles of Philosophy*. I show in what sense it is possible to speak in this connection of a humanist thought, and I proceed to distinguish between the two complementary meanings recognized by Descartes as being a part of the *hominis perfectio*. There is, on the one hand, a remarkable confidence in the natural capacities of the human mind and the conviction that since our faculties are all good by their nature, what is important is to make the best use of them; and on the other hand there is a general project to "elevate our nature to its highest degree of perfection." Still, Descartes rejects, in the wake of Montaigne, the ontological conception of the *hominis perfectio*, which, after the manner of a Raymond Sebond, would make the human being the highest creature in the scale of beings. The main perfection of the human being declares itself for Descartes with respect to the less considerable perfections that are within us: a comparison that is internal to the study of the human being, and that is not carried out with respect to other creatures. Hence it is not between Montaigne and Descartes, between the time of the Renaissance and the Early Modern period, but, within the Renaissance itself, between Sebond and Montaigne, that the break with the image of the human being qua center of the world and goal of creation took place.

René Descartes' opposition to the "ordinary philosophy" (*vulgaris philosophia*) of his time, that is, to philosophy as it was taught in the Schools during the first decades of the seventeenth century—an opposition that he expresses already in the form of his *Rules for the Direction of the Mind*, written during the 1620s—is not the expression of a desire to make his mark in history. He is not interested in his role

---

E. Faye (✉)
Département de Philosophie, Université de Rouen, Normandie, Mont-Saint-Aignan, France
e-mail: emmanuel.faye@univ-rouen.fr

C. Muratori, G. Paganini (eds.), *Early Modern Philosophers and the Renaissance Legacy*, International Archives of the History of Ideas Archives internationales d'histoire des idées 220, DOI 10.1007/978-3-319-32604-7_9

in the history of philosophy. His desire is not so much to break away from his predecessors and his contemporaries as it is to move from probability to certainty and from controversy to science. Nevertheless, his refusal to accept the presupposition of the category of "substance" as it appears at the root of Porphyry's tree, and his refashioning of metaphysics by taking its principle and origin to be the existence of the *mens humana* as apprehended in the "thinking I" were seen, particularly by Hegel—as Cecilia Muratori and Gianni Paganini remind us in the introduction to this volume (but also, in France, by Victor Cousin)—as a new beginning of philosophy and a new era for thought. Indeed, Cousin looks upon the author of *Discourse on the Method* as the one who "puts an end to the risky ventures of the Renaissance."[1]

Contrary to Hegel, several interpreters in the twentieth century, particularly in France, from Étienne Gilson to Jean-Luc Marion, have attempted to connect Descartes' philosophy to possible sources of scholasticism, at the risk of a certain "theologizing" of his thought and of challenging the coherence of his metaphysics.[2] In Germany, it was Martin Heidegger who took his lead from Gilson's thesis, but in doing so he radicalized it; indeed, the author of *Being and Time* went so far as to assert, with no real demonstration establishing it, that "the fundamental ontological concepts of Descartes are directly taken from Suarez, Duns Scotus and Thomas Aquinas." In reality the theologico-scholastic reinterpretation of the *First Philosophy* of Descartes is contradicted by his very clear consciousness of the distinction between metaphysics and theology. That distinction is valid not only for our relation to infinite being, as to whether it comes from natural reason or revelation, but equally for our relation to man. As opposed, for example, to Nicolas Malebranche, Descartes rules out the introduction into philosophy of the question of original sin and predestination. His conception of man rests on a remarkable confidence in the natural capacities of the human mind and its faculties, which are "perfect in their kind." This leads him to formulate, first in the *Meditations*, and then in the *Principles of Philosophy*, a thought of the "perfection of man" (*hominis perfectio*) at once innovative and in kinship with the humanist thought of the Renaissance. The analysis of the Cartesian idea of the *hominis perfectio* can thus make a significant contribution to the elucidation of the question of this volume, which bears on the continuities between Renaissance and Early Modern philosophy.[3]

---

[1] Cousin (1843), 728.

[2] Marion (1986).

[3] That is the study I undertook in *Philosophie et perfection de l'homme. De la Renaissance à Descartes* (Philosophy and Perfection of Man: From the Renaissance to Descartes), devoted to the evolution of the Latin idea of *hominis perfectio* in French thought from Raymond Sebond to René Descartes. Faye (1998). See also Faye (1999).

Descartes, Humanism and the Perfection of Man: A Theme Revisited in a Recent Work

This guiding theme of the perfection of man, which I consider enlightening when it comes to characterizing the humanist thought of Renaissance and Early Modern philosophy, was the object of a manner of debate in a rather recent work on *L'humanisme scientifique de la Renaissance aux Lumières*.[4] Basing my remarks on the outcome of that discussion, I would like to revisit the question of the meaning of the perfection of man according to Descartes, for the purpose of further defining it. The editors Lorenzo Bianchi and Gianni Paganini, in their general introduction to the work, begin by reminding us how, in 1947, Eugenio Garin had advocated a conception of Italian and European humanism that was more philosophical and broader in scope than that of Kristeller, who limited that humanism to "philology." Then, to illustrate the fact that the philosophical renewal of the Renaissance continues in a sense into the seventeenth century in the form of Descartes' intention to "elevate our nature to its highest degree of perfection," Bianchi and Paganini use the following terms[5]:

> When Descartes was composing his *Discourse on the Method* ... he considered, in 1636, entitling it "Project for a Universal Science that can elevate our Nature to the Highest Degree of Perfection." Again, at the end of the Fourth Meditation, he mentioned a similar project, speaking of the "maxima et praecipua hominis perfectio." Thus, he relived, plumb in the middle of the seventeenth century, one of the great ideals of the renascent period, which Montaigne (referring to Socrates) had encapsulated in the very similar expression, "extreme degree of the perfection of man," or in a variant turn of phrase, less developed, but just as emphatic, "summit of human wisdom."[6]

Bianchi and Paganini then stress what differentiates Descartes from Montaigne in the respect:

---

[4] Bianchi and Paganini (2010).

[5] See also the review of the volume published by Frédéric Lelong in the *Bulletin Cartésien*: "This work, dedicated to the flexibility of the concept of humanism, and diametrically opposed to the interpretative perspectives that tended to see a radical break between the classical age and the culture of the Renaissance (M. Foucault is explicitly targeted in the Bianchi/Paganini preface, but one might also think H. Gouhier co-intended) could not but take up a position in relation to the thesis defended by E. Faye [...]. The preface, moreover, explicitly positions itself within that reflection on the *hominis perfectio* of the Renaissance up to and including Descartes." (Lelong (2012), 157).

[6] Bianchi and Paganini (2010), 1–2: ". Nel preparare il suo *Discours de la méthode* [...] Descartes meditava nel marzo 1636 di intitolarlo "progetto di una Scienza universale che possa elevare la nostra natura al suo più alto grado di perfezione." E ancora alla fine della Quarta Meditazione evocava un progetto analogo parlando della 'maxima et praecipua hominis perfectio' Riviveva dunque in piena epoca secentesca uno dei grandi ideali dell'età rinascimentale, quello che Montaigne (riferendosi a Socrate) aveva riassunto nella formula assai prossima: 'extrême degré de la perfection de l'homme', oppure, con una variante meno sviluppata ma altrettando enfatica, "de sommet de la sagesse humaine."

> But what for Montaigne was still the privilege of the individuality of an exceptional case, or at most the fruit of a strictly personal education, became, in the Cartesian discourse, a perspective accessible to all mankind, thanks to the use of a suitable method and appropriate scientific techniques, making it feasible for whoever was possessed of "good sense" or of "natural light."[7]

Cartesian Humanism: A "Philosophical Myth"?

Denis Kambouchner's contribution to the volume essentially addresses this same theme.[8] Let me summarize his comment. Referring to the correspondence between Étienne Gilson and Henri Gouhier on the cogency of describing Descartes as a humanist, he advances his argument by drawing on the *Letter to Voetius* to enrich the debate. Indeed, since we are not in a position to consult the *Treatise on Erudition*, unpublished and now lost, the *De usu librorum* is a remarkable illustration of the Cartesian conception of erudition.[9]

Denis Kambouchner accepts the terms of the discussion on "Cartesian humanism" by Henri Gouhier, once the question of its relation to *studia humanitatis* and erudition—"too negative in tone" in Henri Gouhier, in his view[10]—has been corrected. He accepts "a twofold qualification of Cartesian humanism," which he considers to be "irrefutable." It consists, on the one hand, in "the hopefulness associated with the 'technical self-confidence' of a new science," and on the other the idea of a sovereign good that is "perfectly accessible in this life, with no other condition than a certain resolve of the will" and therefore "the distancing of oneself … from the theme of corrupt nature."[11] Nevertheless, Kambouchner confesses that he feels "a certain discomfort" with respect to the "least questionable aspect of Cartesian humanism" (here the author is thinking of the second above-mentioned qualification) which remains, in his opinion, "purely negative".[12] "If the idea of original sin is removed, in keeping with a decision constitutive of the philosophical genre," "that rigorous abstraction" in his view would not indicate "any passage into the register of pure positivity."[13]

---

[7] Bianchi and Paganini (2010), 1–2: "Ma quello che per Montaigne era ancora il privilegio di un'individualità eccezionale o al più il frutto di un'educazione strettamente personale, diventava invece nel discorso cartesiano una prospettiva accessibile all'intera umanità grazie all'uso di un metodo e di tecniche scientifiche appropriate, in modo da render la praticabile da chiunque fosse dotato di semplice 'buon senso' o di 'lume naturale.'"

[8] Kambouchner (2009).

[9] I had developed this point in a colloquium organized in 1996 on *Descartes and the Renaissance* (See Faye (2000), 16–17), and the interpretation of *De usu librorum* had subsequently been at the center of an initial and brief discussion between Kambouchner and me on the relationship between Descartes and scholasticism, during a session of the *Société française de philosophie* held in 1998 (See Kambouchner (1998), 45–47).

[10] Kambouchner (2009), 345.

[11] Ibid. This second theme is closely connected with my own analyses.

[12] Ibid., 352.

[13] Ibid., 354. I disagree with this last assertion; I will give my reasons below.

From this assertion, Kambouchner deduces another, which in his view follows logically. "Therefore the mention, in a Cartesian context, of the theme of the "perfection of man" requires some caution".[14] The risk, he opines, lies in "imagining as real or given a condition that would essentially be aspirational, or at least only attained at the cost of an entire exercise." He goes on to say:

> This is probably the problem raised by E. Faye's thesis, articulated in the final development of *Med. IV.*, on Descartes' metaphysics as a body of thought on the perfection of man. Noting that the perfection of man no longer has exactly the same meaning as Aristotle's *entelecheia*, the author adds, "The perfection of man henceforth designates not only the ultimate goal, but also, and especially, the primordial truth of our nature."

Indeed, we may consider most of the second section of the present chapter to be a careful and nuanced discussion of the topic of the perfection of man according to René Descartes.

Humanism and "Humanists," from Montaigne to Descartes

But what connects this discussion on the perfection of man according to Descartes with the theme of the article ("Cartesian humanism")? Apparently, it is mainly my theses on the *hominis perfectio* in the work of Descartes, which the author identifies with the defense of that "philosophical myth," that he believes to be representative of that humanism—without making explicit, by the way, what meaning he attaches to the term "myth." Thus, he responds in his own way to the suggestions in the preface to the work by Bianchi and Paganini, which, as we have just seen, calls upon the perfection of man according to Montaigne and Descartes to give meaning to the conception—no longer just literary and philological, but frankly philosophical—of humanism.

I would like to voice a reservation about this. In *Philosophy and Perfection of Man*, it is in an explicit way, sustained by argumentation, that I deny myself the right to use the notion of humanism, which is too anachronistic and polysemous to clarify, without risk of confusion, any specific movement of Renaissance thought.[15] Let me remind the reader here that, in order to clarify on that occasion the reason behind my terminological choices, this methodological refusal, which I do not make bold to impose on anyone else (everyone being perfectly at liberty to use the historiographical categories he or she prefers, provided they are clearly defined), was roundly reproved by Thierry Gontier in an article in the form of a manifesto, published in 2001, and titled "Une catégorie historiographique oblitérée. Humanisme" (Humanism: An Obliterated Historiographical Category). He writes the following.

[14] Ibid.

[15] See Faye (1998), 28–32: "Humanisme," "dignité", and "perfection," in the chapter titled "Le choix des mots" [The choice of words].

> Must we quite simply abandon the term from our vocabulary as historians of philosophy, as Emmanuel Faye, in a very recent article, also suggests, having noted many cases in which this term has been used abusively in recent studies on Montaigne and Descartes?[16]

According to him, I share the responsibility of having "obliterated" the concept of humanism with Heidegger—a concept Gontier considers indispensable in defending the importance of philosophical studies on the Renaissance. An odd collusion indeed, since Heidegger's rejection of humanism comes from a determination to replace the concept of man with the term *Dasein*, and brings no specific view on the Renaissance with it. My refusal to use the term humanism, on the other hand, is more simply motivated by the desire to use, in the history of philosophy, the terms the thinkers of the time themselves used. Hence *hominis perfectio*, and not humanism—a term not created till the end of the eighteenth century, to mean the love of humanity [i.e. mankind] and used in the nineteenth century in the sense of "a thought or doctrine that makes man the ultimate end and supreme value," to repeat the definition quoted by T. Gontier.[17] This thought remains quite distant from the Cartesian conception, far more nuanced, of the place of man in the whole of creation.

While the concept of humanism, which is absent from the vocabulary of the Renaissance and the Early Modern period, is therefore not used in my research, I do emphasize the importance of the term "humanist(s)", which Montaigne uses as a hapax in his essay *Of Prayers.* He understands humanists as being authors of writings "purely human and philosophical, without any mixture of theology." And the context of the quoted passage, which refers, not without humor, to his "fancies merely human," shows that he puts himself in the category of those writers who are not theologians but philosophers and humanists.

It is just as legitimate, in my view, to consider Descartes a humanist in precisely the sense Montaigne gives that word, which I therefore do as well[18] without, however, subscribing to the "philosophical myth" of humanism, if what is meant by that expression is the fact of making man the "ultimate end" of Creation as a whole—before being replaced, during the second half of the twentieth century, by another, no less powerful and still flourishing "myth," largely in the wake of Heidegger's *Letter on Humanism*, the myth of anti-humanism, variously proclaimed by the tenants of what it has become customary to refer to as the *French Theory*—from the theoretical anti-humanism of Louis Althusser to the deletion of man in the Foucault of *The Order of Things.*

---

[16] Gontier (2001), 29 ff. In reality, I do not speak of abuse, but identify and note the extreme diversity of meanings ascribed to the word "humanism," even when applied to Renaissance philosophers. And I do not advocate simply dropping the notion of humanism. On the contrary, I write that "the term humanism is indeed a word that belongs to our time – a word whose meaning philosophy must defend today" (ibid., 30). This "today," written in 1998, is the twentieth century that henceforth belongs to history. A historian of ideas will thus have a lot to say about the word humanism as used in the nineteenth and twentieth centuries. We have only to think of the exemplary work by Toussaint (2008), but this term can be the source of many misunderstandings and much confusion when applied in an inaccurate or insufficiently nuanced manner to Renaissance and Early Modern authors.

[17] Faye (1998), 269.

[18] Ibid., 385 ff. See Montaigne (1978), 323.

The Positivity of the Distinction between Humanists and Theologians, and of the Cartesian Notion of the Greater Perfection of Man

While humanism designates a doctrine yet to be defined, to classify oneself as a humanist, as Montaigne does, is to express a certain mode of philosophical thought and a certain manner of non-theological writing. The author of the *Essays* draws a distinction between philosophers and theologians which, far from being limited to a purely negative determination, will translate, at the level of his concepts, into the "enormous distinction" he recognizes in Book III of his *Essays* between conscience and devotion. This distinction was adopted and amplified by Pierre Charron, and transmitted one way or another from there to Descartes. This humanist and philosophical position was not taken for granted, and was far from being universally shared. On this point we may cite a contemporary of Descartes such as Scipion Dupleix, who in 1510 begins his *Ethics* by evoking the "original sin that we, for our part, carry within us from birth."[19] And to uphold the distinction between conscience and devotion in ethics in the seventeenth century was to run the risk of being taxed with Pelagianism, as evinced in 1647 in the course of Descartes' trials and tribulations with the theologians of Leiden. It therefore reflects a somewhat anachronistic view of the Early Modern mentality to think that the dropping of original sin was at that time obviously "a decision constitutive of the philosophical genre.' The example of the philosophy of Malebranche, and the theses he developed as early as in the introduction to the *Recherche de la vérité*, show quite clearly that this was not the case at that time.

The qualification of the humanist by Montaigne, and my own with respect to Descartes, appears negative only to the extent that it attempts to distinguish the thought of the humanist from the dominant mode of thought, that of the theologians. This is no different from when, in the Platonic dialogues and Socrates' confrontation with the Sophists, we begin by seeing what the philosopher's thought opposed before seeing in what it consists. But that initial opposition, or that distinction, will then make it possible for Montaigne to assert the positivity of conscience—and for Charron to assert the anteriority of nature before grace. Now, there is an analogous movement of thought in Descartes. At the beginning of the Fourth Meditation, he asserts (in an expression that is still negative in form, but not in substance) that it does not seem to him to be possible for God to have given him "a faculty which is not perfect of its kind, or which lacks some perfection which it ought to have." But at the end of the same Meditation, on a remarkable page, long neglected by the commentators, he recognizes very positively, in the capacity that has been given us to acquire the habit of avoiding error, "man's greatest and most important perfection": *maxima et praecipua hominis perfectio.*

[19] Dupleix (1994), 43 (my translation).

A Perfection Given, or Attained?

But let us get to the philosophical substance of this debate. Kambouchner's thesis is that "the philosopher, although having no particular right to use the imperative, should beware of the indicative, which leads one to imagine a condition as real or given that should, essentially, be envisaged as a goal, or at least attained at the cost of a whole exercise".[20] In short, to say that the Cartesian conception of the greatest perfection of man designates not only an ultimate accomplishment but a primary truth of our nature, would amount to forming an imaginary thought appropriate to reinforcing the philosophical myth of Cartesian humanism. But what does Descartes tell us? In what exactly does this greatest and chief perfection of man, on which the whole movement of the thought of the Fourth Meditation culminates, consist?

Descartes' focus here is on what I have received from the infinite being. "Even if I have no power to avoid error in the first way just mentioned," namely the one "which requires a clear perception of everything I have to deliberate on," "I can avoid error in the second way, which depends merely on my remembering to withhold judgment on any occasion when the truth of the matter is not clear."[21] God has given me the natural ability and the freedom to avoid error, not by omniscience, but by mastering my judgment, and it is indeed in this virtue that, according to Descartes, the greatest and chief perfection of man consists.

However, considering the weakness that is in the nature of man and that keeps him from continually keeping the same thought in mind, this perfection that is naturally given to him as a virtue must still be achieved through habit. Hence the need for "attentive and repeated meditation," in order to impress deeply on the memory the resolution taken to judge only those things the truth of which is clearly known.

The long periodic sentence of Descartes, often disconcerting to today's reader, is necessary here because it gathers together into one movement the main faculties of the human *mens* that collaborate in the search for truth: the will's resolve to bridle any precipitous judgment, the help of memory, the understanding's clear conception, and last but not least the attention of the meditative mind. Moreover, there is clearly no contradiction in saying that the chief perfection of man, his capacity to avoid error, is not just a goal, but at once a given of our nature and a habit whose acquisition requires a long and continual exercise of the mind. Thus, there is not just a conditional mood in the Cartesian philosophy of the greatest perfection of man, but also an indicative one, and even, we might say, if not an imperative in the Kantian sense of the term, at least a prescriptive, since what is at stake is the learning of what I must avoid, and what I must do, in order to know the truth. In short, what is involved is how to make the best possible use of the ability I have been given naturally as a human being: my freedom.[22]

---

[20] Kambouchner (2009), 355.

[21] Descartes (1996), 43.

[22] Ibid.: "So today I have learned not only what precautions to take to avoid ever going wrong, but also what to do to arrive at the truth". It is clear that Descartes' words are not merely negative. Though not imperative in the Kantian sense of the term, it is positive and prescriptive.

On the Proper Use of Human Freedom

The twofold dimension, both given and attained, of the *hominis perfectio*, is remarkably illustrated by a text that extends and renders more explicit the concluding passages of the end of the Fourth Meditation, namely articles 37 and 38 of the first part of *Principles of Philosophy.* In the stating of article 37, the difference between the Latin text and the French translation raises some difficulty. The Latin reads: "summam esse hominis perfectionem, quod agat libere, sive per voluntatem". Literally, this translates as "It is the highest perfection of man, that he act freely, or at will." The accent is therefore placed on free, voluntary action—free will and will being one and the same in Descartes. In Abbot Claude Picot's translation, revised by Descartes, we read "que la principale perfection de l'homme est d'avoir un libre arbitre" [that the chief perfection of man is to have a free will]. It is in the possession of a free will in itself that the perfection of man resides. In the one case (the French of 1647), it is what is naturally given to man, his freedom, that constitutes his highest perfection; in the other (the Latin of 1644), it is a certain way of acting that is accentuated. It is not certain, however, that this means we should interpret this difference as an alternative, in which the chief or highest perfection of man would consist either in the fact of having free will (translation of 1647) or in the fact of acting freely or at will (original Latin of 1644). Indeed in comparing these two versions we find the same complementarity between nature and action as the one presented in the Fourth Meditation between natural virtue, received from God, and the habit acquired by human effort. Furthermore, the commentary on article 37 enlightens us on Descartes' thought: it is indeed of a natural capacity proper to the human being as such, namely of our will, that it is question, in that it is a power of acting freely.[23]

The following article and its commentary enlightens us even more, and one can truly assert that we have here one of the most important of statements of Descartes' entire philosophy.

Here is what he writes. "Our falling into error is bad behavior, not the result of a bad nature." He comments in these terms.

> Our falling into error is a defect in how we act, how we use our freedom; it is not a defect in our nature. Whether we judge correctly or incorrectly, our nature remains the same.

Descartes, instead of explaining error and evil as a corruption affecting our nature, as the theologians do, sees the cause of our errors in the bad use of our faculties, which means that we have the ability to correct that use. The whole undertaking of the *Rules for the Direction of the Mind* was already framed in that way of thinking—Descartes showing us how to develop the natural wisdom of the human mind based on the two primary actions of the intelligence, which are intuition and

[23] Indeed, Descartes writes: "It is part of the very nature of the will to have a very broad scope; and it's a supreme perfection in man that he acts voluntarily, i.e. freely." Descartes (2010), 9 [article 37].

deduction. The *Meditations* and the *Principles of Philosophy* extended that way of seeing things to the study of the determinant role of the will in the act of judgment, and therefore in the search for truth.

The Highest Perfection of Man, a First Truth about Our Nature

The Cartesian conception of the perfection of man is therefore of great fecundity, in that it makes the human being more conscious of the capacities within him. On this basis, Descartes can thus elaborate not only a method of thinking, but also a true "project," or program of thought and life. We can see this by the title originally chosen in March 1636 for the future *Discourse on the Method*, namely *Plan for a Universal Science Capable of Raising our Nature to its Highest Degree of Perfection*, as well as by the mention (11 years later, at the end of a Letter/Preface to the French translation of the *Principles of Philosophy*) of the "degree of wisdom" and "perfection of life" to which the truths deduced from the principles of philosophy can lead. In *Principles of Philosophy*, however, this encouragement to pursue the search for truth is philosophically bolstered by the confidence expressed by the philosopher in the natural capacities of the human being to make the best possible use of his or her freedom. This is remarkably expressed in the two texts on the *praecipua* or *summa hominis perfectio*, the analysis of which I initiate. As the ideal of knowledge, wisdom, and life, the highest degree of perfection indicated by Descartes represents, for humankind, an ultimate achievement; but as the recognition of the ability not to err and of the freedom that is in us, our chief and highest perfection constitutes a primary truth of our nature.

To put it differently, all the richness and subtlety of Descartes' conception of perfection, as it applies to the human being, lies in the fact that, on the one hand, he speaks to us, in the plural, of the perfections that are in us. In doing so, he indicates our natural abilities or virtues, those faculties that have been given to us and as such constitute us, beginning with our highest perfection, free will. On the other hand, he speaks to us, in the singular, of a perfection admitting of degrees, to designate a philosophical ideal that we can attain by the best use of our faculties. This latter consists in acting freely (*agere libere, Principia philosophiae*, I, §37) to acquire the habit of not erring (*habitum quaedam non errandi, Meditatio quarta*),[24] and, going back to the original intent of the *Regulae*, in developing the natural light of our reason in the pursuit of truth.

[24] Descartes (1976), AT VII, 62.

Conclusion: Descartes and the Renaissance Philosophers on the Perfection of Man

In *Philosophy and Perfection of Man*, my intention was to propose a nuanced approach to the question of the continuity or discontinuity between Descartes and the philosophers of the Renaissance in respect to their way of conceiving the *hominis perfectio*, emphasizing the elements of continuity. Thus, rather than being attentive in Montaigne only to his declamation on the misery of man as developed in the *Apologie de Raimond Sebond*, as is often done, I showed that it is possible to find certain points of similarity between Montaigne's way of speaking about the "extreme degree of perfection" of Socrates[25] and Descartes' way of bringing out the capacities of human beings as such.[26] Moreover, one could show the similarities between the Cartesian conception of the *hominis perfectio*—both naturally given to man by his free will and *qua* ideal of knowledge and wisdom—and the conception developed by Charles de Bovelles in the *Livre du sage* of the human being who is twice human: by nature and by culture. Thus one could perhaps recognize, in Descartes' project of a *Traité de l'érudition* something like an echo of Bovelles' project of thematizing the passage from *naturalis homo* to *homo studiosus.*

Moreover, a long familiarity with the philosophical works of the Late Middle Ages (the "*artiens*"[27] of the University of Paris) and, even more importantly, with the thinkers of the French Renaissance has heightened my awareness of the theme of the *hominis perfectio*, to the point of prompting me to highlight a motif I believe to be of major importance in Descartes' philosophy,[28] and one that has thus far received too little attention. It is not only the historical question of the continuity or lack of it between the Renaissance and the Early Modern period that is at issue, but also our conception of philosophy itself. And this is so, not only because the perfection of the human being is an important theme in philosophy, but also—as Jacques de Douai wrote during the time of the condemnations of the ethics of the "artiens" in Paris—because "philosophy is the great perfection of man." Hence it is because Bovelles, Montaigne, Charron, and Descartes are philosophers in a full and basic sense that each one of them rethought, each in his own way, yet in a way that was partially common to them all, the motif of the *hominis perfectio.*

Setting out from this general observation, we can delineate a good number of related phenomena. We can point out, for example, as did E. F. Rice, that from Bovelles to Charron, from the beginning of the sixteenth century to the beginning of the seventeenth, the ideal of human wisdom shifts from metaphysics to morality, or from contemplation to action. This is due in large part to the fact that the corpus of

[25] See Faye (1998), 209–211. Also Bianchi and Paganini (2010), 1–2.

[26] See also on this topic the discussion on humanism proposed by Gress (2012).

[27] "Artiens" so called as they were connected (as teachers, students, or more specifically by their doctrine of happiness) with the Faculty of Arts.

[28] The *summa hominis perfectio* is particularly important in *Principles of Philosophy*, of which it takes up articles 37 and 38 of the Part I, containing 76 articles.

authors on which the philosophers draw evolves, the Neo-Platonist references giving way to those of the Stoics and the Skeptics. This contrast between two culturally and conceptually distinct poles of the Renaissance, further coupled in the case of France by the transition from Latin to French, allows us to highlight all the more clearly the fact that, beginning with the *Studium bonae mentis*, Descartes appears as the one who is able to bring about a novel synthesis of these two dimensions, the moral and the metaphysical, of the *hominis perfectio.*

The discontinuity between Sebond and Descartes (to restrict our consideration to the French case), lies more in the cosmological revolution and its anthropological consequences than in a mutation internal to the notion of human wisdom. Indeed, we know how, in the third part of *Principles*, Descartes deduces, from the impossibility of assigning limits to the universe, the impossibility for humans to consider themselves the end of creation. The consequences for man of "this vast idea of the extension of the universe" that we can form on the basis of the consideration of the new physics, are formulated by Descartes in a remarkable letter to Elizabeth of Bohemia of 15 September 1645, in which (like Montaigne) he rejects a certain abusive way of elevating humans above the other creatures, and a certain ontological conception of the *hominis perfectio* that would (a la Raymond Sebond) make the human being the highest creature on the scale of beings. The fact is, while the programmatic writings of Descartes (the original 1636 title, the 1647 Preface/Letter) show a remarkable optimism, his *Meditations* and *Principles of Philosophy* involving the *praecipua* or *summa hominis perfectio* are sober and without insistence. As I have pointed out here we must, in Descartes' opinion, take into account the weakness or *infirmitas* of humans. They have no right to complain that God wanted them to play a role in the world "that is not the chief and most perfect of all" *(quae non est praecipua et maxime perfecta).*[29] They cannot claim to be at the level of the most perfect creatures of the universe. The letter of 15 September 1645 to Elizabeth should thus not be interpreted, in my opinion, as a necessary "precaution" or a corrective to the mention, "in a Cartesian context, of the theme of the 'perfection of man'". [30] Indeed, that letter constitutes, on the contrary, the best illustration of that theme, as I attempted to show (in *Philosophy and Perfection of Man*) by quoting from, and twice commenting extensively on, Descartes' critique of excessive anthropomorphism as elaborated in the letter.[31] Descartes expresses the chief perfection of the human being in relation to less considerable perfections within us. The comparison is internal to the study of the human being: it is not with respect to other creatures. Descartes did in fact benefit, in his own way, from the lessons of *Apology for Raymond Sebond* when he refused to attribute "to other creatures imperfections that

[29] Descartes (1976), AT VII, 61. See also Faye (1998), 331.

[30] See Kambouchner (2009), 85–86.

[31] See Faye (1998), 310–311, 331. On this precise point, D. Kambouchner's analyses and my own are not far from agreeing. What separates us is that I do not believe that Descartes' critique of anthropomorphism is an argument from which it is possible to draw a metaphysical reason for doubting that the perfection of man is a "properly Cartesian" theme.

they do not have."[32] In short, for one who seeks to determine, like Hans Blumenberg in the fourth part of *Légitimté des Temps modernes*, the *limes* or "era threshold" between the middle ages and the Modern period, it will no longer be between the Renaissance and the Early Modern period, but, within the Renaissance itself, between Sebond and Montaigne. That is when the anthropological break with the image of the human being *qua* center of the world and goal of creation took place.

Translated from French by Michael Smith.

Bibliography

Bianchi, Lorenzo, and Gianni Paganini. 2010. *L'umanesimo scientifico dal Rinascimento all'Illuminismo*. Naples: Liguori Editore.

Cousin, Victor. 1843. Vanini, sa vie, ses écrits et sa mort. *Revue des deux mondes décembre* 1843: 673–728.

Descartes, René. 1976. *Oeuvres*, ed. Charles Adam et Paul Tannery. 11 vols. Paris: Vrin, (abbreviated in the essay as AT).

Descartes, René. 1996. *Meditations on First Philosophy*, ed. and trans. John Cottingham. Cambridge: Cambridge University Press.

Descartes, René. 2010. *Principles of Philosophy*. Trans. Jonathan Bennett http://www.ahshistory.com/wp-content/uploads/2012/07/descprin.pdf.

Dupleix, Scipion. 1994. *L'Éthique ou philosophie morale*, ed. Roger Ariew Paris: Fayard.

Faye, Emmanuel. 1998. *Philosophie et perfection de l'homme. De la Renaissance à Descartes*. Paris: Vrin.

Faye, Emmanuel. 1999. Descartes et les philosophes français de la Renaissance. *L'Enseignement philosophique* 49(4):27–71. http://www.appep.net/wp-content/uploads/2012/06/faye01 pdf. Accessed 30 May 2014.

Faye, Emmanuel. 2000. Descartes et la Renaissance: philosophie de l'homme, méthode, métaphysique. In *Descartes et la Renaissance*, ed. Emmanuel Faye, 11–54. Paris: Champion.

Gontier, Thierry. 2001. Une catégorie historiographique oblitérée: l'Humanisme. In *Comment écrire l'hisoire de la philosophie?* ed. Yves-Charles Zarka, 267–281. Paris: PUF.

Gress, Thibaut. 2012. *Descartes et la précarité du monde*. Paris: CNRS-Editions.

Kambouchner, Denis. 1998. Descartes et le problème de la culture. *Bulletin de la Société Française de Philosophie* 92(3). Paris: Vrin.

Kambouchner, Denis. 2009. L'humanisme cartésien: un mythe philosophique? In *Descartes et la philosophie morale*, ed. Denis Kambouchner, 339–360. Paris: Hermann.

Lelong, Frédéric. 2012. Recension de Denis Kambouchner, L'humanisme cartésien: un mythe philosophique ? *Bulletin cartésien XLI. Archives de philosophie* 75(1). http://www.archives-dephilo.com/Bulletins/BC/BC41/BC41Liste.php.

Marion, Jean-Luc. 1986. The Essential Incoherence of Descartes' Definition of Divinity. In *Essays on Descartes' Meditations*, ed. Amélie Oksenberg Rorty, 297–338. Berkeley: University of California Press.

Montaigne, Michel de. 1978. *Essais*, ed. Pierre Villey, and Verdun-Louis Saulnier Paris: PUF.

Toussaint, Stéphane. 2008. *Humanismes et antihumanismes. De Ficin à Heidegger*. Paris: Les Belles Lettres.

[32] We should note that F. Lelong concludes his interesting review of Bianchi and Paganini's volume in the following terms. "That is why this work is characterized by the will to transcend, in the name of a metaphysically "sobered up" anthropology, the contemporary polemic between the enthusiastic use of the term *hominis perfectio* and Heidegger's questioning of humanism as a metaphysics of the subject." (see Lelong (2012)).

Chapter 10
The Return of Campanella: La Forge versus Cureau de la Chambre

Emanuela Scribano

Abstract In his *Traité de l'esprit de l'homme*, Louis de La Forge argues that everything that can be observed in a living body can be explained without resorting to any form of knowledge. La Forge's target, never explicitly mentioned, is Marin Cureau de La Chambre, who in his work as a whole had developed the thesis that animals act through the presence of a form of knowledge that is different from that of the intellect and that can be attributed to the body. In claiming the necessity of a form of knowledge in organic events, Cureau was answering to a problem raised by Campanella in his *De sensu rerum*. La Forge's contention that no knowledge is required to explain nature is addressed against the permanence of Renaissance vitalism in the name of the original inspiration of Cartesian new science.

The physician Louis de La Forge built his entire work around the divulgation, defence and completion of the thought of Descartes. In the course of this endeavour he was, in the name of Descartes, required to refute the notion that knowledge of the mechanisms of the living body is the necessary condition for producing them. At around the same time Arnold Geulincx formulated the principle "Quod nescis quomodo fiat id non facis", that is, an effect can be produced only by he who knows how it is done.[1] Geulincx elaborated this principle within a Cartesian context and it rapidly became an organic element in the arguments supporting occasionalism of Cartesian inspiration. What I wish to demonstrate here is, firstly that La Forge sustained the opposite thesis, using instruments drawn from Cartesian philosophy, and secondly that in doing so La Forge intended to defend Descartes' physiology against a form of vitalism which was fuelling the opposition to Cartesian science in Parisian philosophic and scientific circles.

---

[1] The principle was formulated for the first time in 1663 in the *Disputatio physica* 3, in Geulincx (1965–1968), II, 502–503. Geulincx takes up the principle again in *Ethica*, First Treatise, in Geulincx (1965–1968), III, 30–37, in *Annotationes*, ibid., III, 203–222, in *Metaphysica vera*, ibid., II, 147–157. I have sought to demonstrate the origin of the principle *Quod nescis* and its radical opposition to Cartesian philosophy in Scribano (2011).

E. Scribano (✉)
Dipartimento di Filosofia e Beni Culturali, Ca' Foscari Venezia, Venice, Italy
e-mail: emanuela.scribano@unive.it

C. Muratori, G. Paganini (eds.), *Early Modern Philosophers and the Renaissance Legacy*, International Archives of the History of Ideas Archives internationales d'histoire des idées 220, DOI 10.1007/978-3-319-32604-7_10

The Animals, the Captain and the Ship

Descartes' *L'Homme* was published as a result of the collaboration between Clerselier and La Forge. In 1662 Schuyl had published the Latin translation of Descartes' text under the title *De homine*, supplementing it with a lengthy preface, also in Latin. In 1664, Clerselier published Descartes' text in the original French with a preface and accompanied by the extensive notes by Louis de La Forge, followed by the *Description du corps humain*. Clerselier also included an appendix with the French translation of Schuyl's preface; he was, incidentally, extremely critical of Schuyl's edition.[2]

In their prefaces to the text both Clerselier and – to an even greater extent – Schuyl devoted considerable space to the theory that animals are devoid of thought whereby Descartes ruled out any possible assimilation between man and animals, thus safeguarding as best as possible the immateriality of the mind belonging to man alone. According to the two editors of *L'Homme*, Cartesian physiology offers the strongest argument against animals possessing a soul and hence thought. Indeed it demonstrates that human biological events and instinctive behaviour are accounted for independently from the action of any psychic principle. The acceptance of Cartesian physiology leads to the conclusion that the presence of a mind is not necessary to account for any behaviour on the part of animals, since the mind is not even necessary as a cause for the behaviour that men have in common with them.[3]

The independence of bodily mechanisms from any immaterial principle is also abundantly underscored by La Forge in his notes to *L'Homme*. Simultaneously with the said notes La Forge was also working on the *Traité de l'esprit de l'Homme*, in which he intended to explain not only the functioning of the human body but also the operations of the mind and its union with the body. The *Traité* was published with the date 1666, but was actually printed in November 1665, just a year after the publication of *L'Homme*. In the *Traité*, La Forge refers to the notes to *L'Homme* apropos the study of the human body. He endorses the view that Cartesian physiology offers the best arguments for explaining animal behaviour without having to resort to a mind. Nevertheless, the target of La Forge's contention is different from that against which Clerselier and Schuyl argued, extolling the advantages of the Cartesian theory of animal-machines. Rather than failing to attribute to animals an immaterial soul, he denied that the body is capable of thought.

---

[2] On the text of *L'Homme* see Meschini (2011), 165–204.

[3] Claude Clerselier, in Descartes (1664) Préface, *in fine*. I quote from La Forge (1999), 52: "Comme la grande ressemblance qui est entre les hommes et les bestes, soit dans la conformité de leurs corps, soit dans la conformité de leurs actions, est cause que l'on croit qu'elles agissent par un principe interieur en quelque façon semblable au nostre, c'est à dire, par une Ame qui sent et qui connoist, il me semble que pour combattre ce prejugé [...] un des plus puissants moyens est de faire voire que la pluspart des choses mesmes qui se font en nous, se font sans le ministere de l'ame, et ne sont point connu par elle [...] et partant qu'elles ne laisseroient pas de se faire, quand il n'y auroit en nous aucun principe connoissant [...]."

La Forge devotes three chapters of the *Traité* – III, IV and V – to demonstrating that everything that is immaterial, and only that which is immaterial, can think, and that everything that is material, and only that which is material, is extended, in opposition to the philosophers that attribute thought to matter.[4] Such philosophers include those who, like Hobbes, admit no other substance than that which is material. However, the opinion of those who accept the existence of immaterial substances while maintaining that material substances can also think and that spiritual substances too are extended, is equally erroneous. These philosophers, like the materialists, dispute the very foundation of Cartesian metaphysics, namely the opposition between thought and matter, which is why La Forge sees them as the most formidable opponents. To support their theories they tend to draw on the instinctive behaviour of animals which calls for a form of knowledge, and hence of thought. Since animals do not possess an immaterial soul, the knowledge that their instinctive actions display must belong to the body. A form of knowledge similar to the instinctive knowledge of animals would then be necessary to explain all the phenomena of life.

La Forge defines the distinctive characteristics of the theory he intends to refute. People calling into question the boundary between thought and extension argue that:

(i) Animals act as the result of the presence of a form of knowledge. This knowledge is not intellectual, and therefore can be ascribed to the body without undermining the distinction between spirit and matter.
(ii) Unless some form of knowledge is ascribed to animals one cannot account for their instinctive behaviour.
(iii) Hence, either a living body has a form of knowledge or it is governed by an external Intelligence.
(iv) If a living body were deprived of knowledge, it would be deprived of life too.[5]

The second is the argument La Forge considers the strongest. Indeed, it appears difficult not to attribute to animals feelings and knowledge like those of humans, since animals perform similar actions to humans. To undermine this argument one has to resort to Descartes, demonstrating that everything that can be observed in a living body can be explained without resorting to any form of knowledge. In this, La Forge's argument does not diverge from that sustained by Clerselier and Schuyl, but he does add an ulterior argument. If it is argued that knowledge is necessary to

[4] La Forge (1666), in La Forge (1974), 120: "L'on ne sçauroit [...] sans contradiction attribuer aucune Pensée au Corps, ni aucune Estendüe à L'Esprit."

[5] Ibid.: "nos parties adverses disent trois choses. La premiere, que la connoissance sensitive qu'ils attribuent à quelques corps est d'un genre tout à fait different de celle de l'entendement; La seconde, que sans cette connoissance les actions des brutes ne sçauroient s'expliquer, et partant qu'il faut reconnoistre, ou que les Corps sont capables de quelques pensées, ou establir une Intelligence pour la conduite de chaque animal; La troisième, que suivant nostre opinion, les Corps ne seroient pas seulement privez de la pensée, mais encore de la vie, puis qu'elle se rencontre aussi dans les substances Spirituelles."

explain the behaviour of animals, one cannot stop at the knowledge of the end the animal intends to achieve. The animal also needs to know the physical mechanisms through which the action is produced.

La Forge appropriates the famous paragon between the mind and body and the captain and the ship. This comparison, which Thomas Aquinas considered emblematic of the relationship between mind and body in Platonist theory, was rejected by Descartes in Meditation VI as being unable to explain how bodily movements are transformed into mental sensations.[6] La Forge takes up this analogy and uses it against those who consider that knowledge is necessary to cause bodily movements in a living organism. Those who think in this way commit themselves to using the model of the relationship that exists between the captain and the ship: in effect, the captain uses his knowledge to pilot the ship. But the model has to be followed through to its logical conclusion. It is not enough for the captain to know where he wants to steer the ship for the ship to sail towards its destination. The captain also has to know how the ship works in order to steer it along its course.[7] Consequently, if knowledge were a necessary condition to explain animal behaviour, not only ought animals to be aware of their intentions, but also of the way in which the animal spirits, the nerves and the muscles have to move to pursue such ends.[8] However, in this way we would have to attribute to the animal a much greater knowledge than that which man knows to be necessary to move his body. Indeed, we know that a human being moves his body without his mind knowing how to produce those movements. Hence, knowledge is not required to cause biological events or instinctive actions in either animals or humans. In short, the model of the captain and the ship, used to support the theory that a form of knowledge is required to cause bodily movements, is in itself the best refutation of the theory it is intended to illustrate.

La Forge returns to a more detailed discussion of the ship and captain model later on, in Chap. XV, which is devoted to the union between the human mind and the body. Not only does animal behaviour not require any knowledge in order to be produced, but not even voluntary human actions require it. The will is sufficient for certain effects to be produced in the body without the mind knowing them, precisely because the mind is not connected with the body in the same way a captain is with

---

[6] R. Descartes *Meditationes de prima philosophia*, in Descartes, AT, VII, 81: "Docet etiam natura, per istos sensus doloris, famis, sitis etc., me non tantum adesse meo corpori ut nauta adest navigio [...]." See Manzini (2003).

[7] La Forge (1974), 122: "Seroit-ce assez [...]pour expliquer le mouvement d'un vaisseau qui seroit porté tantost en Syrie, et tantost en Affrique, de dire que le Pilote qui est dedans a dessein d'y aller, et qu'il a connoissance de la route qu'il doit tenir, ne faudroit-il pas outre cela qu'il sçeust parfaitement bien l'usage de tous les instruments du Vaisseau, et qu'il eust l'adresse de s'en bien servir pour agir en vray Pilote et le pouvoir bien conduire; et si par malheur il ignoroit ces choses, ne seroit-on pas obligé de reconnoistre une autre cause du mouvement du Vaisseau, que le seul dessein du Pilote et la connoissance qu'il auroit des chemins si l'on voyoit qu'il suivit fort bien la route?"

[8] Ibid.

his ship.[9] It follows that, even if we were to attribute to animals knowledge on a par with our own, this would be of no use at all in explaining their behaviour.[10]

According to La Forge, the Cartesian rejection of the analogy of the captain and the ship implies that knowledge of the neuromotor processes is never necessary to produce the movements of the body, including in the latter biological events, instinctive behaviour common to both men and animals and voluntary human actions.[11]

La Forge's Adversary

Whom does La Forge have in mind? There was no shortage of people who ascribed knowledge to animals in order to explain their behaviour, but who was it who maintained that the knowledge of the end is necessary and sufficient to explain the behaviour of animals? Who claimed that such knowledge is not intellectual and hence can be ascribed to the body? Who posited the alternative of ascribing to animals either knowledge or the dominion of an external Intelligence? Who put knowledge on a par with life? Finally, who was it that maintained that the body is capable of knowledge and that spiritual substance is extended?

Taken together such tenets unequivocally pinpoint La Forge's target. It was Marin Cureau de La Chambre, the King's doctor. Cureau de La Chambre had been engaged in an important and lengthy dispute with Pierre Chanet on the subject of animal instinct. In 1643 Chanet had published *Considérations sur la Sagesse de Charron*, a refutation of Charron's theory whereby the behaviour of animals displays knowledge and reason. According to Chanet, although knowledge is necessary to explain the instinctive behaviour of animals, such knowledge must comprise an awareness of the bodily mechanisms through which instinctive behaviour is produced in both animals and humans. Not even men possess such knowledge and hence instinctive actions both human and animal have to be traced back to God, the only entity to which the scientific knowledge essential for producing bodily movements can be attributed.[12] Cureau had refuted Chanet's arguments in a brief treatise,

---

[9] Ibid., 223: "Et enfin, ce n'est pas simplement en voulant mouvoir les diverses parties de son Vaisseau qu'il (le Pilote) a la puissance de le faire avancer, et d'en changer la situation; mais c'est par une connoissance distincte qu'il a des instrumens dont il se doit servir, et par l'employ qu'il en fait; au lieu que l'Esprit de l'Homme n'a de sa nature aucune connoissance des moyens necessaires pour mouvoir son corps; et quand mesme il l'auroit, elle luy seroit inutile; la seule Volonté qu'il en a estant suffisante pour cét effet."

[10] Ibid.: "Vous pouvez voir de cecy que les Mechaniques et l'Anatonie, dont la science est tres-utile pour connoistre comment le Corps a la puissance de se mouvoir, sont tres-inutiles pour concevoir comment la pensée de l'Homme a le pouvoir de le faire; et qu'ainsi c'est une chose non seulement inutile, mais mesme ridicule, de vouloir expliquer par elle le mouvement des membres des Bestes."

[11] Sandrine Roux maintains that Descartes' own rejection of the paragon of the captain and his ship already implies the rejection of knowledge as a condition of causality in voluntary movements. See Roux (2015).

[12] Chanet (1643), 64–92.

Quelle est la connoissance des bestes et jusqu'où elle peut aller, published as an appendix to the second volume of the *Charactères des passions* which appeared in 1645. In this work Cureau argued that a minimal knowledge is sufficient to produce animal behaviour, namely the mere awareness of the goal to be achieved, so that this minimal knowledge may be attributed to animals, which are devoid of intellect and operate purely through imagination.[13] Furthermore, the imagination is a material faculty and for this reason is distinct from the intellect.[14]

Therefore, both Cureau and Chanet were convinced that knowledge was necessary to explain instinctive behaviour. The opposition between them hinged on the type of knowledge required and, consequently, on who possessed it. Cureau reduced it to an awareness of the end to be pursued, attributing it to the animal and to a material faculty; Chanet, on the other hand, claimed that a perfect knowledge of the corporeal mechanism was required and hence settled upon God as its repository.

The polemics between Cureau de la Chambre and Chanet on animal intelligence ended in 1647 with Cureau's *Traité de la connoissance des animaux*. Almost twenty years later, however, Cureau resumed the issue in the *Système de l'âme*, published on 27 May 1664. The printing of Descartes' *L'Homme* with La Forge's notes was completed on 12 April 1664, so that it would have been impossible for him to have had Cureau's text in mind as he was drafting these notes. A second edition of Cureau's *Système*, without significant changes, was published in 1665. In this work he argued that the same knowledge required to account for animals' instinctive actions was necessary for all biological phenomena.[15] In this way Cureau expanded and rendered explicit a theory already mentioned in the *Traité de la connaissance*

[13] Cureau de La Chambre (1645b) appended to Cureau de La Chambre (1645a). Chanet replied with Chanet (1646). Cureau responded in turn with Cureau de La Chambre (1647).

[14] Cureau de La Chambre (1989b), 323: "[…] l'imagination est au rang des choses materielles"; Cureau de La Chambre (1989a), 214: "[…] l'Entendement est une puissance separée de la Matiere […] et […] elle est differente de l'*Imagination* qui est dans l'ordre des choses materielles."

[15] Cureau de La Chambre (1664), 173–176: "les actions de l'Ame Vegetative tombent sous la question de l'Instinct […] on donne ce nom-là aux actions qui se font par une obscure et secrette Connoissance […]. Or puisque cette cause est commune à toutes [les choses], il est certain que si nous la pouvons connoistre en quelqu'une, ce sera *la mesme* qui fera agir toutes les autres; et nommément la Vegetative qui est celle qui nous occupe maintenant. Cherchons-la donc dans les Animaux, c'est-à-dire, dans l'Ame Sensitive où il semble qu'elle est plus manifeste, et où l'on en a fait de plus exactes et de plus frequentes observations." Italics mine. See also, 160: "[…] il faut presupposer qu'il y a une Connoissance dans l'Homme, où les Sens ni la Raison n'ont point de part, et qui se remarque principalement dans la faculté Vegetative. Car il est impossible de considerer tant de diverses actions qu'elle fait, et l'ordre et les mesures qu'elle y garde, qu'on ne soit contraint d'avouër qu'il y a quelque connoissance qui regle et qui conduit une si belle oeconomie. Quand il n'y auroit que ce qui se passe dans la premiere conformation du corps, où le nombre, la figure et la situation des parties sont si justes et si regulieres; cela ne seroit-il pas capable de persuader que la cause qui en a la direction, est bien sçavante, et qu'elle fait les choses avec plus de connoissance, que la raison mesme ne pourroit faire, quand elle s'en voudroit mesler toute seule?" For the knowledge that regulates the life of the plants, see 216. Initially inanimate things are compared to vegetables and animals (ibid., 174–175), but later Cureau concludes that the inanimate bodies have only the "ombre de la Connoissace" rather than true knowledge since, unlike living things, they are passive, 217. See also, 222 ff.

des animaux: namely the continuity of nature from minerals to animals, the entire natural world being traversed by different degrees of knowledge.[16]

In his work as a whole Cureau had developed the arguments which La Forge now set himself to refute in the *Traité de l'esprit de l'homme*: animals act through the presence of a form of knowledge that is different from that of the intellect, which can be attributed to the body; animals possess only the knowledge of the end they intend to pursue; without this form of knowledge one could not account for the instinctive behaviour of animals: either a living body has a form of knowledge or it is governed by an external Intelligence[17]; if a living body were deprived of knowledge, it would be deprived of life too.[18] In the *Système*, Cureau himself had argued at length in favour of the extension of the soul, thus in La Forge's eyes putting the finishing touch to his attempt on the distinction underpinning the entire edifice of Cartesian metaphysics.[19] Finally, if the target of the controversy is Cureau de La Chambre La Forge's contention is easier to understand, interested as he was in the absence of *knowledge* in animals rather than the absence of *sensitivity*, which was instead the most shattering aspect of Descartes' theory of animals.

La Forge refutes Cureau's arguments one by one. There is no form of knowledge different from that of the intellect; hence if living bodies had knowledge they would also have to have a non-material mind.[20] If knowledge were an essential prerequisite to explain the movement of the living body it would require a knowledge much greater than simply that of the end to be achieved. Finally, the life of spiritual substances is governed by principles that cannot be reduced to the purely corporeal principles that produce life in animals and plants.[21]

---

[16] See Cureau de La Chambre (1647), chap. 2, *Que la perfection des choses est commencée dans celles qui leur sont inferieures*, 45 ff.

[17] Cureau de La Chambre (1664), 163–164: "Tout le monde voit et admire la sage conduite de cette Faculté (vegetative), et il n'y a personne qui n'advouë qu'elle agit avec un ordre et une justesse merveilleuse qui marque une grande Connoissance. Ce n'est donc pas en cela que consiste la difficulté, c'est de savoir si cette Connoissance est un effect de la Faculté Vegetative, ou si elle part d'une plus noble cause. De la rapporter à la Vegetative il n'y a point d'apparence, puisque personne ne l'a mise au rang des Facultez connoissantes. [...] De façon qu'il faudroit en ce cas recourir à une Cause exterieure et intelligente qui poussast toutes ces choses à faire leurs actions, et qui y mist la regle et la justesse que l'ont y remarque."

[18] Ibid., 217: "tout ce qui est vivant connoist, et [...] tout ce qui connoist est vivant." The argument that La Forge is referring to, whereby life belongs in the first place to a spiritual soul, may perhaps be derived from this passage, 167: "En effet ce sont actions vitales qui font partie de la vie, et toutes les actions de vie doivent estre produites par un principe de vie: or il n'y a point d'autre principe de vie que l'Ame mesme, et par consequent c'est elle seule qui les fait."

[19] Indeed Book V of the *Système de l'âme* is entitled *De l'Extension, Des Parties, De la Figure et de la Grandeur de l'Ame*.

[20] La Forge (1974), 121. La Forge assumes an agnostic attitude as regards the soul of animals. The question that can be answered is not whether or not animals possess a soul, but whether a certain animal behaviour requires a spiritual and knowing soul in order to be produced, which La Forge emphatically denies, ibid. Here La Forge aligns himself with the agnostic position adopted by Descartes in the letter to Henry More of 5 February 1649, in AT V, 276–77.

[21] La Forge (1974), 124.

The Spectre of Campanella

Cureau's work appears to La Forge a stepping stone towards an open materialism. After having refuted those who admit the existence of spiritual and material substance, but deny any distinction attributing thought and extension to both, La Forge devotes a chapter to countering those who admit the existence only of corporeal substances and consider that thought is a property of matter.[22] Here, in passing, La Forge observes that those who maintain that thought is a property or a consequence of the body come close to Campanella's way of thinking and hence to a philosophy that is rejected by all people "de bon sens.[23]

It is probable that, in evoking the spectre of Campanella, La Forge still had Cureau de La Chambre in mind. Cureau had had personal relations with Campanella,[24] and many pages of Cureau's *Système de l'âme* were in effect impregnated with the influence of the *De sensu rerum*. Challenging Aristotle, Campanella had argued that the sensitive soul is material, and Cureau was of the same opinion.[25] Like Campanella, Cureau considered all living phenomena to require some form of knowledge. Like Campanella, Cureau thought this knowledge to be internal to the living body itself. Positing the alternatives of a knowledge internal to the living body or external to it, Campanella had opted for the former since the latter would have impaired divine perfection. He argued that if the works of God are perfect, God must have provided them with the means necessary for their survival, first and foremost with the knowledge of what is beneficial or harmful to self-preservation.[26] Cureau proposed exactly the same alternative as Campanella, applying it to living beings: since living phenomena require knowledge, either the living being itself has

---

[22] Ibid., Chap. 6: "Autre preuve contre ceux qui ne reçoivent que des substances corporelles."

[23] Ibid., 127: "De dire que la Pensée constitue l'essence du Corps, ou qu'elle en soit une suite, on ne le peut sans attribuer la connaissance à tous les corps, ainsi que faisoit Campanella: mais comme personne de bon sens ne suit cette opinion, je ne m'amuse pas aussi à la refuter."

[24] Cureau had been in contact with Campanella on the question of the flooding of the Nile. See *Judicium C.V. Thomae Campanellae De Causa Inundationis Nili allata* in Cureau de La Chambre (1665), 199–212, concerning Cureau de La Chambre (1634). See Firpo (1947), 126–133, and Darmon (1985), 27–29. The direct relations between the two are documented by two passages in letters from Campanella to Pierre Séguier, dated 13 September 1636 and 16 February 1637, that is at the time that Campanella was publishing the second and third editions of *De sensu rerum*, dedicated to Cardinal Richelieu. See Campanella, *Lettere* (2010), 467 and 645–646. The first edition of *De sensu rerum* dated to 1620. See the introduction by G. Ernst to Campanella (2007). I should like to thank Germana Ernst for having informed me about the relevant passages in Campanella's letters.

[25] Campanella (1637), book II, chaps. 7 and 8. On Campanella's thesis about sensibility and conscience, in relation with Descartes' thought, see Paganini (2008), 126–169.

[26] Ibid., 11: "At plurimi Deo tribuunt huiusmodi actus, qui intrinsece in rebus operatur [...] Ego vero respondeo, praedictas opiniones omnes aut perperam declarari, aut errorem continere. Si enim omnia opera Dei perfecta sunt [...] fateri oportet, eas rebus vires ab eo largitas esse, quae ipsarum conservationi sufficiant. Quoniam vero nulla facultas tam necessaria est in tanta rerum varietate, quam cognocendi similia, quibus servamur, et contraria, quibus destruimur, necesse est hanc sentiendi vim innatam esse rebus cunctis."

this knowledge or God has. Like Campanella, Cureau recalls that the Platonists had opted for the second alternative – both quote Avicenna – whereas others had rejected it.[27] The "others" mentioned by Cureau, and with whom he agrees, consider that divine perfection entails attributing creatures enough knowledge to independently guide their own actions. Such "others" undoubtedly include Campanella, and Cureau explicitly sides with him by using the very same argument. The dates allow us to confirm that the fear of a revival of Campanella's animism, which La Forge expressed in 1665, were justified by Cureau's most recent publication.

Cureau's polemic with Chanet, which terminated in 1647, had merely prefigured an extension of knowledge to biological phenomena as a whole, through the allusion to a natural continuity extending from minerals through to man. It was only in the *Système de l'âme* that Cureau revealed his complete adherence to Campanella's animism. Nor was this the only novelty of the *Système*, since in this work Cureau also decided to explicitly attack Cartesian physiology by challenging the theory of the pineal gland.

As a result of his position as the king's physician and a founder member of the Académie des Sciences, Cureau de La Chambre found himself in the thick of the medical debates engaging the most prominent scientists. Among these we should recall a figure whose intellectual career was closely bound up with that of Cureau: the physicist and writer Pierre Petit. In 1660 Petit had published in Paris a treatise entitled *De motu animalium spontaneo liber unus*[28] in which he had utilised some of Cureau's ideas to explain bodily movements. In opposition to Chanet, Cureau argued that the imagination of the animal was equipped with innate images that provided it with the practical knowledge required for instinctive behaviour.[29] Utilising Cureau's account of instinct, Petit set out to explain all biological events and voluntary movements through a form of knowledge transmitted to the parts of the body via images. According to Petit, such images instantly convey knowledge of the agent's intentions from the brain to the limbs; the presence of the images sent from the brain means that the parts of the body involved know what they have to do and therefore can succeed in doing it.[30] An eloquent example of this is the marks

[27] Cureau de La Chambre (1664), 164–5: "Neantmoins il y en a d'autres qui ne peuvent approuver cette opinion. Car quoy qu'elle soit appuyée sur la Bonté et sur la Providence de Dieu … ils croyent qu'elles est injurieuse à sa Toute-puissance et à sa Sagesse, qui a deû donner à ses ouvrages toute la perfection qui leur estoit convenable. De sorte que chaque chose estant parfaite quand elle a la vertu de faire les actions qui luy sont propres; il estoit de la gloire du Createur de luy donner cette vertu, et de ne la rendre pas inutile en faisant de luy-mesme l'action qu'elle doit produire."

[28] Petit (1660).

[29] Cureau de La Chambre (1989b), 344–346.

[30] Petit (1660), 153: "Eadem enim imaginatio, quae in cerebro imperat, in membris exequitur, postquam imperij species per spiritus animales propagata, ad ipsam pervenit. Exempli causa, libet nunc exarare has literas, eodem ipso momento scriptionis species ad eos manus nervos pervenit, qui ad eam actionem comparati sunt, simul quae iis nervis inest imaginatio per speciem acceptam, quid velim, cognoscit: cognitoque consilio spiritus musculis digitorum contentosciet iis motibus, qui ad exarandas has vel illas literas pertinent."

impressed on the foetus, a phenomenon for which, according to Petit, no one has ever provided a satisfactory explanation.

Cureau, in turn, borrowed Petit's account of how the mind can move the body in his *Système de l'âme*. When the imagination wants to move an arm, it forms an image of the movement it wants to produce. This picture spreads like a flash through the parts of the body and joins the natural images that are impressed on the muscles necessary for such movements, which resemble the picture formed by the imagination. The two images come together to move the muscle assigned to that particular movement.[31]

A logical corollary of this explanation of body movements is the rejection by Petit and Cureau of the central role of the brain in perception. In the *De motu* Petit criticised the Cartesian doctrine of the pineal gland, after which Cureau in the *Système* challenged the theory of the pineal gland as extravagant and contrary to experience.[32] Each part of the body has sensitivity and hence a form of knowledge.[33] In order to demonstrate that sensation is independent of the brain, Cureau was one of the first to draw attention to the phenomenon of irritability.[34] The rejection of the centrality of the brain is another aspect that links Cureau to Campanella, who had in his turn asserted that "bones, hair, nerves, blood and spirit, all feel, refuting Aristotle."[35]

Descartes Versus Campanella

With his *Système de l'âme* Cureau joined the fray a month after the publication of *L'Homme* with La Forge's notes. The editors of *L'Homme* had set themselves a challenging commitment. Descartes' unpublished work was called upon to refute the criticisms that since the philosopher's death had begun to cluster around the physiology to be derived from the printed works, the *Discours de la methode*, the *Dioptrique* and the *Passions de l'âme.* This was the gauntlet thrown down by Clerselier and La Forge: to provide Descartes' work with weapons to defend itself

---

[31] Cureau de La Chambre (1664), 489–491.

[32] Ibid., 468: "Je ne veux pas perdre le temps à refuter une opinion qui s'est introduite depuis peu sur ce sujet, parce qu'elle est contraire à l'experience, et n'est pas mesme concevable. Car elle veut que la Glande qui est au milieu du Cerveau se meuve incessament [...]."

[33] Ibid., 208.

[34] Ibid., 209–210: "Il ne faut que remarquer l'*irritation* que la malignité des humeurs donne à la Nature en toutes les parties; les efforts et les mouvemens qu'elle leur fait faire pour chasser ce qui les incommode, comme sont les palpitations, les changemens de pouls, les vomissemens, les diarrhées et mille autres semblables qui se font à l'insceu du Cerveau et de la Faculté Sensitive. Car tout cela montre que la Nature est irritée: et il n'y a rien de si commun en la bouche des Medecins, que cette façon de parler; mais elle ne put estre irritée qu'elle ne sente, et qu'elle ne connoisse ce qui l'offense." Italics mine.

[35] Campanella (1637), book II, chap. XIII, 58: "Ossa, pilos, nervos, sanguinem et spiritum, omnes sentire contra Aristotelem."

against an adversary that might have grown and thrived in the absence of a more thorough physiological text such as *L'Homme*. In his criticism in the *Système de l'âme* of a physiology that placed the brain at the centre of biological and motor phenomena, Cureau went to swell the ranks of scientists such as Petit and Steno's teacher Thomas Bartholin, who had challenged significant aspects of Cartesian physiology.[36] And as if this were not enough, the year following the publication of the *Système de l'âme*, another and even more formidable adversary joined the enemy ranks in the person of the rising star of physiology, Nicolas Steno.

In 1665 Steno gave a famous lecture in Paris at Thévenot's house, in which he anticipated the results of his anatomical research on the brain; these were then published in 1669 in the *Anatomie du cerveau*, in which Steno criticised Descartes' theory of the pineal gland.[37] Both Cureau and Petit were probably in the audience at Thévenot's house, together with a group of Cartesians of strict observance.[38] In 1669 the publisher of the *Anatomie du cerveau* actually dedicated it to Cureau de La Chambre, whereas Steno's teacher Bartholin had associated Petit and Steno as supporters of the notion of sensitivity spread throughout the body.[39] La Forge had explicitly challenged Bartholin in his notes to *L'Homme*.[40] Nevertheless, far from extinguishing the anti-Cartesian fire, the edition of *L'Homme* with La Forge's notes appears to have poured oil on the flames. This led to the decision to tackle one of the philosophical cornerstones of the enemy camp: the knowledge attributed to the living body in order to explain its movements. This is exactly what La Forge proposed to do in the *Traité de l'esprit de l'homme*.

---

[36] On Thomas Bartholin see Porter (1963), 99–125. A thorough review of the editions of Bartholin's *Anatomia reformata* between 1641 and 1674 is to be found in Meschini (1998), 75–80. See also Trevisani (1992), 223.

[37] Franco A. Meschini has convincingly argued that the target of the *Discours sur l'anatomie du cerveau* was not only Descartes but also – or more importantly – La Forge's notes to *L'Homme*. See Meschini (1998), 85–98.

[38] A list of those who probably attended Steno's famous lecture is to be found in Meschini (1998), 22–23. It is probable that writers close to Descartes such as Géraud de Cordemoy, Jaques Rohault and Claude Clerselier were also present.

[39] Bartholinus (1673), 477: "*P. Petitus* non cerebrum tantum imaginationis esse sedem, sed eandem in omnes corporis nervos continuari liberaliter concedit. Non multum dissimilis est *Stenonius* cui animales operationes omnes non soli cerebro, sed spinali quoque medullae tanquam primae scaturigini, adscribuntur."

[40] La Forge (1999), 308. In the *Traité*, La Forge was to insist on the central role of the brain in sensing, seeking to explain the error of those who denied it. See La Forge (1974), 221. The pineal gland's movements causing sensations in the mind do not allow us to grasp their true causes, i.e. movements of the brain, "mais elles nous representent l'action de l'objet, ou comme dans l'objet mesme, et hors du corps, ou du moins dans l'extremité de quelqu'un de nos Membres." Perceiving sensations in their remote origin (which would be ineffective if they did not reach the brain) is at the origin of the error that ascribes a sensitive faculty to the parts of our body: "nous avons attribué la faculté de sentir aux parties de nostre Corps, ou du moins nous avons cru que l'Ame l'exerçoit dans les organes exterieurs, dautant que les pensées des Sens nous representent l'action des objets, comme dans nos Membres exterieurs, et non pas comme dans le cerveau."

The Captain and the Musician

An understanding of the context of the battle for hegemony in the field of physiological studies allows us to more fully grasp the logic behind the arguments La Forge uses against those who maintain that biological and instinctive phenomena can be explained purely through knowledge of bodily mechanisms. As we have seen, La Forge reasons that if knowledge were necessary to explain animal movement, then it would have to be the perfect knowledge that the captain needs in order to be able to steer his ship. In this way, not only did La Forge use Descartes against those who attributed knowledge to animals, but also challenged the devaluation of knowledge necessary to explain such behaviour. This devaluation was implemented by the writers such as Petit and Cureau who exerted themselves to attribute a form of knowledge to bodies, and was essential to their strategy.

Both Cureau and Petit attempted to reply to a problem raised by Campanella in *De sensu rerum*. According to Campanella, everything requires knowledge in order to perform the functions aimed at self-preservation. But how can the human mind move the body when it is ignorant even of its anatomy? "I am surprised that man is so ingenious and that his mind can guide his body, even without knowing how it does so."[41] It was a quandary that Campanella was unable to fully resolve, and Cureau shared this difficulty.[42] The difference was that, like Petit, Cureau felt he had an answer to the problem that had tormented Campanella: how can the mind move the body if it does not know the mechanisms whereby the movement is produced? Both Petit and Cureau strove to show that the knowledge required to move the body was not the perfect knowledge of the body's mechanisms. According to Petit, "it is not necessary for the imagination to understand all the relations of the movements and which muscles are required by each movement." Who can possibly know how all the different muscles function? It is enough to know the purpose, what one wishes to do, and immediately the spirits linked with the imagination move in the manner proper to each function.[43] Already in the controversy with Chanet Cureau had argued that animals produce their instinctive actions purely through the

[41] Campanella (1637), 95: "Admiror equidem hominem tanto praeditum ingenio, animamque eius regentem corpus, nec tamen ipsam scire, qua ratione regat. Fiunt intra nos tot concoctiones, separationes, aggregationes, nutrificationes, assimilationes, nec tamen intelligere possumus, quomodo fiunt; et quidem nos ipsi, qui animae sumus, hos actus operamur, nec tamen nostra opera, nec operationes nostras scimus."

[42] Cureau de La Chambre (1664), 488: "Car c'est une chose merveilleuse que l'Ame ne sçait point qu'elle ait des muscles, ni combien elle en a, ni quel usage ils peuvent avoir; et neantmoins quand elle veut remüer un membre, de plusieurs dont il est composé et qui font de mouvemens contraires, elle choisit si justement ceux qui sont propres à l'action qu'elle veut faire, qu'elle ne prend jamais l'un pour l'autre."

[43] Petit (1660), 153–155: "Porro id praestare non est cujuslibet notionis, sed ejus tantum, quae practica est, hoc est, induta circumstantiis boni, vel mali, item loci, temporis, aliisque ejusmodi, quibus ad agendum determinamur. […] Atque haec sufficit cognitio ad moderandos partium motus: neque enim necesse est comprehensas haberi imaginatione omnes movendi rationes, et qui ad quosque motus musculi faciant. Quotusquisque enim novit musculorum usus et differentias, aut

knowledge of the end they wish to achieve.[44] In the *Système de l'âme* the theory that a scientific knowledge of physical mechanisms is not necessary to produce bodily movement is taken up again and expanded. Through the natural or acquired images that are stored in the memory, the imagination of the animal knows which movements it intends to produce in the limbs without being aware of how the muscles have to act to produce such movements. The image of the movement that the animal wishes to achieve spreads through the body, joining up with the images similar to that movement that are inscribed only in the muscles capable of generating it. This explains how the animal is able to produce a specific movement despite being unaware of which muscles are required for it. The same thing happens with a harpsichord player. He is aware of which sound he wants to produce and which keys he has to play although he is unaware of the mechanism whereby the struck key produces precisely that sound.[45] It is not the analogy of the captain and his ship but that of the musician and the harpsichord that best illustrates the relation between the mind and the movement of the body.

La Forge is well aware that this strategy can provide a powerful argument in support of the theory that knowledge is a necessary condition for producing animal movement. If in order to produce physical movements it is necessary and sufficient to know the proposed intention and which parts of the body are to be moved, without knowing the physiological mechanisms behind such movement, this strengthens the notion of an animal knowledge that can be assimilated to that of man. This is why La Forge insists that, if knowledge is indeed required to move the body, then this must be the perfect knowledge that allows the captain to steer his ship, and not that of someone who obeys orders without knowing why or who plays a musical instrument without understanding its mechanics.[46] In short, the knowledge that is claimed to be necessary to move the body must be of the kind that Chanet and not Cureau referred to. Chanet, however, drew from it an argument for attributing such knowledge to God. La Forge, on the other hand, concludes that *no* knowledge is required to move a body, as demonstrated by Descartes. Not only is the mind not a captain, but bodily movement is not traced back to *any* captain at all, not even the supreme captain, God.

qua ratione membra moveantur? Sufficit, id quod agendum est, finem inquam non ignorari: mox enim conjuncti imaginationi spiritus moventur, ut unicuique functioni consentaneum est."

[44] Cureau de La Chambre (1989b), 344–346.

[45] Cureau de La Chambre (1664), 493: "[…] il en est comme d'un Homme qui jouë du Clavessin: il connoist bien les accords qu'il veut faire, et sçait les touches qu'il doit abattre; mais il ne void et ne connoist point les sautereaux qui remüent les chordes; quoy que les touches qu'il a abatuës, fassent mouvoir les sautereaux. L'Imagination sçait aussi les mouvemens qu'il faut donner aux membres; les Images qu'elle forme sont les touches qui esbranlent les Images naturelles qui sont dans les Muscles; et les Muscles sont comme les sautereaux qui font le mouvement des membres."

[46] La Forge (1974), 122, see note 7. Malebranche too took up a stance opposed to Cureau's thesis, claiming that there is no knowledge inferior to the scientific and hence no instinctive knowledge. See Malebranche (1972), 24.

Conclusion

The challenge of Campanella's animism, regenerated through the pages of Cureau, drove La Forge to take up a stance that placed him way out on the side lines from what was to become mainstream occasionalism, marshalling under the banner of "Quod nescis quomodo fiat id non facis" as an authentically Cartesian principle. This is a point that needs to be underscored. Although La Forge was an early subscriber to a form of occasionalism,[47] he did not use the "Quod nescis" principle to call the mind-body interaction into question. Indeed, La Forge's rejection of the need for knowledge to produce bodily movement was made in the name of Descartes himself, of his physiology and his refusal to compare the mind to a captain.

Nevertheless, from the very start the position adopted by La Forge the better to attack Cureau's vitalism was parallelled by the alternative that was to furnish grist to the mill of occasionalism. In his preface to the Latin edition of Descartes' text, Florentin Schuyl had rejected the theory that animals had to be attributed a mind and hence knowledge, since the knowledge necessary to produce animal movement was the prerogative of the divine mind. Schuyl quoted the motto "Opus naturae est opus intelligentiae", and referred to God and not animals the knowledge necessary to account for their behaviour: "the knowledge that allows animals to act does not belong to the animals, but to the author of nature whose wisdom is celebrated by all creatures."[48] Schuyl felt that Descartes' comparison of the body to a machine, while ruling out that the machine possessed an intelligence, also implied that an intelligence was nevertheless indispensable to explain its functioning. Schuyl too was entrapped by the alternative posed by Campanella and taken up again by Cureau: a form of knowledge is indispensable to explain the regularity of nature, and this knowledge is either in nature as Campanella and also Cureau were convinced, or in God, as – according to Schuyl – Descartes believed.

La Forge's conviction that *no knowledge* was required to explain nature continued to be marginal and disparaged, and with it the path chosen to stifle the revival of Campanella's vitalism in the name of loyalty to Descartes.

Translation from the Italian Aelmuire Helen Cleary.

Acknowledgment I would like to thank Matteo Favaretti Camposampiero for his careful reading and insightful remarks.

---

[47] On La Forge's occasionalism see Bardout (2002) and Nadler (2011), 104–114.

[48] Descartes, (1662), 8r: "Tantam autem affinitatem Nobis cum Bestiis non intercedere, neque etiam illas tantae dignitatis esse constabit, uti existimo, si probatum fuerit, Cognitionem, qua aguntur bestiae, non illarum esse, sed ipsius Authoris Naturae, cujus sapientiam omnes creaturae celebrant: juxta decantatum illud: *Opus naturae est opus intelligentiae*. Secundum hanc providentiam gravia deorsum, levia sursum feruntur: totumque hujus Mundi horologium tam ordinate circumagitur. Haec Tulipa, licet omni propria cognitione destituta, folia sua matutino Soli explicat, quae, ne à nocturno frigore semini fiat injuria, vesperi colligit, atque constringit."

Bibliography

Bardout, Jean-Christophe. 2002. Occasionalism: La Forge, Cordemoy, Geulincx. In *A Companion to Early Modern Philosophy*, ed. Steven Nadler. Boston: Blackwell.

Bartholinus, Thomas. 1673. *Anatome ex omnium veterum recentiorumque observationibus in primis Institutionibus b.m. parentis Caspari Barholini, ad circulationem Harvejanam, et vasa lymphatica quartum renovata*. Lugduni Batavorum: ex officina Hackiana.

Campanella, Tommaso. 1637. *De sensu rerum et magia*. Parisiis: apud Dyonisium Bechet.

Campanella, Tommaso. 2007. *Del senso delle cose e della magia*. Rome/Bari: Laterza.

Campanella, Tommaso. 2010. *Lettere*, ed. Germana Ernst. Florence: Olschki.

Chanet, Pierre. 1643. *Considérations sur la Sagesse de Charron en deux parties*. Paris: C. Le Groult.

Chanet, Pierre. 1646. *De l'instinct et de la connoissance des animaux, avec l'examen de ce que M. de la Chambre a escrit sur cette matière*. La Rochelle: T. de Gouy.

Cureau de La Chambre, Marin. 1634. *Nouvelles pensées sur les causes de la lumière, du débordement du Nil et de l'amour d'inclination*. Paris: P. Rocolet.

Cureau de La Chambre, Marin. 1645a. *Les charactères des passions, vol. II, où il est traité de la nature et des effets des passions courageuses*. Paris: P. Rocolet.

Cureau de La Chambre, Marin. 1645b. *Quelle est la connoissance des bestes et jusqu'où elle peut aller* [appended to Cureau de La Chambre (1645a)]. Modern edition in Cureau de La Chambre (1989b).

Cureau de La Chambre, Marin. 1647. *Traité de la connoissance des animaux où tout ce qui a esté dict pour ou contre le raisonnement des bestes est examiné*. Paris: P. Rocolet.

Cureau de La Chambre, Marin. 1664. *Le Système de l'âme*. Paris: Jacques d'Allin.

Cureau de La Chambre, Marin. 1665. *Discours sur les causes du débordement du Nil*. Paris: J. D'Allin.

Cureau de La Chambre, Marin. 1989a. *Traité de la connoissance des animaux où tout ce qui a esté dict Pour, et Contre le raisonnement des bestes est examiné*. Paris: Fayard.

Cureau de La Chambre, Marin. 1989b. *Quelle est la connoissance des bestes et jusqu'où elle peut aller*. In Cureau de La Chambre (1989a).

Darmon, Albert. 1985. *Les corps immatériels. Esprits et images dans l'oeuvre de Marin Cureau de La Chambre (1594–1669)*. Paris: Vrin.

Descartes, René. 1662. *De Homine, figuris et latinitate donatus a Florentio Schuyl*. Lugduni Batavorum: apud Petrum Leffen & Franciscum Moyardum.

Descartes, René. 1664. *L'Homme de René Descartes et un Traitté de la formation du foetus du mesme autheur, avec les remarques de Louys de La Forge* [...]. Publié par Claude Clerselier, et suivi de la version de la préface que M. Schuyl a mise. Paris: Jaques Le Gras.

Descartes, René. *Oeuvres*, ed. Charles Adam et Paul Tannery. Paris: Vrin (abbreviated as AT).

Firpo, Luigi. 1947. *Ricerche campanelliane*. Firenze: Sansoni.

Geulincx, Arnold. 1965–1968. *Opera Philosophica*, recognovit J.P.N. Land, 3 vols. The Hague: Martinus Nijhoff, 1891–1893, repr. with some additions in *Sämtliche Schriften*, ed. Herman Jan De Vleeschauwer. Stuttgart-Bad Canstatt: Frommann.

La Forge, Louis de. 1974. *Traité de l'esprit de l'homme, de ses facultez et fonctions, et de son union avec le corps. Suivant les Principes de René Descartes*. In L. de La Forge, *Oeuvres philosophiques*, ed. P. Clair. Paris: PUF.

La Forge, Louis de. 1999. *L'Homme de René Descartes*. Paris: Fayard.

Malebranche, Nicolas. 1972. *Conversations chrétiennes*. In N. Malebranche, *Oeuvres complètes*, IV, ed. A. Robinet. Paris: Vrin.

Manzini, Fréderic. 2003. Comme un pilote en son navire. *Bulletin Cartésien* XXXI, *Archives de philosophie* 66:163–169.

Meschini, Franco Aurelio. 1998. *Neurofisiologia cartesiana*. Florence: Olschki.

Meschini, Franco Aurelio. 2011. Note per un'edizione critica de *L'Homme* di Descartes. In *Le opere dei filosofi e degli scienziati. Filosofia e scienza tra testo, libro e biblioteche*, ed. Franco Aurelio Meschini, 165–204. Firenze: Olschki.
Nadler, Steven. 2011. *Occasionalism*. Oxford/New York: Oxford University Press.
Paganini, Gianni. 2008. *Skepsis. Le débat des Modernes sur le scepticisme. Montaigne, Le Vayer, Campanella, Hobbes, Descartes, Bayle*. Paris: Vrin.
Petit, Pierre. 1660. *De motu animalium spontaneo liber unus. In quo partim Aristotelis de hujus motus principio sententia illustratur, partim nova musculorum motus ratio indagatur*. Parisiis: E Typographia Claudii Cramoisy.
Porter, Ian Herbert. 1963. Thomas Bartholin (1616–1680) and Niels Stensen (1638–1686). Master and Pupil. *Medical History* 7: 99–125.
Roux, Sandrine. 2015. L'union cartésienne à la lumière du problème du "défaut de connaissance". *Journal of the History of Philosophy* 53(2): 207–219.
Scribano, Emanuela. 2011. Quod nescis quomodo fiat id non facis. Occasionalism Against Descartes? *Rinascimento* 51: 63–86.
Trevisani, Francesco. 1992. *Descartes in Germania*. Milano: Franco Angeli.

Chapter 11
From Animal Happiness to Human Unhappiness: Cardano, Vanini, *Theophrastus Redivivus* (1659)

Cecilia Muratori

Abstract The topic of the distinction between man and animal as discussed in *Theophrastus redivivus* (1659) is a noteworthy example of the engagement with Renaissance sources in the seventeenth century. This essay argues that it displays how conceptual continuities intertwined with significant interpretative shifts. In dealing with the specific question of human and animal happiness, the anonymous author carefully selects and brings together passages from Renaissance philosophers – especially Cardano and Vanini – but inserts them in a completely new frame, ultimately employing the Renaissance roots of this philosophical problem in order to develop his own original view. Thus Cardano's reflections on the animals' capability of attaining happiness and Vanini's doubts about man's qualitative distinction from animals are woven together in order to point to a conclusion which is in fact a subversion of the sources used to reach it: *Theophrastus redivivus* shows that man has lost his assigned place in the economy of the universe altogether. The legacy of the Renaissance debate on the animals' happiness therefore changed form even while it persisted as a crucial point of reference from the fifteenth to the seventeenth century.

"To live happily is to live according to nature": with this firm statement, the anonymous author of *Theophrastus redivivus* opens the discussion in the sixth treatise, which concludes the long manuscript written in France in the mid seventeenth century.[1] Right from the beginning, the definition of the "vita secundum naturam"[2] is developed through detailed and far-reaching comparisons between man and the animals, pivoting

All translations are my own unless otherwise stated.

[1] TR, 783. On *Theophrastus*' treatment of the relation between the "life according to nature" and the pursuit of happiness, see Canziani (1981).

[2] On "vita secundum naturam", with special regard to the topic of the foundation of society, see Bianchi (1988), 115.

C. Muratori (✉)
School of Modern Languages and Cultures, University of Warwick, Coventry, UK
e-mail: c.muratori@warwick.ac.uk

C. Muratori, G. Paganini (eds.), *Early Modern Philosophers and the Renaissance Legacy*, International Archives of the History of Ideas Archives internationales d'histoire des idées 220, DOI 10.1007/978-3-319-32604-7_11

around these key questions: what does it mean, for man and for all other living beings, to live happily and according to nature? How does man attain this happiness, and are there relevant differences between man and the animals in this respect? In the background hovers of course a crucial question: can animals be happy at all or is happiness an exclusively human aim, as Aristotle had already established?

As is well known, the backbone of the *Theophastus redivivus* is constituted by its use of sources – some openly acknowledged, some tacitly integrated in the text, many strongly adapted and abridged. Two of the main authors to whom *Theophrastus* recurs in the discussion of the meaning of a (happy) life according to nature, are Girolamo Cardano and Giulio Cesare Vanini. Indeed, both Cardano and Vanini are placed on the right hand side of the author himself on the title page of the manuscript in two Wien codices.[3]

In this essay I aim to show how *Theophrastus redivivus* employs quotations and main ideas drawn from Cardano and Vanini with specific reference to the topic of the happy life that the creatures, human and animal, can lead according to nature.[4] I argue that reconstructing the unfolding of this topic with attention to the adaptation of Cardano and Vanini's own views on this same matter, can provide a fitting case study for the reception history of Renaissance philosophy into Early Modern philosophy.[5] In particular, although the significance of the debate about the happiness of life changes, the debate itself persists from the mid-sixteenth to the mid-seventeenth century. But the main reason for focussing on these three authors in the context of a study of the 'Renaissance legacy in Early Modern Philosophy' lies in their connections with regard to the ways in which texts and ideas were transmitted and readapted. Their methods of dealing with their sources effectively show that subtle shifts and complex changes of scenery might better describe this phase than assuming a gap between Renaissance and Early Modern period.

It is not by chance that all three authors had and even recently have been accused of plagiarism because of their ways of dealing with sources and traditions. Scaliger had already reproached Cardano for his more or less unacknowledged use of other philosophers, especially Pomponazzi.[6] In the case of Vanini the accusation of plagiarism and lack of originality continued almost until the present day: Luigi Corvaglia brought to light the intricate pattern of undisclosed quotations (many from Cardano) in Vanini's two surviving works, *De admirandis* and *Amphitheatrum aeternae providentiae*, and on this basis dismissed the author as a mere plagiarist.[7]

---

[3] TR, vol. 1 (unnumbered pages: reproductions inserted before the *Proœmium*).

[4] This essay is a companion piece to Muratori (forthcoming). While in the present work I analyse the topic of the man-animal distinction with special attention to the question regarding the pursuit of happiness, in the latter I investigate the practical, ethical consequences of *Theophrastus*' critique of anthropocentrism, bringing attention to the often neglected aspect of food choice in the anonymous text.

[5] On the careful and selective use of Vanini in *Theophrastus redivivus* see Paganini (1998). On *Theophrastus* as reader of Cardano see Canziani (1985).

[6] Maclean (2003), 196. On Cardano and Pomponazzi see Ingegno (1980), 1–78, and Paganini (1985).

[7] Corvaglia (1933–1934) and Corvaglia (1991); contrast the role assigned to Vanini in the libertine tradition by Spini (1983), 125–143. For a clear presentation of the 'plagiarism affair' see Raimondi (2010), 60–61.

Similarly, *Theophrastus redivivus* (who quotes not only from Cardano and Vanini, but also – further complicating the picture – from Vanini's Cardano) has been often considered as a none too original mixture of sources, and the author has even been described as a "a man of an earlier generation in outlook",[8] hopelessly far from the 'new philosophy' of the seventeenth century.

Tracing the development of one particular topic dealt with by Cardano, Vanini and ultimately *Theophrastus redivivus*, can thus reveal important shifts but in the context of a certain continuity. This will point to a different way of interpreting the problem of reception and appropriation beyond accusation of plagiarism and lack of originality. Yet these are side-effects of bringing attention to the question about the (happy) place of animals in nature: the main point, however, is the change that occurs in understanding the relationship between man and the animals, the animals' capability of attaining happiness and man's prospect of a happy life, from Cardano, to Vanini, to *Theophrastus redivivus*.

Readapting Cardano's Happy Animals

Echoing a famous passage of Cardano's *De subtilitate* (without declaring the source), *Theophrastus redivivus* states in the chapter on "what it is to live according to nature and happily (*feliciter*)" that living beings differ from each other only inasmuch as they belong to different species and thus "adorn the world" in various ways:

> Nature gave equally to all of them [the living creatures] similar desires, similar reason and similar concern for their own preservation. Neither did it determine that there be any differentiation among all living creatures with regard to the condition, but only with regard to the species. [...] They [the animals] differ from each other according to species in order to adorn the world, just like plants and stones [...].[9]

In the following chapter *Theophrastus* even mentions directly the example that Cardano himself had used in order to argue that no animal whatsoever was created to be useful to human beings, not even the bothersome fly:

> You might ask: what is then the use of the fly, to mention a most base and highly insolent animal? I reply that that animal exists to perpetuate its own species, and for itself, and to adorn the world, and is provided with everything it needs not just for life but for a happy life [*vita beata*]. It is made for its own sake, not to molest man.[10]

---

[8] Spink (1960), 60–71.

[9] TR, 802–803.

[10] *De subtilitate* in Cardano (1663), III, 550a. Cardano's example of the fly is studied in detail by Guido Giglioni, who places it in the context of Cardano's natural philosophy, and his treatment of the order of nature: Giglioni (2014), 247–248. (I thank the author for having allowed me to read his work before publication). See also the section on 'The Soul and the Order of Nature' in Giglioni (2013). On Cardano's view of the relations between the creatures, beyond anthropocentric patterns, see Giglioni (2002), 115–116.

This famous passage about the happy fly is contained in the eleventh book of Cardano's *De subtilitate*, dedicated to the relationship between man and the rest of the creation, and to the shape of man's body (*De hominis necessitate et forma*). The statement about the fly can be considered as summing up clearly two main elements of Cardano's approach. First and foremost, Cardano aims at showing that the structure of the world is not arranged around human beings as its raison d'être: provocatively, he does so by using the example of a creature which not only is of no apparent use to human beings, but even torments them. Indeed from this interpretation of nature that attempts to take into account points of view and modes of existence other than the human one, derives the second aspect of Cardano's argument: the extension of the conception of 'happiness' to animals, too – and this is the aspect that the author of *Theophrastus redivivus* places at the centre of his reinterpretation of the content and role of that *felicitas naturalis*, which pertains to all creatures.

The crucial element for understanding both Cardano's argument and its reception through *Theophrastus* is the meaning that Cardano attributes to happiness, when he says that even flies can lead a happy life, a *vita beata*. In these pages of *De subtilitate* Cardano uses the terms *beatus* and *felix* quite freely, and roughly as synonyms.[11] The *beatitudo* he ascribed to all creatures seems to be linked to very practical aspects: primarily being or not being oppressed by other creatures. Thus a fly is happy if it can lead its life relatively freely, without being killed (or eaten) by other creatures, such as man. To the objection that if being free and not oppressed is a requirement for a happy life, surely small tame creatures like hares must live the most miserable lives, being hunted and eaten by predators, Cardano answers by insisting on the fact that in principle each animal's life is directed to its own happiness, and indeed "many hares are unhappy, but not all: indeed some of them never saw a man, or a dog, and did not suffer from being chased".[12] This basic definition of a happy life does not directly involve the connection between happiness and virtue, which played a pivotal role in the aftermath of the Aristotelian tradition: the possibility of attaining *eudaimonia* would set human beings apart from all other creatures.[13] In attributing the potentiality of happiness to the fly, Cardano seems therefore to have changed the content of what happiness means, as compared to this tradition. Yet it is important to underline that this regards only what I have called the

[11] On the happiness of animals, and especially of the elephant, see *De subtilitate* in Cardano (1663), III, 530b. On man's achieving happiness (using both terms, *felix* and *beatus*) see also *De rerum varietate* in Cardano (1663), III, 149b.

[12] *De subtilitate* in Cardano (1663), III, 550a.

[13] Aristotle (2011), 17–18 (1099b11-34) "it appears that even if happiness is not god sent but comes to be present through virtue and a certain learning or practice, it is among the best divine things. For the prize of virtue or its end appears to be best and to be something divine and blessed. [...] happiness was said to be a certain sort of activity of soul in accord with virtue. [...] It is to be expected, then, that we do not say that either a cow of a horse or any other animal is at all happy, for none of them are able to share in such an activity. It is natural, then, that we call neither ox nor horse nor any other of the animals happy". Julia Annas has pointed out that in the Aristotelian corpus one finds the idea that every creature contributes in its own way to the stability of the whole (eco)system: Annas (1993), 156.

'basic definition of happiness', because elsewhere Cardano does insert elements of differentiation between man's and the animals' ways of being happy.[14]

Cardano continues his argument by stating that "the common good had to be placed before the inconvenience of a few":[15] this is thus the key to understanding the relationships of all the creatures – of man and the fly, or of the hare and the dog. All in all, divine *sapientia* formed everything to be the best it could be.[16] As to the relation between man and fly, this means that man should not expect the fly to be of any use to him, but also that he should not consider that the insect was made to torment him either. Moreover, widening the horizon to consider other areas of the earth and climates helps us to realize that man and the fly are not always linked together in a relationship necessarily beneficial or necessarily harmful: "not all flies are vexatious to humans, but some always remain in wooded areas, and they don't exist in some regions, such as Lapland, and there are few of them in the Western Indies".[17] Rather, they pursue their own lives, which must be considered from the point of view of the whole, and not of man's limited perspective.

The author of *Theophrastus redivivus* often recurs to Cardano in order to challenge the view, loaded with Christian biblical interpretations, of man's superiority to the animals. In the second treatise he quotes openly from Cardano's *De subtilitate*: "To state that many species of snakes, says Cardano, that are harmful to man are made for man's sake, is utterly delirious".[18] But at this point Cardano had then inserted a rhetorical "What then?",[19] upon which he had reintroduced a new kind of hierarchy of the creatures. Nature shaped all creatures according to a principle of general purposiveness, pursuing the idea of usefulness to the whole the living beings, which can be catalogued according to their degree of complexity (from minerals, through plants and animals to man), are linked together primarily by being sources of food for each other.[20] Thus Cardano argues that man can make use of other creatures for his own benefit, without having to explain those creatures' existence exclusively in terms of human subsistence. Introducing this hierarchy does not

[14] On the close relation between happiness (as *felicitas* and as *beautitudo*) and securing one's progeny, see *De utilitate* in Cardano (1663), II, 252. On reaching happiness (*beate vivere*) within human society see *Proxeneta sive De prudentia civili*, ibid., I, 365a. On the *ratio vivendi* in *Proxeneta* see Ingegno (1980), 339ff. On the relation between happiness and rationality, and thus on the highest form of happiness, that of the *sapiens*, see ibid., 327.

[15] *De subtilitate* in Cardano (1663), III, 550a.

[16] Ibid.

[17] Ibid. The passage is quoted and commented upon by Giglioni (2014), 248, where he traces back the main addressee of Cardano's critique to Aristotelian metaphysics.

[18] TR, 216–217.

[19] *De subtilitate* in Cardano (1663), III, 549b: "Ergo species rerum ipsae, aut propter se factae sunt, aut hominem. Tot autem genera serpentum quae homini exitio sunt, propter hominem facta dicere, insanientis prorsus est, tum venena mortifera. Quid igitur?"

[20] Ibid., III, 549–550.

weaken the core of Cardano's argument: that nature is not oriented towards man as its sole centre.[21]

Cardano's distinction between man and the animals certainly involves further aspects, which also have an effect on the consequences to be drawn with regard to the type of happiness that animals and humans can reach, as well as to the unique complexity of human happiness. A crucial factor in human happiness is the knowledge of the Divine, which is firmly linked to the specificity of man's soul.[22] But, leaving these issues aside and keeping the focus on the dissolution of a clear teleological order of nature, the main consequence of this reasoning is that Cardano can at the same time argue against an anthropocentric view of the world, deriving from the acknowledgment of the plurality of life forms, and state that a hierarchy still remains, from the less to the most perfect being.

Theophrastus redivivus, on the other hand, inflects the quotation from Cardano differently, by framing it with the following remark: "And surely if god gave to men this kind of dominion, he had to grant also the powers with which to defy rebels, for without these powers no dominion can last".[23] *Theophrastus* introduces a doubt about man's 'practical' superiority: is man really stronger than the animals? And as a consequence can he be considered capable of making the best possible use of natural resources, supposedly given to him for that purpose? Thus the basis of Cardano's argument is ultimately unsettled, and this in two respects. The overpowering strength of human beings is indeed far from being an undisputable fact – actually, as *Theophrastus* points out, man is constantly hunted, chased and attacked by other creatures. But if this is true, nothing remains of the conception according to which man was created superior, at least with regard to his potential in making use of nature. *Theophrastus* clinches this adaptation of Cardano with the words:

> Indeed everyone understands how foreign to truth and reason it is to say that man rules over the other living beings and dominates them, since he always has to be on guard that he is not infested, devoured or that he doesn't suffer some harm, and it was for this that the society of men was first established.[24]

According to the anonymous author, man's lack of superiority, and indeed his radical inferiority to other animals, does not emerge from subtle reasoning on the human condition, but rather from unbiased observation of his relationships with the other creatures: it is ultimately a matter of being torn apart and eaten (or not) by other creatures. *Theophrastus redivivus* states directly: "What is said about man's dominion over the other animals is highly false and it surpasses all lies. For it is obvious from daily experience that no right or power of command over them was

---

[21] For a discussion of the opening pages of *De subtilitate*, Book XI, with special attention to the implications of this idea of a hierarchy in nature, see Giglioni (2014), especially the "Introduction".

[22] On human happiness, deriving from the knowledge of the Divine (leading to the conclusion that the happiest life is that spent in contemplation), see *Theonoston IV* in Cardano (1663), II, 433–436; on happiness as being conscious of being happy, see ibid., 454a: "nam omnis felicitas et miseria consistit in existimando se beatum aut miserum".

[23] TR, 217.

[24] Ibid.

granted to man".[25] The animals' power consists in sheer bodily strength (such as the wolves' or the lion's) and man cannot do anything but seek protection against the beasts through association with other human beings.

Of course depriving man of any sort of dominion over the animals implies abandoning the biblical tradition according to which man's place in the creation is defined by his power over all other creatures.[26] *Theophrastus* marks its distance from this tradition by repeatedly using Ecclesiastes as its principal biblical source, thus underlining the essential similarity between man and the animals. For instance, in the first treatise of *Theophrastus redivivus* we read: "what has man more than a beast of burden? – asks Solomon".[27] And again in the sixth: "as Solomon says: the destruction of man and of the beasts is one and the same, and equal is the condition of both".[28] Moreover, *Theophrastus* mocks the biblical representations both of man's superiority, as imparted by God, and of the shift in the relationship between man and animal following the Fall. Not only is it unclear why the animals should have been damned too, if they did not sin, but it is the logic of the whole story that *Theophrastus* finds wanting. With regard to the animals' usefulness for man, *Theophrastus* even asks why fish were not extinguished by the flood like all other animals, if they were all created only for man's sake.[29] Furthermore, the Biblical story even proves that an animal, the snake, actually dominated man, thus being superior to both man and the woman.[30]

From Cardano, *Theophrastus* derives the radical view that no animal at all was made for man's sake, while at the same time he changes significantly the overall conception of nature's aim and order, which in Cardano's case framed the resulting anti-anthropocentrism within firm limits. But what happened, in *Theophrastus*' readaptation, to the second part of Cardano's argument from which we started? How does *Theophrastus* deal with the connection asserted by Cardano between the plurality of life forms and the possibility that the animals might live a happy life (*vita beata*)?

In *De natura* Cardano asks directly how it is possible to conceive that the aim of the creatures is to live happily, despite the plurality of life forms, which seems necessarily to produce conflicting aims: in particular, the animals' striving towards a *vita beata* seems to be diametrically opposed to man's own pursuit, which often

[25] Ibid., 216. See also ibid., 218: "Homo dominatur agno et vorat illum, lupus utrumque saepissime. Igitur imaginarium et fictitium est imperium quod in belluas homines habere dicuntur [...]."

[26] Ibid., 217–218. Far from viewing the man-animal relation in terms of (at least possible) harmony, *Theophrastus* states: "nunc enim, sicut et ante redemptionis tempus, inter belluas et homines similis discordia viget" (218).

[27] Ibid., 125.

[28] Ibid., 805. See also ibid., 242.

[29] Ibid., 290: "quid enim peccaverant bruta? Si, quia propter hominem deus illa creaverat, extincta fuere, quare non etiam pisces extincti fuerunt, qui hominum quoque causa creati fuerant?" I consider *Theophrastus*' critique of the Biblical narrations, with particular regard to the problem of diet and eating, in Muratori (forthcoming).

[30] TR, 218 (reference is to Genesis 3:1).

involves subjugating the animals.[31] Therefore, if we consider this plurality, there is no convergence of the creatures' aims into a single one, and moreover Cardano underlines yet again that chance also plays an important role: the rain does not fall for the sake of the wheat, for instance, but rather because the vapours attracted by the heat of the sun are condensed.[32] As a consequence of this view of nature, the conception of a "happy life" cannot be univocal, but its meaning will vary at least from one species to another: "If one were to say that the first aim is to live happily, certainly this one aim will be such only in name, because the freedom of the horse, and man's ownership of the horse and use of it are contrary to each other."[33]

Living happily seems here to mean two things: first, living according to one's own nature, without restrictions imposed by others (like for the horse not to be oppressed by man), and to have at one's own disposal what one needs to do so (like for man to be able to use the horse). It is noteworthy that this same logic can be applied to the relationship between man and the animals, and also to the dynamics of human social interaction. Indeed, Cardano continues as follows: "And for the king it will be happy [*beatum*] to reign, but not so for those oppressed by the power of the king."[34] If these are then the basic requirements for a happy life, it is clear that, at least in the first instance, both animals and man are capable of attaining it, if the circumstances allow it. The fact that man retained a strong position in the hierarchy of living beings, due to his outstanding capabilities, meant that his chances of attaining the kind of basic happiness he shares with other creatures would be significantly higher than those of certain animals (the horse, which man subdues, or the hare, which man – and the wolf – eat).

In *Theophrastus redivivus* man not only has no dominion over the animals as a divine investiture: using the criteria derived from Cardano, the anonymous author concludes that his chances to achieve this happiness appear to be weaker than those of the animals, by which he is constantly attacked and subdued – notwithstanding the narratives man himself created to support the opposite view. In addition to integrating Cardano in his own new context, *Theophrastus* also recurs to Vanini's treatment of the same topic, that is the question regarding the supposed superiority of man over the animals, which Vanini had in his turn developed, especially in *De admirandis*, by introducing *ad hoc* interpretations of Cardano.

---

31 Cardano (1663), II, 292b: "An vero in universi ordine finis sit dubitatione dignum. Primum quia res cum sint infinitae ad unum finem reduci non possint".

32 Ibid.: "Et plura etiam casu accidere cum sit concessum ea autem non ob finem, finis horum nullus erit et quoniam sic experimur: nam calore solis attracti vapores cum in unum coëant congregatur pluvia, non ergo frumenti causa pluvia decidit neque eam immisit Iuppiter".

33 Ibid., 293a.

34 Ibid.

Reinterpreting Vanini's Beastly Men

The first occurrence of Vanini's name in *Theophrastus redivivus* is indirectly connected to the topic of happiness. In chapter five of the first treatise, *Theophrastus* ridicules belief in a providence that governs "our things",[35] which is ultimately the basis for believing in the existence of the gods: therefore if it is dispelled, belief in the gods will succumb as well. Here *Theophrastus* refers to a passage in Vanini's *Amphitheatrum* with the aim of attacking the theologians' view that torments in the afterlife will redress any crime left unpunished during one's lifetime:

> To the objection against providence according to which surely many crimes are left unpunished, the theologians customarily respond that god's providence directs everything to the future life, and that punishments will then be given for the evils, and so a reward for the good deeds as well, according to what each will have done. But Vanini says that the atheists considered the punishments of hell as a delirium of the ancients [...].[36]

Vanini's text contained several argumentative twists: first he had distanced himself from the views of the theologians, then he had introduced the doubts of the atheists, which were finally dismissed while retaining the suspension of opinion regarding the existence of a world of the dead. This *exercitatio* ended with a remark on the limits of natural reasoning and a reference to Cardano:

> since we can't demonstrate with natural reasoning that in the other life a specific place for punishment is assigned, and much less that demons exist, if we don't grant the existence of female demons as well, to speak in the manner of Cardano at the end of the book *De rerum varietate* and also in the book *De subtilitate*, for this reason we have to use a different refutation.[37]

Theophrastus, on the other hand, leads the reader to the conclusion that all suffering, punishment or reward belongs to the world of the living, and not to a hypothetical world of the dead, and readapts the passage as follows:

> we cannot demonstrate by any natural reasoning that in the other life a specific place for punishment is assigned, and much less that demons exist. Therefore if a place of punishment is not constituted anywhere, there can't be a place of rewards either. And thus crimes will remain unpunished, and good deeds will remain unrewarded, therefore there is no providence and as a consequence there are no gods.[38]

Theophrastus continues by pointing out ironically that the theologians picture god to be *beatissimus* and *felicissimus*[39]: with reference to Epicurus' doctrine, it is suggested that this kind of *beatitudo* is directly connected to the divine detachment from all worldly affairs. The life of man, by contrast, appears far from this kind of happiness. He rather seems to share with the animals the possibility to achieve a

---

[35] TR, 94.

[36] Ibid., 106. As Gianni Paganini points out in Paganini (1998), 257, this is the only quotation from *Amphitheatrum* in the whole text.

[37] Vanini (2010), 446 (*Amphitheatrum aeternae providentiae[...]*, exercitatio 10).

[38] TR, 107.

[39] Ibid.

"natural happiness" (*felicitas naturalis*), which resembles the *vita beata* that Cardano had ascribed to the fly, namely a life that follows what is beneficial for the creatures and avoids what is harmful to them: "This natural happiness does not pertain only to man, but in equal measure to all living beings: to those, I say, that we call brutes, this kind of happiness is due just as it is to men."[40] It is true that in the last chapter of the text the emphasis lies on a different type of happiness than this natural one, shared equally with the animals: it is the happiness of the wise man, which stems directly from *Theophrastus*' reflections on directing life according to nature.[41] A link is apparent between this type of happiness and the *beatitudo* that Epicurus had ascribed to the gods: in both cases, happiness is equivalent to lack of involvement – in the case of the former with society, in case of the latter with the world of the mortals.[42] Thus even a figure like Timon the misanthrope could be viewed as an example of a man who lived *beate*, without contacts with other humans.[43] Indeed the "Peroratio operis ad sapientes saeculi" ends with a series of four words which are deeply interconnected, and define the happiness of the detached wise man: *gaudium*, *pax*, *tranquillitas*, *felicitas*.[44]

But leaving aside the case of the wise man, and returning to the man-animal distinction in the context of the pursuit of happinness, it is worth noting that *Theophrastus*' reference to Vanini (and Cardano) on the non-existence of demons echoes a passage from *De admirandis*, not quoted explicitly – a passage which once again connects the topic of happiness with man's relation to the animals. In dialogue 50, the two main characters, Alexander and Iulius Caesar, begin a discussion about man's "aim" ("de hominis fine"), "which is God".[45] Referring to the theme of the "misery of man", Alexander states: "Man is full of so many and such miseries that if it did not contravene Christian religion [...], I would dare to say: if demons exist, they are punished for their crimes into human bodies."[46] Iulius Caesar's answer steers the discussion towards a very practical aspect that connects the life of animals with that of humans: generation.

> The desire for perpetuation is present in all animals, and for this reason they strive for perpetuity through their offspring and reputation. But very few are attracted by the love of the real eternity. Therefore hardly anyone desires to die, even if he is highly miserable.[47]

The fact that humans do not desire death, even when they are most unhappy, suggests that they act like animals, which desire preservation more than anything else: humans are not different from animals, but rather they belong to the group of the animals. These statements about the equality of life aims had already been prepared

---

[40] Ibid., 802.

[41] Ibid.

[42] Solitude is thus the essential basis for achieving this kind of *felicitas*: ibid., 884.

[43] Ibid., 886, and 887.

[44] On the relation of the wise man and the people see Bianchi (1988), chapter 3.

[45] Vanini (2010), 1344 (*De admirandis*, III, 50).

[46] Ibid.

[47] Ibid.

in the previous dialogue, when Iulius Caesar had demolished the main barrier dividing man from the animals, that is the possession of reason: "What in us is called reason, we call in animals instinct of Nature".[48] *Theophrastus* uses the same terminology (without, this time, spelling out Vanini's name) in the chapter of the sixth treatise entitled: "man does not differ from the animals". Here he asks: "Indeed what is instinct other than reason? And what is reason other than instinct? With a different name the very same thing is meant and expressed."[49]

If reason does not mark a qualitative distinction between man and the animals, then the lives of humans and animals are pulled closely together, and indeed Vanini dedicates a relevant portion of *De admirandis* to discussing questions regarding the generation of all creatures, a topic which he approaches by drawing crucial ideas from Cardano.[50] In the dialogue on the generation of man, immediately following that on the generation of bees, the two interlocutors discuss the theory attributed to Cardano that man can be generated from putrid matter, just like some other species of animals.[51] Staging this conversation, Vanini adapts a passage from *De subtilitate* in order to introduce the theory that man not only could have originated from putrid matter, but even from "the putrid matter of monkeys, pigs and frogs", thus suggesting that man may even derive from the animals.[52] This dependence on, or derivation from the brutes with regard to generation is expanded in the following part of the dialogue to include a different kind of dependence, one that *Theophrastus redivivus* clearly considers to be key to subverting the Christian view of man as the centre of the cosmos: man is physically subdued to the animals, and not the other way round. This is the crucial point where Vanini's adaptation of Cardano becomes clearly apparent: from Cardano's treatment of spontaneous generation as a natural phenomenon, Vanini draws man's submission to the animals, constantly using a dialogical form as a rhetorical device:

A.: You reason acutely. But I thought that man had been created to rule over the rest of the animals.
I.C.: You dare to say that man dominated over the basilisk?
A.: Indeed man kills the basilisk, not to speak of dominating him.

---

[48] Vanini (2010), 1328 (*De admirandis*, III, 49).

[49] TR, 825. On *Theophrastus*' critique of reason as the point of distinction between man and the animals, and on the assimilation of reason and imagination: Paganini (1981), 74–75, 79. On the role of the man-animal comparison within the libertine tradition (with Vanini as an important source of inspiration for the critique of anthropocentrism) see Gregory (1981), 34–37.

[50] See especially Vanini (2010), 1108–1120 (*De admirandis*, III, 30: On the generation of fish); 1144–1156 (*De admirandis*, III, 34: On the generation of birds); 1158–1160 (*De admirandis*, III, 36: On the generation of bees); 1162–1166 (*De admirandis*, III, 37: On the generation of man).

[51] See the notes by the editors, Raimondi and Carparelli, in Vanini (2010), 1729. On Cardano's view on spontaneous generation, also with reference to Scaliger's critique of it, see Gliozzi (1977), 316–319.

[52] On the topic of generation from corruption (with special reference to Bruno) see Papi (1968), 3–6. I have considered Vanini's discussion of spontaneous generation in the context of a broader Renaissance debate on the dangerous implications of this type of reproduction in Muratori (2013).

I.C.: And the basilisk man. Therefore the dominion is equal. In truth the sages believe that there is dominion where there is servitude or obedience. But what sort of man establishes a state with basilisks, bees, swallows, whales or eagles? Actually if man seizes beasts, more often he is seized by them. […] In fact if man kills, he is also killed, if he eats, he is eaten.[53]

Cardano had already emphasised that man is involved, like all other creatures, in a battle for survival in nature, in which equal interests clash. Yet Vanini goes a step further in stressing not only the equality of intentions that drive man and the animals' actions, but even man's extreme difficulty in finding his way in a world in which the creatures rebel against him: not only can man not easily subdue them, but the whole of nature seems to have turned against man, trying to annihilate him.

In a passage already quoted, *Theophrastus redivivus* had underlined that no dominion can last if one is not endowed with enough strength to resist rebels. Indeed Vanini had even presented the interpretation of the Fall as a case of rebellion of nature against humans: "A. But this rebellion of the animals against humans happened after the sin".[54] The inferiority of man's strength compared to the animals' is reinforced in *Theophrastus redivivus* through a twist in the reception of the theory of spontaneous generation, filtered through Cardano and Vanini. *Theophrastus* underlines that while small animals and plants regularly reproduce via spontaneous generation, thus ensuring their survival, man's way of generating is far less successful: he should therefore be considered "the most arrogant of all animals", since he is in fact inferior even to the smallest and most insignificant of them, and even to some plants, whose endurance on earth is guaranteed by their efficient way of reproducing themselves.[55]

By intertwining and interpreting its sources, *Theophrastus* in fact subverted the whole framework of the debate: starting with Cardano's discussion of the happy animals, *Theophrastus* concluded that man's chances of reaching happiness are in fact lower than the animals'. At the same time *Theophrastus* brought Vanini's discussion of man's bestial origins to its extremes, ultimately setting man apart even from the group of the animals. This reversal is exemplified in the following statement in the chapter on "what it is to live according to nature and happily", from which we started:

All are equal with regard to nature, therefore all living beings participate in natural happiness [*felicitas naturalis*]; indeed, given that men are completely thrown into disorder by the laws and opinions they set up, they can be said to be much unhappier than the rest of the animals.[56]

[53] Vanini (2010), 1164–1166 (*De admirandis*, III, 37).

[54] Ibid., 1166.

[55] TR, 178.

[56] Ibid., 803.

Conclusion

Theophrastus does not simply let the radical core of Cardano and Vanini's arguments emerge, as if it were just a matter of disclosing what had been present but veiled in the Renaissance authors, primarily because of the role of censorship.[57] It is rather a re-reading that introduces selected ideas into a new context. *Theophrastus* recombines the sources, and consciously draws its own conclusion: that not only is man the most unhappy of all animals, but, more radically still, he has no place assigned in the economy of the universe.

Theophrastus asserts that there is no absolute good or bad in nature,[58] but that the perspectives of the creatures define what is positive and what is negative: indeed even the extreme case of animals eating members of their own species – an example derived from Cardano's remark on certain species of fish in *De subtilitate*[59] – is not an aberration, something "bad according to nature".[60] And this same reasoning can be used in the case of humans as well, not only of fish: even cannibalism can thus be conceived as a practice which in itself is not unnatural ("contra naturam"). But in *Theophrastus redivivus* this openness to the plurality of perspectives in nature does not lead to including man in the group of the animals: it rather marks the exclusion of this hopelessly unhappy creature from the rest of the living beings. The chapter entitled "man does not differ from all other animals", close to the conclusion of the manuscript, in fact achieves the exactly opposite outcome: man is very much different from all animals, and not simply because of his lower 'happiness expectation' but rather because he is the only creature to have no orientation whatsoever in the world.

Radicalising Vanini's statement about the lack of qualitative distinction between reason and instinct, *Theophrastus redivivus* lists at length all the practical capabilities that different species of animals possess and man lacks. In the first instance, 'instinct' enables the animals to successfully preserve themselves: their natural inclination towards things that are beneficial to them, and aversion to what could damage them, support the striving for self-preservation, which is present in all creatures, humans and beasts alike. One might think of Campanella's treatment of the animals' capabilities, to which *Theophrastus* indeed often refers.[61] Yet – and this is the crucial point – *Theophrastus* uses the description of the animals' extraordinary forms of behaviour as a means to argue that the affirmation of the animals' superiority to man with regard to the external senses actually involves asserting their superiority with regard to the internal ones as well. The fact that the animals

---

[57] On the creation of Cardano's fame as a free-thinker, also with reference to the seventeenth-century reception of his works (including Vanini) see Maclean's observations, based principally on the detailed study of Cardano's psychology: Maclean (2009).

[58] See especially TR, 793–797.

[59] Ibid., 795.

[60] Ibid., 796. This aspect is discussed by Laursen (2014).

[61] TR, 807.

are capable of doing things "of which man knows nothing",[62] not only means that they have more acute senses, but that they must even have an advantage on the level of imagination and judgement. All in all, if the animals feel more acutely, than they must think more acutely as well (*Theophrastus* uses both the terms *intellectus* and *ratiocinatio* in this context).[63]

Once again, *Theophrastus* goes even beyond the demolition of all distinction between man and the animals: not only are the animals 'equal' to man, since rationality no longer represents a barrier between the two, but they are even better equipped than him, since they perceive the world more sharply, and therefore are more capable of reflecting on their sensations. The debate on the place of animals in the world, and on the animals' inclusion in the prospect of a happy life, traditionally considered properly human, has tilted into a vision of nature in which man has lost his specific place, surrounded by animals which on the contrary show all their ability in the pursuit of life – and of a happy one, too.

While this conclusion emerges as the result of *Theophrastus*' adaptation of sources, it is also achieved through stressing certain continuities with the authors he uses. *Theophrastus* constructs a solid foundation for 'encountering' his chosen Renaissance conversation partners. This foundation consists in the interest in the topic of the animals' happiness, developed to its paradoxical conclusion, but it also involves stylistic aspects. *Theophrastus* frames his subversive arguments, often readapted from his sources, by stating that he simply aims at displaying atheistic reasoning to better embrace true religion. This is a rhetorical method which Vanini had also employed, but through more complex intertwining of opposing views.[64]

Most importantly *Theophrastus* himself has achieved the creation of a continuity by constructing a tradition of thinking about man's relationship with the animals as characterized by the radical equality of all living beings – so radical that in fact no equality is visible any more after *Theophrastus*' intervention. Yet the distance from the authors used in the anonymous manuscript is achieved through quoting and commenting, selecting and adapting, including acknowledged and unacknowledged sources.

For this reason, *Theophrastus* is a case of reception history which bridges the gap between supposedly fixed periods – from the Renaissance into the Early Modern age. But at the same time the red thread we followed – the debate on the happiness of animals – proved that a gap did in fact open precisely within *Theophrastus*' engagement with 'Renaissance sources', to the point that the question about the place of animals in the world has ultimately lost its importance. It is the place of man that has been dissolved in the process, and at the same time the natural happiness of animals appeared to the anonymous author writing almost a hundred years

[62] Ibid.

[63] Ibid., 809.

[64] On the rhetorical devices by which an author is ultimately made more available thanks to discussion of his atheistic views and their apparent condemnation, see Mulsow (2001), 67 and 77 (especially on La Croze's treatment of Vanini).

after Cardano to be out of the reach of humans. This duality – the continuity and the shifts – is visible only by reconstructing specific paths within the vast material constituted by *Theophrastus redivivus* together with its sources: it is the legacy of the animals' happiness, in this case, that changes form in persisting as a crucial point of debate from the fifteenth to the seventeenth century.

Bibliography

Anon. 1981–1982. *Theophrastus redivivus*, 2 vols, ed. Gianni Paganini and Guido Canziani. Florence: La Nuova Italia. (abbreviated in this essay as TR).

Annas, Julia. 1993. *The Morality of Happiness*. Oxford: Oxford University Press.

Aristotle. 2011. *Nicomachean Ethics*, ed. Robert C. Bartlett and Susan D. Collins. Chicago: University of Chicago Press.

Bianchi, Lorenzo. 1988. *Tradizione libertina e critica storica*, 115. Milan: FrancoAngeli.

Canziani, Guido. 1981. La critica della civiltà nel *Theophrastus redivivus* II. Ordine naturale e legalità civile. In *Ricerche su letteratura libertina e letteratura clandestina nel Seicento*, ed. Tullio Gregory et al., 83–118. Florence: La Nuova Italia.

Canziani, Guido. 1985. Une encyclopédie naturaliste de la Renaissance devant la critique libertine du XVIIe siècle: le *Theophrastus redivivus* lecteur de Cardan. *Dix-septième siècle* 37: 379–406.

Cardano, Girolamo. 1663. *Opera omnia*. Lyon: Jean-Antoine Huguetan and Marc-Antoine Ravaud.

Corvaglia, Luigi. 1933–1934. *Le opere di Giulio Cesare Vanini e le loro fonti*, vol. I–II. Milan/Genoa/Rome/Naples: Società Editrice Dante Alighieri.

Corvaglia, Luigi. 1991. *Le opere di Giulio Cesare Vanini*, vol. III.1. (*Da Omero a Cardano*), ed. Maria Corvaglia Aprile and Gino Pisanò. Galatina: Congedo Editore.

Giglioni, Guido. 2002. Medicina e metafisica della vita animale in Cardano. *Bruniana & Campanelliana* 1: 113–158.

Giglioni, Guido. 2013. Girolamo Cardano. In *The Stanford Encyclopedia of Philosophy*. http://plato.stanford.edu/entries/cardano/#SouOrdNat. Accessed 3 Mar 2014.

Giglioni, Guido. 2014. Humans, Elephants, Diamonds and Gold: Patterns of Intentional Design in Girolamo Cardano's Natural Philosophy. *Gesnerus. Swiss Journal of the History of Medicine and Sciences* 71(2): 237–257.

Gliozzi, Giuliano. 1977. *Adamo e il nuovo mondo. La nascita dell'antropologia come ideologia coloniale: dalle genealogie bibliche alle teorie razziali (1500-1700)* Florence: La Nuova Italia.

Gregory, Tullio. 1981. Il libertinismo della prima metà del Seicento: stato attuale degli studi e prospettive di ricerca. In *Ricerche su letteratura libertina e letteratura clandestina nel Seicento*, ed. Tullio Gregory et al., 3–47. Florence: La Nuova Italia.

Ingegno, Alfonso. 1980. *Saggio sulla filosofia di Cardano*. Florence: La Nuova Italia.

Laursen, John Christian. 2014. Cynicism in the *Theophrastus redivivus*. In *Entre la Renaissance et les Lumières, Le Theophrastus redivivus (1659)*, ed. Nicole Gengoux and P. F. Moreau, 47–64. Paris: Honoré Champion.

Maclean, Ian. 2003. Cardano on the Immortality of the Soul. In *Cardano e la tradizione dei saperi*, ed. Marialuisa Baldi and Guido Canziani, 191–207. Milan: FrancoAngeli.

Maclean, Ian. 2009 [2008]. Cardano's Eclectic Psychology and Its Critique by Julius Caesar Scaliger. In *Transformations of the Soul: Aristotelean Psychology 1250-1650*, ed. Dominik Perler, 170–195 (Special Offprint of *Vivarium* 46:3). Leiden: Brill.

Mulsow, Martin. 2001. *Die drei Ringe: Toleranz und clandestine Gelehrtsamkeit bei Mathurin Veyssière La Croze (1661-1739)*. Tübingen: Niemeyer.

Muratori, Cecilia. 2013. The Earth's Perilous Fertility: Telesio on Spontaneous Generation and the Continuity of Living Beings. In *The Animal Soul and the Human Mind: Renaissance Debates*, ed. Cecilia Muratori, 131–151. Pisa-Rome: Fabrizio Serra (*Bruniana & Campanelliana:* Supplementi XXXVI, Studi 15).

Muratori, Cecilia. Forthcoming. Food for "Free Thought": Diet and Libertinism in *Theophrastus Redivivus* and its Sources. In *Les philosophes et la libre pensée (XVIIe-XVIIIe siècles)*, ed. Lorenzo Bianchi and Gianni Paganini, 181–198. Paris: Honoré Champion (Series "Libre pensée et littérature clandestine").

Paganini, Gianni. 1981. La critica della civiltà nel *Theophrastus redivivus*. I, Natura e cultura. In *Ricerche su letteratura libertina e letteratura clendestina nel Seicento*, ed. Tullio Gregory et al., 49–82. Florence: La Nuova Italia.

Paganini, Gianni. 1985. L'anthropologie naturaliste d'un esprit fort. Thèmes et problèmes pomponaciens dans le *Theophrastus redivivus*. *XVIIe siècle* 38: 349–378.

Paganini, Gianni. 1998. Le *Theophrastus redivivus* et Vanini: une lecture selective. In *Kairos. Revue de Philosophie* 12 (Monographic issue dedicated to Vanini), ed. Jean-Pierre Cavaillé and Didier Foucault, 255–274. Toulouse: Presses Universitaires du Mirail.

Papi, Fulvio. 1968. *Antropologia e civiltà nel pensiero di Giordano Bruno*. Florence: La Nuova Italia.

Raimondi, Francesco Paolo. 2010. Monografia introduttiva. In *Tutte le opere*, ed. Giulio Cesare Vanini and Francesco Paolo Raimondi, 7–293. Milan: Bompiani.

Spini, Giorgio. 1983. *Ricerca dei libertini. La teoria dell'impostura delle religioni nel seicento italiano*. Florence: La Nuova Italia (second edition).

Spink, John Stephenson. 1960. *French Free-Thought from Gassendi to Voltaire*. London: Athlone Press.

Vanini, Giulio Cesare. 2010. *Tutte le opere*, ed. Francesco Paolo Raimondi and Mario Carparelli. Milan: Bompiani.

Part IV
Moral and Political Theory

Chapter 12
Ethics, Politics, and Friendship in Bacon's *Essays* (1625): Between Past and Future

Annalisa Ceron

Abstract This chapter sheds light on the final version of Bacon's *Essays*, an early modern advice book that weaves together different and often contrasting Renaissance lines of thought. When offering his *Counsels, civil and moral* Bacon was in fact deeply influenced by Machiavelli's pessimistic view of man and combined it not only with the new Tacitean humanism that laid emphasis on private and personal interests, but also with Ciceronian ideas and Machiavellian republican arguments that prioritised public duty. He thereby continuously oscillated between common and personal good, generating ambiguities and ambivalences that should be neither emphasised nor minimised but instead related to his view of moral philosophy as a therapy that needed to be grounded on a realistic diagnosis of human nature in order to heal the mind of its perturbations and misleading tendencies. In *Of Friendship* the connection between Bacon's advice and the doctrine of the idols is clearer than anywhere else. Moreover this essay developed a very interesting reflection on friendship that is suspended between past and future. On the one hand, Bacon was the heir to the authors of the fifteenth-century 'mirrors for princes', who used the language of friendship to describe the counselors of the prince. On the other, he conceived friendship in terms that would be familiar to us today: a private and intimate relationship of mutual affection between people committed to taking care of one another. Bacon's view of friendship confirms that Renaissance ways of thinking continued to be far-reaching and were inseparable from new, more modern, conceptions.

A. Ceron (✉)
Dipartimento di Filosofia, Università degli Studi, Milan, Italy
e-mail: annalisa.ceron@unimi.it

C. Muratori, G. Paganini (eds.), *Early Modern Philosophers and the Renaissance Legacy*, International Archives of the History of Ideas Archives internationales d'histoire des idées 220, DOI 10.1007/978-3-319-32604-7_12

The Intricacy of Bacon's Civil and Moral Counsels

From the time of Voltaire to the present day Francis Bacon has generally been considered the father of modern science[1] and, on account of the key role he played in the development of literature, also the father of the English essay.[2] Like many of his other writings, Bacon's *Essays* was a work in progress, to which he devoted more than twenty years of effort. His view of human knowledge as an open and perpetually incomplete series of investigations, his distrust of philosophy as promoting a speculative approach, and above all his suspicion of moral systems insufficiently grounded in empirical data, may help to explain his habit of continually revising his work.

The first edition of the *Essays* was published in 1597, soon after Bacon had learned that an unauthorised version had been printed without his knowledge. As suggested by the dedicatory letter to his brother, this first endeavour to systematise his observations on human nature targeted a restricted readership. Moreover, the ten essays included in the first edition were written in an aphoristic style that Bacon himself said made them appear like "short dark oracles". He thus began to revise them after the publication of *The Advancement of Learning*, not only by trying to remould them into "a clear and perspicuous exposition",[3] but also by adding 28 new compositions, which displayed his growing interest in the ethical, political and historical influences on human behaviour.

The second edition appeared in 1621 and was dedicated to the Prince of Wales, rather than to the heir to the throne, who had died suddenly. Although this edition won Bacon national and international renown,[4] he was still not satisfied with the work and so chose to revise it once more. The third and final edition—to which I will refer in the course of my analysis—was published in 1625 and comprised 58 essays, 20 totally new and the others comprehensively modified. As the dedicatory letter to King James I's favourite, George Villiers, made clear, Bacon's interest in political matters (and perhaps his concern for gaining political appointments) had not faded after his fall from public office in 1621.[5] The political meaning of Bacon's

[1] See Voltaire (2010) and, for instance, Agassi (2013).

[2] For example Lee (2008), 600–609.

[3] Bacon (1857–1874), vol. VIII, 323; henceforth abbreviated as *Works*.

[4] The 1612 collection was reprinted in 1613, 1614 and 1624. The first anonymous Italian translation of Bacon's *Essays* was published in London in 1617 and reprinted both in England and in Italy before he died. Another Italian translation by Andrea Cioli appeared in 1619 and the first French translation was published in the same year. Moreover, in a famous letter written in the autumn of 1625 to Fulgenzio Micanzio, Bacon announced his desire for a Latin translation of the *Essays*. For further information, see Vickers' introduction in Bacon (1999), xxiii–xxv.

[5] Bacon was expelled from public office after the House of Lords convicted him for bribery. In the Jacobean court the distinction between bribes and gifts from followers was often ambiguous. Moreover the Lord Chancellor become gradually involved in several conflicts with judges of other courts, constitutionals battles between the parliament and the king, and clashes due to his association with the corrupt network of patronage of the favourite of the King, George Villiers. On Bacon's impeachment see Levy Peck (1990), 50–52 and 86–190; Jardin and Stewart (1998), 444–475.

Essays was openly declared in the subtitle added to the last collection: the *Essays* was intended to offer *Counsels, Civil and Moral*. Accordingly, it cannot be thought of as merely a commonplace book, but instead belongs to the heterogeneous genre of Renaissance advice books. As I will show in the course of my analysis, the *Essays* echoed the style of advice books written for princes, in particular the fifteenth-century "mirrors for princes" and Machiavelli's *Prince*, and followed the conventions of the Italian books of manners that were fashionable in Elizabethan and Jacobean England: Castiglione's *The Courtier*, Guazzo's *Civil Conversation* and Della Casa's *Galateo*.[6] Therefore, in order to avoid misunderstanding Bacon's ideas, it is important to consider the last version of *The Essays* as a handbook of advice on how to conduct private and public life.

The essays of the 1625 edition were less aphoristic than those of the earlier versions and were also sustained by a far more discursive structure. Even so, they were no less disjointed than those of the past: not only did Bacon fail to connect the essays together, but he also constructed the majority of them around an assortment of oppositions, associations and metaphors. Moreover, even in the few essays that were more systematically organised, he seldom subdivided his themes into topics or developed a continuous flow of argument. Instead, he set out to study the same subject from different points of view, to weigh up the influence of affections, dispositions, characters and tempers over men from within an Aristotelian framework, according to which virtues were seen as habits rather than correct judgements of reason. However, he did not align his ideas with Aristotle's doctrine of the *mesotes*, but instead examined the influence of fortune and external circumstances on men's inner dispositions without attempting to find a balancing force and, in order to highlight the pros and cons of certain forms of behaviour in specific situations, he overlapped quotations from the Bible and the classics. Furthermore, in preference to drawing general conclusions and passing judgement on the correct form of moral conduct in order to leave his readers with a specific lesson, he invited them to form their own opinions about the questions under discussion. This is one reason why his essays are sometimes deemed to be ambiguous and ambivalent.

This equivocal and indecisive outlook can more easily be explained if we compare *The Essays* to Bacon's scientific writings. In an old but nevertheless crucial article, Robert Crane demonstrated that Bacon wrote the second and third edition of his *Essays* in order to fulfil the scientific programme he had previously outlined in *The Advancement of Learning*.[7] To be more precise, the main subjects of Bacon's

---

Wootton (1999) accounts for Bacon's references to gift-giving and patronage in the different versions of his *Essays*.

[6] Castiglione's *The Courtier* was published in 1561 in an English translation by Thomas Hoby; Guazzo's *Civil Conversation* was published in 1586 in an English translation by George Pettie and Bartholomew Yonge; Della Casa's *Galateo* was printed in 1576 in an English translation by Robert Patterson. Even though the influence of those works is generally emphasised by Vickers and other editors of Bacon's *Essays*, there are no specific studies dedicated to this aspect. For a general framework see Burke (1993) and Wyatt (2005), 179–184.

[7] Crane (1923). For Bacon's scientific method as a medicine of the mind see Corneanu (2011), 14–45.

Essays correspond in large measure to the second part of moral knowledge in *The Advancement of Learning*, which Bacon called "regiment" or "culture" or "medicine of the mind". This part of moral knowledge prescribes directives about how to "subdue, apply and accommodate" the will to the good, and consists of an investigative branch, whose goal is the discovery of the perturbations and disorders of men, and an operative branch, whose goal is the health of the mind.[8] In his description of those branches of moral knowledge, Bacon revised the Stoic tradition of thought, according to which moral philosophy is *cultura animi* (Cic., *Tusc*. III, 6) or *medicina animi* (Sen., *Ep*. XV, 1–3), and thus helps men to master their passions.[9] As Ian Box has recently pointed out, the pervading presence of medical metaphors in *The Advancement of Learning* and *The Essays* indicates that Bacon's moral philosophy was intended as a type of psychological therapy which he believed should be grounded in empirical evidence and oriented towards gradual changes rather than definitive transformations in the human mind.[10] Moreover, according to the divisions defined in *The Advancement of Learning*,[11] moral knowledge cannot be separated from civil knowledge, and this implies three kinds of wisdom: that of conversation, in the wider sense of social intercourse (as in Guazzo's *Civil Conversation*); that of negotiation or business; and that of government or state. This subdivision is congruous with the main topics of Bacon's *Essays*, and once more highlights the practical aims of his advice.

Bacon's *Essays* achieve their practical goals thanks to their rhetorical structure. According to Stanley Fish, in fact, the scientific aspect of Bacon's *Essays* resides in the experience of reading them, because they lead to "a more self–conscious scrutiny of one's mental furniture".[12] Furthermore, as Quentin Skinner has shown, Bacon's idea of moral knowledge was deeply indebted to Cicero's *scientia civilis*, according to which the power of rhetoric is necessary in order to instil love for virtue and justice in men.[13] The fact that Bacon's ethics cannot be separated from rhetoric has also been emphasised by Brian Vickers, who has recently made clear that reason does not exert absolute power over other mental faculties and is involved in a constant *psycomachia* against passions and affections: for Bacon, the power of eloquence contracts a "confederacy" with imagination, capable of overcoming passions and affections.[14]

---

[8] *Works*, vol. VI, 310–311 and 342.

[9] On this Stoic tradition of thought see Nussbaum (1994) and Hadot (2002), in which the Stoic notion of the cultivation of the mind is analysed as a 'way of life'.

[10] Box (1996), 260–282.

[11] In the *De augmentis scientiarum*, moral knowledge is included in the *doctrina animae humanae* and separated from the *doctrina civilis*, but Bacon himself underlined that his divisions were branches of the same tree.

[12] Fish (1971), 56.

[13] Skinner (1996), 215–244; according to Skinner, Hobbes may have been one of the Latin translators of Bacon's *Essays*.

[14] Vickers (1996), 200–223.

To enhance the power of reason, and thus to heal the minds of his readers, Bacon weaves together many contrasting lines of thought, a variety already made clear by the title of the collection. As is well known, Bacon took the term *Essays* from Montaigne's *Essais.* These appeared in print for the first time in 1580, one year after Bacon had concluded his mission in France as companion to the English ambassador, although he probably learnt of Montaigne's masterpiece through his brother Anthony, who lived in Bordeaux and became a correspondent with the father of French essay writing.[15] Although John Florio's English translation of Montaigne's *Essais* was published in 1603, Bacon referred specifically to Montaigne's works only in the last edition of his own essays. Moreover, to explain the choice of the title in the dedicatory letters of the second edition he pointed out that the "word is late, but the thing is ancient" and cited Seneca's "dispersed mediations", namely the *Moral Letters to Lucillius*, as his true source of inspiration.[16] What is more, Martin Dzelzainis has recently argued that the subtitle added to the final version of the *Essays* alluded to Guicciardini's *Ricordi*, whose first edition was published as *Più consigli et avverimenti* by Jacopo Corbinelli in 1576.[17] Bacon in fact mentioned Guicciardini in *Of Empire*, and was deeply influenced not only by his aphoristic style, but also by his idea that accidents prevail over virtues in human history and especially by his pessimistic conception of man.[18] Nonetheless, in the wake of Paolo Rossi's study,[19] scholars have usually focused on Machiavelli's influence to explain Bacon's realistic and disenchanted description of human nature. Indeed, in *The Advancement of Learning* Bacon himself praised the author of the *Prince* for having described what men do, not what they ought to do.[20]

Machiavelli's scandalous lesson seems to have been a determining factor in Bacon's work, especially in the essays in which he dealt with characteristics and actions traditionally considered to be vices, such as simulation, envy, cunning and ambition, and argued that they were useful and necessary aspects of political life. Still, in order to explain the emphasis on those qualities, we must understand that Bacon's *Essays* contained a new kind of humanism, which looked to Tacitus rather than Cicero for its ancient leading guide. According to Richard Tuck,[21] this new humanism was a combination of moral scepticism, Senecan stoicism and reason of state, and it saw ethics and politics as an interaction of personal interests rather than an exercise of civic virtues. This new perspective on moral and political questions spread throughout Europe thanks to Giovanni Botero's *Della ragion di stato* and Justus Lipsius's political works, including his three related treatises on the greatness of the Roman Empire. Botero and Lipsius's influence was immediately evident in

---

[15] Kenneth (1991).

[16] *Works*, vol. XI, 340.

[17] Dzelzainis (2000).

[18] *Essays*,10–12, 80–83. For Guicciardini's influence on Bacon's *Essays* see Lepri and Severini (2011).

[19] Rossi, (1957), 234–255.

[20] *Works*, vol. VI, 327.

[21] Tuck (1993), 105–110.

Bacon's *Of the True Greatness of Kingdom and Estates*, although, as Makku Peltonen has demonstrated, in this essay Bacon challenged the theories of the new humanists using Machiavellian arguments.[22] In fact, unlike Lipsius and Botero, he believed that civic greatness relied on valiant men and not on the wealth of the state, and therefore argued for a citizen's militia. In this context Machiavelli was praised as a republican thinker who defended the active involvement of citizens in political life, not as an unidealistic writer who disclosed the reality of human behaviour.[23]

It is important to bear in mind that Machiavelli's influence on Bacon's *Essays* included, so to speak, an anthropological side and a republican side. On the one hand, Bacon painted a disenchanted and pessimistic portrait of human nature in line with the critique of traditional virtues elaborated in Machiavelli's *Prince*. On the other, he made use of the republican theses developed in Machiavelli's *Discourses* to emphasise the engagement of citizens in civic life. Moreover, the anthropological and the republican sides of Machiavelli's influence were frequently set in opposition to one another. Bacon, in fact, did not limit himself to advocating the primacy of an active participatory life, but went as far as to say that the public good is of greater value than individual good.

In the passages of *The Advancement of Learning* that Bacon used to describe the first part of moral knowledge, which he called the "platform of good", he claimed that good had a double nature. Since everything in the world is a separate and independent entity, it tends to preserve its autonomous existence, which is the individual good or the self-good. Yet everything in the world is also "a part or member of a greater body", and is thus oriented towards the "conservation of a more general form", namely the common good, the good of communion or the good of others. According to Bacon the latter was greater than the former because it was grounded in the "summary laws of nature", the appetites impressed upon matter at the moment of Creation that drove all natural objects to unite with others. And, as stated by this view, since those appetites were more deeply imprinted on human beings, the conservation of the duty to the public good ought to be much more precious for men than the conservation of their private and individual good.[24] Bacon's insistence in the *Essays* on the moral superiority of the common good is drawn partly on the Christian doctrine about the pre-eminence of charity, partly on Machiavelli's republican arguments, and partly on the tradition of Ciceronian humanism, which extolled the supremacy of the *vita activa* over the *vita contemplativa*.[25]

In the work, Bacon's emphasis on the common good and active life coexists with the importance he placed on self-love, cunning, ambition and other Machiavellian human qualities useful in political life. The coexistence is difficult and gives rise to further ambiguities and ambivalences, which come to light in many of his other essays.

---

[22] Peltonen (1992); the republican side of Machiavelli's influence is less accentuated in Zagorin (1998), 129–174.

[23] For a detailed analysis of Bacon's ideas about territorial expansion, his opposition to Botero's theses, and his affinity with Machiavelli's arguments see the chapter by Sara Miglietti.

[24] *Works*, vol. VI, 311–313.

[25] For the influence of Ciceronian humanism on Bacon's *Essays* see Vickers (1984).

In *Of Fortune*, for instance, Bacon stated that man should have "a little of the fool, and not too much of honest", in order to be "*faber fortunae suae*".[26] In *Of Goodness*, he compared Christian charity to Greek *philanthropia* and exalted it as the greatest moral virtue, without which man is "no better than a kind of vermin". Nonetheless, he also realised that man can only be "so good, that is good for nothing" and added that "misanthropi" are the "fittest timber to make great politiques of", because they are "like to knee timber, that is good for ships, that are ordained to be tossed".[27] In *Of Cunning*, he showed that cunning is not completely at odds with wisdom and is useful especially in royal courts and places in which baseness looms large.[28] Conversely, in *Of Wisdom for a Man's Self* he rejected any form of selfishness, which destroys the public good, and compared self-love to "the wisdom of rats, that will be sure to leave a house, somewhat before it fall", the wisdom of the fox, "that thrusts out the badger, who […] made room for him", and the wisdom of crocodiles, "that shed tears when they would devour".[29] In *Of Envy*, he argued that private vice could serve as a public virtue because envy is a sort of social "ostracism" that keeps overly powerful men "within bounds". On the other hand, love is a private virtue of little public value: when examining private morality from a civil perspective in *Of Marriage*, Bacon considered the domestic values he later celebrated in the *New Atlantis* to be "impediments to great enterprise".[30]

Thus as should be clear, Bacon continuously oscillated from individual to common good, from self–love to the love of others, from private interest to public duty, from new humanism and Machiavelli's realistic view of human nature to Ciceronian humanism and Machiavelli's republican arguments. If we consider the *Essays* as the "good advice from Satan's kingdom", as William Blake suggested and many scholars repeated,[31] we focus only on the negative pole of the author's analysis. But for Bacon men were neither devils nor angels: they were simply men, rational and finite beings corrupted by original sin. If it is true that they can act like foxes, rats and crocodiles, there is no doubt that they can also achieve the highest level of humanity through their commitment to political life. Bacon's *Essays* underlined the misery of the human condition without removing its dignity. Using the medical language that Bacon often adopted in his works, we might say that men are patients who are capable of healing. We have already seen that *The Essays* set forth Bacon's idea that moral philosophy is the medicine of the mind. What I now wish to emphasise is that the best therapy depends on accurate diagnosis, or rather—in this case—on Bacon's realistic and disillusioned description of human nature.

---

[26] *Essays*, 93.

[27] *Essays*, 28–29. The sentence is explicitly, albeit not faithfully, related to Machiavelli's thesis on Christian religion; for sake of brevity in this chapter it is not possible to delve into this aspect of Machiavelli's influence.

[28] *Essays*, 53.

[29] *Essays*, 55.

[30] *Essays*, 16.

[31] On Blake's opinion and its diffusion in twentieth-century scholarship, see Melchionda (1979), 5–10.

Bacon's View of Friendship: Therapeutic Effects and Secrets of Power

Of Friendship offers an intriguing vantage point from which to see how the conflicting and contrasting threads of ideas woven through *The Essays* were meant to cure the mind of Bacon's readers. In the last edition of 1625, at the request of his friend Tobie Matthew, Bacon totally rewrote the essay on friendship included in the previous collection. The new essay was intended as an *encomium amicitiae*, but Bacon described the precious benefits of friendship in realistic and pragmatic terms. Like other Renaissance thinkers, he continued to be absorbed by classical concepts of friendship while nuancing and adapting them "according to changing circumstances, audience, experiences, and ideas".[32] In order to try and understand his point of view, it is worth noting that in the 1592 edition Bacon had dealt with friendship and love in the same essay and, although in the second and third editions he instead dedicated separate essays to the two topics, their theoretic connection is still evident. In *Of Love* Bacon provided a Stoic account of human sociability explaining that "there is in man's nature a secret inclination and motion towards love of others, which if it be not spent upon some one or a few, doth naturally spread itself toward many, and makes men humane and charitable".[33] Thus, if love of others is directed towards only a few people, it becomes friendship (*philia*), while if it is spread amongst many, it changes into charity or philanthropy (*philantropia*). Since Bacon saw friendship as a restricted expression of the natural love of mankind, in *Of Friendship* he compared it to that "natural and secret aversion towards society" that turns men into animals. He in fact opened his essay by alluding to Aristotle's famous thesis (*Pol.* 1255a), according to which whoever delights in solitude is either a wild beast or a god. For Bacon, man is by nature a social being to such an extent that solitude and the contemplative life can be justified only on the condition that the "higher" reflections that result from it are of benefit to mankind. When he explained the meaning of the Latin saying from Erasmus's *Adagia*, according to which *magna civitas* is *magna solitudo*, he did not deny that friendship is the distinctive sign of human sociability; on the contrary, he celebrated friendship as a dyadic, intimate and close tie, without which men could neither live well nor be happy.[34] If it were not conceived in this way, friendship could not play the crucial role that Bacon ascribed to it for its therapeutic effect on the human mind.[35]

The description of the first fruit of friendship makes it clear that friends are the best medicine of the heart, whose curative effects are stronger than sarsaparilla,

[32] James and Kent (2009), 111; this study analyses Bacon's *Of friendship* (and Montaigne's *De l'amitié*) in the light of Burke (1999), Hyatte (1994), Langer (1994), and other relevant contributions on the Renaissance conceptions of friendship.

[33] *Essays*, 23.

[34] *Essays*, 59.

[35] For friendship as part and parcel of Bacon's medicine of the mind, see also Giglioni (2011).

steel and the other remedies suggested by Paracelsus.[36] Bacon did not demonstrate the validity of his theory on the basis of a logical set of arguments, but adopted a rhetorical approach supported by a large number of allusions and quotations. By evoking Hippocrates and Galen's theory that illness is the result of a bad mixture of bodily fluids, he argued that friendship ensured a healthy balance of passions and decreased the pressure on the heart. To explain how friendship can undermine physical and psychological health, he cited Pythagoras's obscure saying *Cor ne edito*, explaining that men would become "cannibals" eating their own heart, were it not for friends with whom to share their lives.[37] Since friendship redoubled joys and halved grief, its effects were compared to the value of the Alchemist's stone.[38] In this way, Bacon brought the Stoic tradition of the *medicina animi* into harmony with the Renaissance tradition of medicine and alchemy.

The benefits of friendship are as dramatic as the Alchemist's stone because it results in "a sort of civil shrift" that is based on moral, rather than religious, confession. To convey the idea more clearly, we might say that friendship consists of a frank discussion and an honest interchange of ideas that allow men to communicate their own self to others. When Bacon described the second benefit of friendship, he saw the act of communicating one's self as a process that enabled a man to organise his ideas in a much more rational, ordered, and efficient way than could be done by indulging in silent meditation. Hence he ironically suggested that, in the absence of friends, people would resort to conversing with statues and pictures. However, the process he described can only be successfully completed by a man seeking the response of a true friend who can give him sound and faithful advice. Since Cicero observed that true friends are expected to give advice freely and administrate admonition even with severity (*Lael.* 42), this comes as no surprise. Nevertheless, Bacon offered an original revision of this commonplace idea of friendship, going on to say that the counsels of a true friend free a man's mind of darkness and confusion of thought. Indeed, he argued that the light emanating from the advice of our friends is drier and purer than that which comes from our personal opinions and judgements, because the latter are inevitably contaminated and disturbed by passions and habits.

To make the matter clearer, Bacon quoted the "enigma" by Heraclitus according to which "dry light is ever the best" because it corresponds to the noblest and most rational part of the human soul, which is made of fire rather than water. The quotation enables us to relate Bacon's reflection on friendship to certain crucial passages of *The Advancement of Learning* and *The Novum Organon* that referred to Heraclitus and his metaphor of the dry light. Right at the beginning of *The Advancement of Learning*'s dedicatory, Bacon explained that knowledge is *lumen siccum* (dry light) that turns into *lumen madidum* or *maceratum* (wet and soaked light) if "steeped in the humours of the affections". Moreover, at the end of his analysis of the branches of science, he emphasised that rational knowledge could be so dry that it might

[36] *Essays*, 60.

[37] *Essays*, 62. Pythagoras's saying is quoted from Plutarch's *The Education of Children*.

[38] *Essays*, 62.

parch "men's watery and soft nature".[39] In *The Novum Organon*, in the final version of his famous doctrine of the idols of the mind, he stated that "the human understanding is no dry light, but receives an infusion from the will and affections", an infusion which tinges and infects the human mind. It is not possible to dwell here on Bacon's complex doctrine of the natural tendencies of the human mind, which prevent men from achieving a full and accurate understanding of nature, but for this analysis it is worth emphasising that Bacon mentioned Heraclitus when outlining the second group of idols. These are the idols of the individual man, whose mind is like a cave or den, full of prejudices and false beliefs that stem partly from the natural agitation of human understanding, partly from personal inclinations and inner dispositions, and partly from education and conversation with others. In this context, Bacon's allusion to Heraclitus made plain that the idols of the cave cause men to draw into themselves and thus induce them to "look for sciences in their own lesser world" rather than in the "greater and common world".[40] If we compare *Of Friendship* to *The Novum Organon* and *The Advancement of Learning*, it is not difficult to grasp that the honest and frank counsels of a true friend are the best remedy against our deceptive beliefs, fallacious opinions and misleading judgements: in Bacon's view, friendship connects the inner world to reality and brings light into the cave of the human mind.

In order to become the dry light of our mind, friends must dispense advice without flattery. Although in Cicero's *On Friendship* adulation is repeatedly criticised and friends were expected to give advice freely and honestly, Bacon appeared to have drawn his inspiration from Plutarch's *How to Tell a Flatterer from a Friend*.[41] In fact, like Plutarch, he argued that flatterers gain ascendancy over others because people are naturally disposed to love themselves: self-love (*philautia*) clears the way for self-flattery, which in turn leads men to welcome flatterers with open arms. Moreover, like Plutarch, Bacon claimed that frankness, or freedom of speech (*parrhesia*), was the main feature that distinguished friends from flatterers.

Before Bacon, the advice books of humanists like Erasmus and Castiglione accentuated the link between self-flattery and flattery from others in Plutarchan terms, but Bacon's considerations were slightly different. Unlike Erasmus and Castiglione, he did not claim that princes and powerful men develop a false self-image when they are charmed by flatterers, and his emphasis on self-flattery might have instead been connected to the tendency of the human mind to worship its own idols. What is more, Bacon did not limit himself to simply describing the counselors of the prince as *parrhesiastes*, but instead provided a more complete study of their characteristics.[42]

---

[39] *Works*, vol. VI, 95 and 260.

[40] *Works*, vol. VI, 77.

[41] According to Colclough (2005), 40–45, at the request of King Henry VIII, Thomas Elyot made an English version of Erasmus's Latin translation, but Bacon probably referred to the English translation by Philemon Holland (1603).

[42] See *Essays*, 63; Erasmus (1995), 56–57; Castiglione (1967), 261–265. Bacon's indebtedness to Plutarch is examined in Achilleos (2010), which refers to Erasmus and Castiglione's works, but insists on the similarities.

Bacon's description of counselors as friends of the prince was given in a long passage that he placed between his analysis of the first and the second fruits of friendship to explain why it played a key role in political life. His reflections on the political meaning of friendship were at once ambivalent and ambiguous: on the one hand, he argued that princes—whether "weak and passionate" or the "the wisest and the most politiques that ever reigned"—require the free and frank advice of a true friend; while, on the other, he suggested that friends might endanger a prince's greatness and safety. To avoid this risk, he explained which friends are best for a prince to choose. Many scholars have shown that Bacon's analysis of the most suitable friends for a prince reflects the importance of the system of royal patronage in the Elizabethan and Jacobean era. Indeed, despite the fact that patronage dealt with hierarchical relationships based on the exchange of honours and benefits, it required trust and involved personal feelings to such an extent that it was generally conceived as a form of close friendship.[43] This interpretative framework offers important viewpoints from which to gain an understanding of Bacon's idea of friendship, but it fails to recognise that the theory behind his observations on the political meaning of friendship was rooted in the humanistic tradition of the fifteenth-century mirrors for princes. He was primarily the heir to humanists like Giovanni Tinto Vicini of Fabriano, Bartolomeo Platina and Francesco Patrizi of Siena, who devoted entire chapters or long sections of their works to advising their princes on how to choose true friends. In giving their advice, they used the Ciceronian language of *vera amicitia* while placing the notion of it in a new political space—that of the Renaissance courts—in an attempt to indicate the moral and intellectual aristocracy over which the prince should aim to exert and administer his power.[44]

Like Patrizi in the *De regno* (1481–1484), Bacon stressed that friendship required equality, and was aware that subjects cannot be equal to their prince, yet he also argued that equality is the consequence rather than the basis of friendship: the disparity in power and wealth between the prince and his subjects can easily be reduced if the prince decides to "raise some persons to be, as it were, companions and almost equals to themselves".[45] He included servants among those people, echoing the famous passage from the Gospel of St John in which Christ calls his followers friends rather than servants. However, to explain exactly why Bacon tended to treat the words "friends" and "servants" as synonyms, we need to refer to Vicini's *De institutione regiminis dignitatis* (before 1406), in which Cicero's distinction between *vera amicitia* and *vulgaris amicitia* (*Lael.* 21) becomes one between true friends, who serve as counselors to the prince, and ordinary or common friends. Vicini included in the latter category not only domestics and retainers (*servi et domestici*)

[43] LaBreche (2010) offers a critical discussion of recent studies on friendship and patronage in the Jacobean era.

[44] For a close examination of friendship in fifteenth-century mirrors for princes see Ceron (2011), 215–479.

[45] See *Essays*, 60 and Patrizi (1531), 337. As shown by Warren (1950), Patrizi's *De regno* was a crucial source for English mirrors for princes.

but also supporters and followers (*sequaces et clients*),[46] and thereby showed that friendship had been theoretically connected to hierarchical relationships and forms of patronage since the beginning of the fifteenth century. In light of this, we should not be surprised that Bacon's *Of Followers and Friends* explained how great men should choose their followers. Like Vicini, in this essay Bacon focused on followers, who were friends inferior to their patrons, and claimed that "there is little friendship in the world, and least of all between equals".[47] By contrast, in *Of Friendship* he called attention to true friendship between equals, which, although rare, did exist.

In attempting to describe the true friends of the prince, Bacon refused the modern name "favourites or *privados*" on the grounds that a prince should not choose his counselors "as it were matter of grace and conversation", clearly implying the qualities of the perfect courtier outlined by Castiglione. Moreover, he made a critical allusion to the role of favourites in the Jacobean court. From the end of the sixteenth century the semi-institutional practice of the "*privanza*", based entirely on the monarch's favour, had spread from Spain to all of Europe and, as far as England was concerned, George Villiers's rise to power was generally thought to epitomise how the favourite was the king's creature and obtained a monopoly over royal patronage and the royal counsel.[48] Furthermore, since the king sacrificed Bacon to protect Villiers, Eleanora Lupini has argued that Bacon's fall from public office was the main reason why he deplored the role of favourites in the Jacobean court and compared them with members of the Privy Council in the Elizabethan court, not only in *The Essays* but also in other political works.[49] Nonetheless, in *Of Friendship* Bacon juxtaposed the favourites with the *participes curarum*, which, according to Roman historians, was the name given by Tiberius to his favourite Sejanus when he assisted in the rule of Rome.[50] The ancient name also appears in *Of Honour and Reputation*, the essay in which Bacon specified the public offices to which citizens might aspire: the lowest level of honour was that of "gratiosi or favoriti", who were expected to comfort and console the prince, while the highest level was that of *participes curarum*, subjects "upon whom princes discharge the greatest weight of their affairs". When Bacon added that these were the "right hand" of the prince, he made use of the same organicistic metaphor employed by Platina in the *De principe* (1471) to highlight the political function of the friends of the prince.[51]

Though Bacon, like Platina and other fifteenth-century humanists, set out to identify the counselors and friends of the prince, his reflections on the political

[46] Vicini (1977), 59 and 63.

[47] *Essays*, 100–102.

[48] See Elliott and Brockliss (1999), 1–12 and 54–71.

[49] Lupini (2010).

[50] *Essays*, 180; Vickers refers to Xylander's Latin translation of Dio Cassius' work since in Tacitus's *Annales* the term was *socius laborum*.

[51] See *Essays*, 122 and Platina (1979), 69. While Platina drew the metaphor from Dio Chrysostom's orations, in the Plutarchan chapter on the diversity between flatterers and friends Elyot borrowed it from Aristotle's *Politics*: Elyot (1966),110–113.

meaning of friendship were more pessimistic and disenchanted. This side of his analysis clearly resurfaced in the list of historical evidence that concluded the section on the political meaning of friendship. In line with Tacitus, Bacon also mentioned the altar dedicated by the Senate to "the great dearness of friendship" between Tiberius and Sejanus. His decision not to refer to the tragic fall of the Emperor's favourite was more eloquent than words might have been, since the silence led readers to recall that Sejanus was executed after Tiberius had become suspicious of the great power he had given him.[52] Bacon's pessimistic and disenchanted reflections were even more notable in *Of Counsel*. In this essay Bacon defended the function of the counselors of princes without hesitation, going as far as to say that "the wisest princes need not think it any diminution to their greatness, nor a derogation to their sufficiency, to rely upon counsel".[53] Moreover, he restated that the prime condition for the giving of advice was that of allowing the counselors, still described as *parrhesiastes*, to speak freely. Bacon once more revealed "a secret of power", which this time did not come from Tacitus but from the "monstrous fable" of Hesiod's *Theogony*, which he had previously included in the *De sapientia veterum*: when his wife Metis was pregnant, Jupiter ate her—in the metaphor, he represents the king and she the adviser. Jupiter then became pregnant, and the resulting child, Pallas Armed, symbolising "decrees and final directions", sprang from his head.[54]

The fable was ambiguous and ambivalent, even though Bacon insisted on the marriage between "sovereignty" and counsel. On the one hand, it showed that kings cannot avoid appropriating the counsels formed in the womb of their counselors, and, in doing so, they create in the subjects the illusion that they were the real authors of their decisions. On the other hand, it revealed that kings needed the generative and creative power of advisers in order to deliberate and make resolutions. Bacon used the fable to defend the power of the king, while at the same time revealing how it was contingent on his advisers' involvement. Bacon's ambivalences and ambiguities may be excused if we accept that he was not a supporter of the absolute monarchy,[55] but nevertheless the suspicion remains that power—both the power of the king and that of his friends—depended on a subtle and cunning trick. Bacon did not overtly invite either the kings or the counselors to hide the truth but, if anything, he highlighted the therapeutic effect of a form of double simulation: while the king enhanced his power by pretending to be the author of his own decisions, his advisers avoided the fate of Sejanus by pretending to be less creative than they actually were. In chapter XXIII of *The Prince*, Machiavelli claimed that the prince had no other way to avoid flatterers than to allow his counselors to tell the truth, but also pointed out that respect was diminished if the prince offered the right to tell such truths to

[52] A well-known drama by Ben Johnson (*Sejanus. His Fall*, 1603), which has been read as a topical reference to the fall of the former royal favourite Robert Devereux, rendered famous the tragic death of Sejanus. Around twenty years later, George Villiers became the target of the literary criticism of royal favourites: see Keenan (2011).

[53] *Essays*, 46–47.

[54] *Essays*, 47.

[55] On the republican inclinations of Bacon's political thought see Peltonen (1996).

just anyone. Then how is it possible to give frank advice to the prince without reducing his power? Bacon answered the question that Machiavelli had raised by revealing a secret of power based on a trick.[56] Once we understand Bacon's trick, it is hard to see how the prince's counselors could act as *parrhesiastes*, but the subterfuge he suggested might have been a side effect of the therapy required to preserve the common good.

Bacon's description of the third and final benefit of friendship showed that advice from friends can also be useful in business. While making clear that friends help in carrying out plans and in fulfilling desires, he completely redefined the Aristotelian and Ciceronian trope of the friend as another self (*Et. Nich.*, IX, 1966a; *Lael.*, XXI, 80), arguing instead that friends allow men to transcend the temporal and physical limits that nature places on their finite bodies.[57] This sort of corporeal redoubling of the self can be seen as the opposite of the *confusion des volontés*, experienced by Montaigne. As Montaigne himself explained in the *De l'amitié* when describing his perfect and extraordinary friendship with La Boétie, he became a part of La Boétie, who in turn "lost" himself in Montaigne's will: the two men shared one soul in two bodies, so absolute was their union.[58] According to Montaigne, friends reflect each other in order to then find the real image of their self in a shared soul. On the other hand, for Bacon friends helped each other to fulfil their duties and business. Both Montaigne and Bacon extolled the existential meaning of friendship but did so in different ways: whereas Montaigne placed friendship at the highest level of interior life and at the foundation of subjectivity, Bacon praised its therapeutic effect on the human mind and its utility in ordinary life.

Final Remarks: Between Past and Future

Of Friendship offers an intriguing viewpoint from which to evaluate the intricacy of Bacon's *Essays*, since the main currents of thought that he linked together to form his moral and civil advice emerge clearly in this work: he continuously referred to ancient thinkers, not only to Aristotle and Cicero, but also to Heraclitus, Plutarch, and Hesiod; he combined the Stoic tradition of *medicina animi* with the Renaissance tradition of medicine and alchemy; like the Ciceronian humanists, he upheld the superiority of the active over the contemplative life and the commitment to political engagement, but he also exposed the secret and dark side of power in keeping with the new humanism and the lessons of Machiavelli. As in his other essays, in *Of Friendship* the lines of Renaissance thinking that Bacon wove together generated ambivalences and ambiguities that should neither be exaggerated nor understated.

[56] See Machiavelli (2005), 81. Both Holcomb (1995) and Solomon (1998), 103–160 offer a close reading of Bacon's fable and insist on Bacon's trick, but they make no direct references to Machiavelli's *Prince*.

[57] See *Essays*, 64–65.

[58] Montaigne (1965), I: 188–192. On the absolute union of the wills in Montaigne's *De l'amitié* see Starobinski (1982), 52–70.

Rather, they should be accepted and placed within the interpretative framework that sees his *Essays* as part of a therapy based on a disenchanted account of human nature and aimed at relieving the human mind of its perturbations and misleading tendencies.

The connection between Bacon's civil and moral counsels and the doctrine of the idols of the mind is more clearly articulated in *Of Friendship* than anywhere else. When making his claim that the advice of friends illuminates the cave of our mind, he extolled the existential meaning of friendship and its importance in life's private and personal sphere. To understand how friendship came to be seen as a close and intimate relationship, more vital to inner life than to political life, scholars have generally focused on Montaigne's *De l'amitié*.[59] Bacon's legacy to the history of friendship is still to be examined in depth and it is not possible to dwell on a comparison between Bacon's and Montaigne's conception of friendship here. What I would like to emphasise is that Bacon's view of friendship is particularly interesting precisely because it is suspended between the past and the future. Bacon looked to the past since he was the heir to the fifteenth-century humanists who advised the prince on how to choose his friends. However, he revised the humanist perception that the counselors of the prince were always his frank and honest friends, shedding light on a secret aspect of power that had not previously been uncovered. Unlike Montaigne, Bacon chose to emphasise the political meaning of friendship and showed that friendship continued to be considered a political relationship that had a key role in the management of power. Yet at the same time, Bacon placed friendship in the private and intimate sphere of life, believing that the advice of friends not only helps to illuminate the human mind, but also to face up to life. Unlike Montaigne, when extolling the existential meaning of friendship, he did not insist on the extraordinary friendship that leads friends to share one soul in two bodies, but highlighted the importance of ordinary friendship in everyday life. Bacon's ideas were more forward-thinking, since his view of friendship as a civil confession useful in the attempt to live happily and fruitfully in the course of our everyday activities is more similar to our present-day understanding, which sees friendship as a relationship of mutual affection between two (or more) people, who take care of each other and help each other to live their lives.

Being suspended between past and future, Bacon suggests that the political sphere of advice, within which the authors of the fifteenth-century mirrors for princes based their reflections on the friends of princes, had started to morph into a personal and private sphere, where the civil confession between friends was deemed to be the best form of medicine with which to face life. If this is true, we may suppose that friendship will be useful for inner life rather than political life when the connection between the government of the self and the government of the other, which Bacon closely and clearly tied together in *Of Friendship*, is loosened. What is certain is that Bacon's view of friendship can be considered as a case study with which to understand that Renaissance ideas continued to be far-reaching and sometimes inseparable from new, original, and more modern conceptions.

[59] See, for instance, Von Heyking and Avramenko (2008) and Daumas (2011).

Bibliography

Achilleos, Stella. 2010. The Discourses of Friendship and Parrhesia in Francis Bacon's *The Essays or Counsels, Civill and Morall*. In *Friendship in the Middle Ages and Early Modern Age*, ed. Albrecht Classen and Marilyn Sandidge, 543–674. Berlin: De Gruyter.

Agassi, Joseph. 2013. *The Very Idea of Modern Science: Francis Bacon and Robert Boyle*. Dordrecht: Springer.

Bacon, Francis. 1857–1874. *Works*, ed. James Spedding, Robert Ellis, Douglas Heath, London: Longmans. (Abbreviated as *Works*).

Bacon, Francis. 1999. *The Essays or Counsels Civil and Moral*, ed. Brian Vickers. Oxford: Oxford University Press. (Abbreviated as *Essays*).

Box, Ian. 1996. Bacon's Moral Philosophy. In *The Cambridge Companion to Bacon*, ed. Makku Peltonen, 260–282. Cambridge: Cambridge University Press.

Burke, Peter. 1993. *The Art of Conversation*. Cambridge: Polity Press.

Burke, Peter. 1999. Humanism and Friendship in Sixteenth-Century Europe. In *Friendship in Medieval Europe*, ed. Julian Haseldine, 262–274. Stroud: Sutton.

Castiglione, Baldassarre. 1967. *The Book of the Courtier*. Engl. trans. Thomas Hoby. New York: AMS Press.

Ceron, Annalisa. 2011. *L'amicizia civile e gli amici del principe*. Macerata: EUM.

Colclough, David. 2005. *Freedom of Speech in Early Stuart England*. Cambridge: Cambridge University Press.

Corneanu, Sorana. 2011. *Regimens of the Mind: Boyle, Locke and the Cultura animi Tradition*. Chicago: University of Chicago Press.

Crane, Robert. 1923. The Relation of Bacon's *Essays* to His Program for *The Advancement of Learning*. *Shelling Anniversary Papers* 1: 87–105.

Daumas, Maurice. 2011. *Des trésors d'amitié: de la Renaissance aux Lumières*. Paris: Colin.

Dzelzainis, Martin. 2000. Bacon's *Of Simulation and Dissimulation*. In *A Companion to English Renaissance Literature and Culture*, 233–240. Oxford: Blackwell.

Elliott, John H., and Laurence W.B. Brockliss (eds.). 1999. *The World of the Favourite*. New Haven: Yale University Press.

Elyot, Thomas. 1966. *The Book Named the Governor*. London/New York: Dent-Dutton.

Erasmus. 1995. *The Education of a Christian Prince*, ed. Lisa Jardin. Cambridge: Cambridge University Press.

Fish, Stanley. 1971. Georgics of the Mind: The Experience of Bacon's *Essays*. *Critical Quarterly* 13: 45–70.

Giglioni, Guido. 2011. *Francesco Bacone*. Rome: Carocci.

Hadot, Pierre. 2002. *Exercices spirituel et philosophie antique*. Paris: Michel.

Holcomb, Christopher. 1995. Kings and Counselors: The Politics of Francis Bacon's Rhetorical Theory. *Philological Quarterly* 74: 221–227.

Holland, Philemon. 1603. *The Philosophie, Commonlie Called: The Morals Written by the Learned Philosopher Plutarch of Chaeronea*, translated out of Greeke into English and conferred with the Latine translations and the French. London: Arnold Hatfield.

Hyatte, Reginald. 1994. *The Arts of Friendship: The Idealization of Friendship in Medieval and Early Renaissance Literature*. Leiden: Brill.

James, Carolyn, and Bill Kent. 2009. Renaissance Friendships: Traditional Truth, New and Dissenting Voices. In *Friendship: A History*, ed. Barbara Caine, 111–164. London: Equinox.

Jardin, Lisa, and Alan Stewart. 1998. *Hostage to Fortune: The Troubled Life of Francis Bacon*. London: Phoenix.

Keenan, Siobhan. 2011. Staging Roman History, Stuart Politics, and the Duke of Buckingham: The Example of *The Emperor's Favourite*. *Early Theather* 14(2): 63–103.

Kenneth, Alan. 1991. 'Mountaigny Saith Prettily': Bacon's French and the *Essays*. *Publications of the Modern Language Association* 106: 71–78.

LaBreche, Benjamin. 2010. Patronage, Friendship, and Sincerity in Bacon and Spenser. *Studies in English Literature* 50: 83–108.

Langer, Ullrich. 1994. *Perfect Friendship: Studies in Literature and Moral Philosophy from Boccaccio to Corneille*. Genève: Librairie Droz.

Lee, John. 2008. The English Renaissance essays. In *A Companion to English Renaissance Literature and Culture*, ed. Michael Hattaway, 600–609. Oxford: Blackwell.

Lepri, Valentina, and Maria Elena Severini. 2011. *Viaggio e metamorfosi di un testo: i Ricordi di Francesco Guicciardini tra XVI e XVII secolo*. Genève: Droz.

Levy Peck, Linda. 1990. *Court Patronage and Corruption in Early Stuart England*. London: Routledge.

Lupini, Eleanora. 2010. La figura del favorito regio nelle Lettere di consiglio e negli *Essays* di Francesco Bacone. *Historia Constitucional* 11: 339–367.

Machiavelli, Niccolò. 2005. *The Prince*. Engl. trans. Peter Bondanella. Oxford: Oxford University Press.

Melchionda, Mario. 1979. *Gli Essays di Francesco Bacone*. Florence: Olschki.

Montaigne. 1965. *Les Essais*, ed. Pierre Villey. Paris: Presse Universitaire de France.

Nussbaum, Martha. 1994. *The Therapy of Desires*. Princeton: Princeton University Press.

Patrizi, Francesco. 1531. *De regno et regis institutione*. Parisiis. Galeotus a Prato.

Peltonen, Makku. 1992. Politics and Science. Francis Bacon and the True Greatness of the States. *Historical Journal* 25: 279–305.

Peltonen, Makku. 1996. Bacon's Political Philosophy. In *The Cambridge Companion to Bacon*, ed. Makku Peltonen, 283–311. Cambridge: Cambridge University Press.

Platina, Bartolomeo. 1979. *De principe*, ed. Giacomo Ferraù. Palermo: Il Vespro.

Rossi, Paolo. 1957. *Francesco Bacone: dalla magia alla scienza*. Rome/Bari: Laterza.

Skinner, Quentin. 1996. *Reason and Rhetoric in the Philosophy of Hobbes*. Cambridge: Cambridge University Press.

Solomon, Robert J. 1998. *Objectivity in the Making: Francis Bacon and Politics of Inquiry*. Baltimore: John Hopkins University Press.

Starobinski, Jean. 1982. *Montaigne en mouvement*. Paris: Gallimard.

Tuck, Richard. 1993. *Philosophy and Government*. Cambridge: Cambridge University Press.

Vicini, Giovanni Tinto. 1977. *De institutione regiminis dignitatis*, ed. Pasquale Smiraglia. Rome: Edizioni di Storia e Letteratura.

Vickers, Brian. 1984. Bacon's So-called "Utilitarianism". In *Francis Bacon: terminologia e fortune nel XVII secolo*, ed. Marta Fattori, 281–313. Rome: Edizioni dell'Ateneo.

Vickers, Brian. 1996. Bacon and Rhetoric. In *The Cambridge Companion to Bacon*, ed. Makku Peltonen, 200–223. Cambridge: Cambridge University Press.

Voltaire. 2010. *Lettres philosophique*, ed. Olivier Ferret and Anthony McKenna. Paris: Garnier.

Von Heyking, John, and Richard Avramenko. 2008. *Friendship and Politics: Essays in Political Thought*. Notre Dame: University of Notre Dame Press.

Warren, Leslie. 1950. Patrizi's *De regno et regis institutione* and the Plan of Elyot's *The Boke Named the Governor. The Journal of English and Germanic Philology* 49: 67–77.

Wootton, David. 1999. Francis Bacon: Your Flexible Friend. In *The World of the Favourite*, 184–204. New Haven: Yale University Press.

Wyatt, Michael. 2005. *The Italian Encounter with Tudor England*, 179–184. Cambridge: Cambridge University Press.

Zagorin, Perez. 1998. *Francis Bacon*. Princeton: Princeton University Press.

Chapter 13
Thomas Hobbes Against the Aristotelian Account of the Virtues and His Renaissance Source Lorenzo Valla

Gianni Paganini

Abstract This chapter objects to the "ethicist" interpretation of Hobbes's theory of morals, considering whether and how a more historical and contextual approach could confirm or disconfirm this sort of reading Hobbes. In this connection, it will be shown that knowledge of Hobbes's Renaissance sources, first of all Valla, can help us to avoid not only historical but also philosophical misunderstandings, such as dismissing Hobbes's objections to the Aristotelian theory of virtues. For his scientific approach to ethics that excludes the doctrine of *mesótes*, for his stressing the value of pleasure and self-preservation, for his criticism of the classic and Renaissance concept of "glory", Hobbes reveals himself to have been influenced much more by Valla's similar topics than by Aristotle's approach, as Leo Strauss in the past and more recently Boonin-Vail and Ewin thought.

Hobbes as a "Virtue Ethicist"? Some Controversial Interpretations

Over the last sixty years almost every possible attempt has been made to construe Hobbes's moral theory as a consistent philosophical position. Even a quick look at the main interpretations is enough to register how wide is their range and how difficult it is to encompass the entire spectrum of Hobbes's work. There is general consensus on only a couple of theses: that Hobbes had a genuine moral theory and not only a theory of politics or right; and that he was not at all moral objectivist, denying that goodness can inhere in the object themselves. Even starting from there, scholars are split into different factions. Gauthier's position, according to which Hobbes's moral system is nothing more than a system of common or universal prudence, is now on the decline, since the supporters of this claim have withdrawn

G. Paganini (✉)
Dipartimento di Studi Umanistici, Università del Piemonte Orientale (Vercelli), Vercelli, Italy

Centro di ricerca della Accademia dei Lincei, Rome, Italy
e-mail: gianenrico.paganini@uniupo.it

C. Muratori, G. Paganini (eds.), *Early Modern Philosophers and the Renaissance Legacy*, International Archives of the History of Ideas Archives internationales d'histoire des idées 220, DOI 10.1007/978-3-319-32604-7_13

to a milder version. In other words, they see Hobbes as attempting to ground morality in prudence with recourse to an indirect approach, claiming "that an agent best promotes his or her well-being by not always directly seeking to promote it".[1] On this view, in the long run an agent of this kind would fare better than people immediately engaged with their current self-interest.[2]

However, even this minimal trace of morality still remains open to contrasting interpretations. It makes a big difference whether you read Hobbes as a moral contractarian, for whom nothing is by nature good or bad, but some actions are rendered so by a human and social convention; or if you understand him to be a rule egoist, or a consequentialist mainly interested in the effects of adopting some general rule, rather than in the effects of a single action.[3] On the other side, there were serious attempts to disentangle Hobbesian morality from his allegedly egoistic psychology and to reconnect it to some sort of divine-command theory. This interpretation, inaugurated by A. E. Taylor and renewed by Hood and more recently by A. P. Martinich, is highly controversial, conflicting with Hobbes's repeated claim that ethics or "moral science" actually belongs to the knowledge of human passions and not to a branch of religion or theology, which for Hobbes are by no means sciences: he states clearly in *De corpore* that they are not parts of philosophy, since they depend not on reason but on legitimate authority. They are basically matters of politics and decision.[4]

Another interpretation that aims at preserving the autonomy of morals without falling into a pure deontology, be that theological or not, has been recently developed by the supporters of virtue ethics. They claim that Hobbes would have been interested in actions as well, and even more, in an agent's character. According to Boonin-Vail, who is one of the main advocates of this thesis, "there is an abundance of textual evidence to support the claim that Hobbes's moral philosophy is ultimately concerned with what makes just people, rather than with what makes just actions".[5] This particular reading also has the ambition of reconciling Taylor's deontologist approach with Kavka's consequentialist one. In other words, from the virtue ethicist point of view, Taylor would be right in stressing that the moral agent ought to consider respect and love for justice as a rule, yet Kavka in his turn has fair

---

[1] Boonin-Vail (1994), 63.

[2] It is obvious that this "indirect" approach to the concern for prudence and self-preservation paves the way to the rehabilitation of morality as a regulation separated from the strictly prudential calculation of the agent's immediate interests. These two interpretations (the prudential one and the rule egoism) largely overlap and they seem best to fit the features of Hobbes's anthropology, considering that the moral psychology of man as self-interested was the basis of his normative ethics.

[3] For the interpretation focused on the effect of a general rule cf. Kavka (1986). This approach was intended as the most convincing attempt to reconcile the demands of the customary moral theory with the prevailing requirement of prudential and instrumentalist rationality.

[4] Cf. Hobbes (1999), 281–283 (*De Corpore*, XXVI, 1). For the transition from a previous and broader definition of philosophy, stated in *De motu, loco et tempore*, to another narrower view that excludes theology, cf. Paganini (2013a, b) and Paganini (2015).

[5] Boonin-Vail (1994), 109.

reason to claim that the agent is required to emulate the virtuous character because doing so is good for him or her. According to the ethicist reading, one of Kavka's main mistakes would consist in making rules of conduct, and not character or virtuous conduct, the ultimate object of moral evaluation.[6]

In this chapter I shall concentrate on this "ethicist" interpretation of Hobbes's theory of morals, considering whether and how a more historical and contextual approach could confirm or disconfirm this sort of reading Hobbes as a virtue ethicist. In this connection, I am going to show how knowledge of Hobbes's Renaissance sources, first of all Valla, can help us to avoid not only historical but also philosophical misunderstandings, such as dismissing Hobbes's objections to the Aristotelian theory of virtues, or considering them as insignificant for the main thesis, as Boonin-Vail and his followers tend to do.

In fact, these Hobbesian virtue ethicists (such as Boonin-Vail and Ewin[7] who preceded him in almost the same path) consider historical research as consisting of purely textual analysis, with little regard for the broad context. As to the sources of Hobbes, they usually go straight back to Aristotle, establishing some sort of direct comparison between the two thinkers, separated by nearly twenty centuries. Furthermore, their standpoint is explicitly influenced by modern twentieth-century theory, which since the pioneering works of Elizabeth Anscombe and Philippa Foot, has posited a great divide between ancient and modern ethics: the latter supposedly characterized by a focus on actions and rules, whereas the ancients stressed the issue of the agent and his character.

Even at a glance, Hobbes's case seems to be enough to disconfirm this view, not least because the English philosopher underlined both the relevance of rules in acting and the agent's intentions. Laws of nature, which are the same as laws of morals, are meant to operate *in foro interno*, involving not directly the action, but just the intention or disposition to act according to the laws, if a condition of reciprocity is afforded by the other partners. This is the reason why the flourishing of these intentions into actual "manners", which in their turn are called "virtues" or "vices" when they are good or bad respectively, is made possible only by civil society that strengthens and protects these "habits".[8] This to say that Hobbes was not hostile to the topic of virtues, as conventional wisdom would claim. On the other side, that of the ancients, even Aristotle who was supposedly the supporter of virtues against rule-based morality, was in reality no less careful about the quality of actions and their conformity to normative patterns in the form of rules.

---

[6] Ibid., 113. By contrast, it is not a mistake (for Booning-Vail) to view "beneficial consequences as Hobbes's agent's criterion for doing the evaluation" (ibid.). Another different way of disentangling Hobbes's morals from the "standard philosophical interpretation" (based on the topics of self-interest and self-preservation) is proposed by Lloyd (1992), esp. 6–47. Although original and worthy of a careful consideration, this new interpretation is not directly connected with the issue of virtue; therefore it remains out of the scope of this chapter.

[7] Ewin (1991), esp. chapter 7, 163–194.

[8] OL II, 116 (*De Homine*, XIII, 8).

A more appropriate view should focus on another watershed between the ancients and the moderns. The latter, being heir of Christian monotheism, base duties and obligation on the binding strength of the lawgiver, starting from the supreme one, whereas ancient ethics, and above all the ethics of Aristotle, were in a sense self-contained, not needing an external authority in order to be binding. Hobbes could make exception in this case too, however, being a thinker who faced in two directions, inheriting from both traditions. On the one side, he saw the law of nature as the simple dictate of reason, even though he remarked that from this point of view the word "law" is just metaphorical, as there is no authority in the natural state to enforce this dictate as a law. This is the reason why he tried to strengthen the dictate of reason by declaring that it is tantamount to the natural word of God: in this connection, reason is not just reasoning or counsel, but also "law" in the proper sense of this word, deriving from the authority of the supreme lawgiver.[9]

Hobbes's "Science of Good and Evil" Against Aristotelian Ethics

It is true that the historical scholarship on Hobbes inspired by the topic of "virtue ethics" is sometimes more subtle than it appears, and recognizes, for example, the need to integrate diverse factors, such as the rule of law in addition to habits and characters. Nevertheless, this incorporation is always made with the Aristotelian theory still dominant, and in this direction Boonin-Vail and Ewin go much further than Leo Strauss who had already recognized the importance of the early Hobbes's studies on *Rhetoric* about the nature and list of the human passions.[10] Boonin-Vail portrays Hobbes as very close to Aristotle, and much closer than one could expect when reading his strong polemics against "Aristotelity". This proximity, for Boonin-Vail, regards not only the descriptive materials drawn from Aristotle, as Strauss thought, but also and foremost the very structure of ethics, which Strauss did not believe.

At first glance, the thesis is highly controversial, if only because it requires diminution of the main contrasts between Hobbes and Aristotle. Thus, the scorn that Hobbes pours on the whole of Aristotle's philosophy, including natural philosophy, metaphysics, ethics and politics,[11] is attributed by Boonin-Vail partly to a kind of rivalry, since Hobbes seemed to have the ambition of replacing in the University the teaching of Aristotelianism by his own philosophy, and more seriously because the English philosopher "associated Aristotle's normative views with the doctrine of

[9] For a balanced evaluation of these topics in Hobbes and their current interpretations, see Curley (1998).

[10] See Strauss (1952). For a more recent study of this topic, stressing the importance of Descartes's moral theory for Hobbes, see Pacchi (1987).

[11] See Hobbes (1994), 456–457, among the many passages that attack the Aristotelian philosophy and more generally "the School of the Grecians" (*Leviathan* XLVI, 11).

natural inequality".[12] The second possible explanation of this contempt would regard rather the degenerated forms of Aristotelianism than the original: 'Hobbes's attacks on Aristotle are often attacks on the Church's manipulative uses of Aristotle rather than the doctrines of Aristotle himself".[13]

However suggestive they may be, these partial justifications cannot efface the fact that the main passage where Hobbes criticizes Aristotelian virtue ethics is not associated either with the mistake of inequality or the decay of scholasticism; rather the critique targets Aristotle himself, contrary to what Boonin-Vail claims.

The objections of Hobbes to Aristotle's theory of virtue are well known; yet they still deserve at least one full quotation, because the motivations of Hobbes's critique are often neglected by the supporters of virtue ethics. Hobbes's attack concerns the very heart of the Aristotelian doctrine, which claims that the virtuous good is placed in a "mediocrity" (moderation) of passions, and that vices are placed at the opposite extremes of a spectrum, in which virtue represents the middle point. Already in his first great work, *Elements of Law*, Hobbes criticizes the "mediocrity"-thesis not only for its inability to describe the true nature of virtues, but also and above all due to the inability showed by its supporters to find the real reason for which one thing is called virtue and another vice. The text of *Elements of Law* reads as follows:

> As for the common opinion, that virtue consisteth in mediocrity, and vice in extremes, I see no ground for it, nor can find any such mediocrity. Courage may be virtue, when the daring is extreme, if the cause be good; and extreme fear no vice when the danger is extreme. To give a man more than his due, is no injustice, though it be to give him less; and in gifts it is not the sum that maketh liberality, but the reason. And so in all other virtues and vices. I know that this doctrine of mediocrity is Aristotle's, but his opinions concerning virtue and vice, are no other than those which were received then, and are still by the generality of men unstudied; and therefore not very likely to be accurate.[14]

This criticism, which is constantly repeated in all Hobbes's subsequent works, does not simply derive from a misunderstanding of Aristotle, as the virtue ethicists would claim; neither is the consequence of an infelicitous emphasis placed by Hobbes on a secondary point of the doctrine, as Boonin-Vail thinks.[15] Far from

---

[12] Boonin-Vail (1994), 180. For a deeper examination of Hobbes's objections to the Aristotelian philosophy, see Wolfers (1991). This book is focused on metaphysics (chapters 1–2) and political theory (chapter 3); neither ethics nor the theory of virtues are considered as such by this author.

[13] Boonin-Vail (1994), 181.

[14] Hobbes (1889), 94 (*Elements of Law* I, xvii, 14). The criticism of the mediocrity thesis can be found also in *Leviathan* XV, 40. See Hobbes (1994), 100: "Now the Science of Virtue and Vice, is Morall Philosophie; and therefore the true Doctrine of the Laws of Nature, is the true Morall Philosophie. But the Writers of Morall Philosophie, though they acknowledge the same Vertues and Vices; Yet not seeing wherein consisted their Goodnesse; nor that they come to be praised as the means of peaceable, sociable, and comfortable living ; place them in a mediocrity of passions : as if not the Cause, but the Degree of daring, made Fortitude; or not the Cause, but the Quantity of a gift, made Liberality". And again in Hobbes (1990), 44 (*Behemoth*) "It is not the Much or Little that makes an action virtuous, but the cause; nor Much or Little that makes an action vicious, but its being unconformable to the laws in such men as are subject to the law, or its being unconformable to equity or charity in all men whatsoever".

[15] Boonin-Vail (1994), 182, 88.

being a rhetorical device, Hobbes's attack is the preliminary move in a clever strategy aimed at framing morals into a "science of the good and evil". This implies a redefinition of virtue that is closely connected to Hobbes's effort to avoid the apparent subjectivism involved by the other notion he often puts forward, according to which the good is something that always is *bonum sibi*, that is in relation to the agent. If Hobbes had simply followed this path, it would have been impossible to him to get over the subjective and current preferences of the agents, which means that the foundation of a "science" of morals would have been simply impossible to discover.

However, in the fundamental chapter of the *Elements of Law* (I, XVII, 14) where the laws of nature are listed, Hobbes takes another path, which would enable him to grasp good and evil as they are "in reason", and not "in passion" as by contrast men usually judge. For everyone, this different criterion is "the whole way to his preservation", and accordingly the fulfilling of these laws "that tend to our preservation" is called "good in reason", and the contrary bad. From "actions" this objective moral qualification extends to "habits", then to "dispositions" or "intentions", and finally to "virtues" in which all these good forms of behaviour result.[16] The virtues listed by Hobbes are far from being original: they include justice, equity, gratitude, temperance, and prudence, largely overlapping with the list given in Aristotle's *Nicomachean Ethics*. By contrast, the "reason", which is also the "cause", of these virtues is peculiar to Hobbes's system, in that it refers to the criterion of self-preservation. Also original is the declared ambition of transforming ethics into a "science of justice", as the Preface proclaims. Both features, the criterion and the scientific program, mark a sharp contrast with regard to Aristotle: self-preservation does not play any significant function in the latter's ethics, which notoriously stops at the level of prudence, *phronesis*, and does not engage in elaborating any moral *episteme*. By contrast, this epistemic claim is constantly reaffirmed by Hobbes, even where – as in *Leviathan* – the polemical reference to Aristotle is superseded by a vague mention of the "Writers of Morall Philosophie" who are blamed for their misunderstanding of the true nature and "cause" of virtues and vices.[17]

As is well known, one of the main ambitions of Hobbes is to establish "the science of what is *good* and *evil*" and, to this end, to find out some sort of scientific test for fixing the moral categories. He certainly wanted to be considered as the inventor of "the science of Virtue and Vice". Moreover, he was convinced that this science would be the core of "moral philosophy"[18]; and this latter in turn largely overlaps with the laws of nature, even though there are some laws that regard private life only and not social relations. From this point of view, Hobbes reaffirms in *Leviathan* what he had already said before in the *Elements* and *De cive*: that the Aristotelian

---

[16] Hobbes (1889), 94. It is notable that the beginning of the section is extremely 'subjectivist': "Every man by natural passion, calleth that good which pleaseth him for the present…", according to the general definition of good in Hobbes. Cf. Hobbes (1889), 29 (*Elements of Law* I, VII, 3): "Every man, for his own part, calleth that which pleaseth, and is delightful to himself, GOOD."

[17] Hobbes (1994), 100 (*Leviathan* XV, 40).

[18] Ibid.

definition of virtue as consisting "in a mediocrity of passions" cannot afford the scientific test for moral categories he is looking for. In order to be "scientific", the theory of virtues must indicate the cause of them, according to Hobbes's general definition of science, including true philosophy, which consists in finding out the possible causes of the observed effects, or vice versa.

We might follow Quentin Skinner in his perfectly clear analysis of Hobbes's "science of virtue and vice".[19] The main points of this doctrine are basically three: (1) laws of nature prescribe what natural reason dictates for preserving our lives; (2) maintaining peace is the first means to this end. Therefore, social virtues like modesty, equity, trust, humanity, and especially keeping contracts and covenants recommend actions aimed at peace, which is the first and foremost condition of anyone's self-preservation. (3) The "cause" by which we call these behaviours "virtues" is definitely that they conduce to peace and thus ensure the preservation of our lives.

In conclusion, Hobbes's attack on the theory of virtue as "*mediocrity* between two extremes, while the vices are to be found in the extremes *themselves*",[20] is too central and constant in the whole of his work to be dismissed as a minor and secondary point of his moral philosophy. As *Leviathan* clearly claims, because they followed Aristotle, "the writers of moral philosophy" made a mistake that was not insignificant. Looking at the "degree" of the passions or at the "quantity" of the objects (for example, in the case of "gifts" and "liberality"), in order to fix the "middle point" between the two extremes, they missed the true "cause" of virtues, "not seeing wherein consisted their goodness". Even though their list of virtues is nearly the same as the one also presented by Hobbes, those "writers" who are criticised in *Leviathan* placed them in "a mediocrity of passions", because they did not realise that virtues are in reality "the means of peaceable, sociable, and comfortable living".[21]

These critical observations made by Hobbes require that we have to address another issue: is that criticism original or does it rely on some antecedents?

Grotius and the Renaissance Rhetorical Background

So far, research on Hobbes's sources about this specific topic has pointed in two different directions: Grotius, on one hand and classical and Renaissance rhetoric on the other. We shall see later that at least another third direction is much more relevant and also promising for understanding the context and the meaning of the Hobbesian theory.

---

[19] I am here summarizing Skinner (1996), 316–326, esp. 320ff.

[20] Hobbes (1983), 120 (*De cive* III, xxxii). For many parallel passages both in *Elements* and *De cive*, see Skinner (1996), 316–326.

[21] Hobbes (1994), 100 (*Leviathan* XV, 40).

Besides the attack on the mediocrity thesis, Grotius's *Prolegomena* have some points in common with Hobbes, as R. Tuck has remarked.[22] First of all, there is Grotius's appeal to "libertas philosophandi", which is at the same time an appeal to "the liberty of the old Christians" against "Aristotle's tyranny".[23] This criticism goes in the same direction as Hobbes's polemics against "Aristotelity", which is blamed for having perverted both philosophy and religion. The manipulative use of Aristotle made by the Church was aimed at establishing absolute power over consciences by promoting "absurdities" to the rank of dogmas, as the fourth part of *Leviathan* insists. The second point is more specific: discussing the cases of liberality and meanness, veracity and justice, Grotius already demonstrated that these virtues do not consist in any graduation of affects, between too much and too little, but in an objective criterion.[24]

However, despite these strong affinities, Grotius misses the key point that Hobbes's criticism dwells upon. Even though self-preservation plays a significant role in *De jure belli ac pacis*,[25] Grotius does not invoke it in the context of the "cause" of virtues, which he treats as more formal than substantial. Even for the foremost and most social virtue, justice, Grotius considers that its motive or cause does not lie in being functional to preservation or peace, as in Hobbes, but in a more traditional and formal criterion: "in forbearing from someone else's good" ("quae tota in alieni abstinentia posita est").[26]

The second possible source was pointed out by Q. Skinner, who showed that Hobbes's criticism of the mediocrity thesis has to be understood against the background of the oratorical technique of "rhetoric redescription": if virtue is the middle of a spectrum going from one vice to the opposite, orators, philosophers, politicians, and theologians might exploit the proximity between virtue and its opposites, transforming virtue into a vice and vice versa, and following for this purpose not reason, but passions and whatever interests were at stake.[27] The awareness of the dangers related to this sophistical and demagogic use of the Aristotelian theory by politicians and rhetoricians certainly had a great impact on Hobbes's depiction of the ideological manipulation afforded by the theory of virtue as a middle point between two extremes. However and without diminishing the importance of this connection with the reflection on the power of eloquence, it seems that it is still worth looking for Hobbes's possible sources on the side of technical philosophy, and specifically in the field of the theory of virtues.

---

[22] See Tuck (1993), 200, 220, 279, 296, 303–304, 306, 311, 347.

[23] Grotius (1712), xxiii–xxiv ("Prolegomena" § 42).

[24] Ibid., xxiv–xxix (§§ 43–45).

[25] See Tuck (1993), 173–177, 199–200.

[26] See Grotius (1712), iv–xiii ("Prolegomena" §§ 5–22) against any utilitarian foundation of human sociability, which Grotius identifies with political skepticism. For Grotius's general theory of "right" ("jus") and justice ("iustitia") see Grotius (1712), 1–24 (*De Jure*, I, i).

[27] This is the general thesis of Skinner (1996): see esp. chapter 10 ("Hobbes's practice of Rhetoric"), 376–425. On the mediocrity thesis in the classical tradition and Hobbes's criticism see Skinner (1996), 153–8, 323–6.

The origin of Hobbes's attack on the notion of virtue as "mediocrity", I will now argue, lies in the Renaissance neo-Epicurean tradition as it was represented by one of its first and most notable supporter in the modern age: Lorenzo Valla. I have already pointed out the relevance of Valla's humanism for some central topics of Hobbes's thought: its influence is easily recognizable in Hobbes's criticism of separated essences and in his discovery of their linguistic origins (the reference is to *Dialectica*),[28] in the humanistic reading of *persona* that can be found in *Leviathan* and in the theological writings referring to the problem of the Trinity (Valla's main authoritative texts were in this case *Elegantiae* and again *Dialectica*),[29] and finally with regard to the topic of anticlericalism, concerning the temporal power of the papacy, along the lines traced in *De falso credita et ementita Constantini donatione*.[30] It is extremely probable that this, Valla's most famous work, inspired Hobbes's depiction of the "Kingdom of Darkness", as well as some of the harshly satirical parts of *Historia ecclesiastica*. For this dismantlement of the Aristotelian theory of virtue another work of Valla was decisive: *De voluptate*, one of the great classics, perhaps still the most famous in the seventeenth century, in the field of humanistic moral philosophy.

Thomas Hobbes and Lorenzo Valla's Neo-Epicureanism

I will first summarize Valla's position on the main issue (the attack on the "mediocrity" thesis), which has a central place in *De voluptate* and represents at the same time an original application of the Epicurean doctrine to a topic that ancient Epicureanism did not discuss at all.[31] In the third book, after Antonius, speaking as almost a "reborn Epicurus" has introduced the main tenets of the Epicurean moral doctrine, Nicolaus[32] (in the subsequent version, probably published in Milan as

---

[28] See Paganini (2007).

[29] See Paganini (2003).

[30] This aspect is treated in the context of Hobbes's connections with the heritage of Valla in Paganini (1999). Some aspects of this long tradition starting from ancient Epicureanism and reaching seventeenth-century philosophy, through Valla, Gassendi and Hobbes, are considered in Sarasohn (1996): on Hobbes esp. 118–141. The influence of the Epicurean tradition is underlined in Springborg 2003; and Springborg 2004, which emphasized politics. A broader spectrum of Epicurean and neo-Epicurean topics is treated in her commentary on Hobbes's *Historia Ecclesiastica*. See Hobbes (2008).

[31] The best and most up-to-date treatment of Valla's moral philosophy can now be found in Ebbersmeyer (2010), 222–255; cf. also Nauta (2009), 152–190. On the specific topic of Valla's criticism to the Aristotelian doctrine of virtue as mediocrity, see Laffranchi (1999), 271–277.

[32] I am quoting *De voluptate* from the edition of Valla's *Opera Omnia* (Valla 1540), that had wide diffusion in the early modern period. The text contained in this edition is in fact an expansion of the first draft (*De voluptate* 1431, printed in the Paris edition Josse Bade 1512), and indeed the setting and the characters of the dialogue are the same. In the second version, written in Pavia between 1431 and 1433, Valla changed the title (*De vero falsoque bono*), the setting (Pavia), and the characters, while also expanding and reworking the text.

De vero falsoque bono, Antonius's role was taken by Antonius from Rho) arrives in order to settle the controversy between Antonius and the Stoic Leonardus (in the next version, respectively Vegius and Cato). Nicolaus explains that the Aristotelian mediocrity theory has many problematic points: it involves the supposition that vices are at least twice as numerous than virtues, which is not a sure thing; and it fails from a logical and epistemological standpoint, because virtue can be set in the middle only at the cost of elaborate classifications that contrast with basic moral intuitions.[33] However, according to Valla, Aristotle's worst mistake was conceptual rather than descriptive, because the Stagirite, in order to maintain the thesis of mediocrity, was obliged to put the opposites into one single nomenclature, mixing up different things into the same species.[34] The examples of this mistaken view represented by Nicolaus are very close to those that will later be suggested by Hobbes.[35]

The affinities between Hobbes and Valla go much further in the rejection of the mediocrity thesis and extend to its illustration. These affinities regard three other main points and constitute a whole web of references to neo-Epicurean humanism: (a) a scientific approach to ethics; (b) the grounds of moral psychology; (c) the topic of self-preservation.

Let's look at these aspects separately.

(a) *A scientific approach to ethics*. The very point of any ethical evaluation, according both to Valla and Hobbes, is the "reason", and not the "middle" position, occupied by a virtue or behaviour. Both Valla and Hobbes touch on the examples of justice, courage and liberality, so as to show that in all these cases it is not the deal, but the "cause" that makes certain forms of behaviour virtuous or vicious. The point, in Valla as well as in Hobbes, lies in the "reasons", as the text of *Elements* clearly states: "Courage may be virtue, when the daring is extreme, if the cause is good; and extreme fear no vice when the danger is extreme".[36] It is noteworthy that Valla, too, is well aware of the rhetorically harmful possibilities opened up by what he calls "virtutum vitiorumque vicinitas".[37] However, he thinks that the possible mistakes involved by this narrowness can be avoided first by discarding ambiguous nomenclature, and secondly by drawing attention to single actions, one by one, apart from any artificial classificatory schemes. It must be emphasized that this position is not tantamount to a sort of nominalism

---

[33] Cf. Valla (1540), 966: "Sed mihi evidentissima ratione videtur probari posse singulas virtutes singulis vitiis adversa fronte consistere, falsoque illud dici hinc et hinc excessum ac defectum, in medio esse virtutem, quae mediocritas quaedam dicitur inter nimium et parum, incassumque disputari utrum duorum contrariorum sit magis contrarium".

[34] Ibid., 967.

[35] In Hobbes (1889), 94 (Elements of Law I, xvii, 14), the example of "courage" is found at the end of a review of the main virtues (justice, equity, gratitude, temperance, prudence) and underpins a strong attack on the Aristotelian position: "As for the common opinion, that virtue consisteth in mediocrity, and vices in extremes, I see no ground for it, nor can find any such mediocrity" (ibid.). Cf. Valla's anti-Aristotelian treatment of the virtue of courage in Valla (1540), 966.

[36] Hobbes (1889), 94.

[37] Cf. Valla (1540), 971.

or ethical empiricism, as it might appear at first glance. On the contrary, and despite all the emphasis put on the actual features of every single action, Valla does not stop at the level of pure description, since he thinks that what is decisive for imputing the name of virtue to same behaviour is looking for its "science and cause" ("scientia vel causa"): "Quod eo audacius dixerim, quod virtutes et vitia in ea ratione dinoscuntur, quia sunt in infimo, aut in medio, aut in summo. Non enim gradu ponderamus haec, aut modo, sed scientia vel causa".[38] And one page later Valla stresses the same concept, against the Aristotelian mediocrity thesis: "Nec earum [i.e. the passions] parvitate, mediocritate, magnitudine, et ut dixi, mensura, sed ratione ac scientia".[39] It is not quantity (excess or paucity) that makes virtue or vice, but according to the motive that we call an action virtuous or vicious: "In summa, omnia et recte fieri possunt et prave".[40]

As we have noted, Valla also speaks of "ratione ac scientia" to tell the difference between a positive and a negative behaviour. Even if these notions ("reason", "science", "cause") are not yet fully developed in a real "science of virtue", as will later be the case in Hobbes, these strong requirements invoked by Valla bestow on the moral discourse, at least in principle, a claim to scientificity that is to be found neither in the Aristotelian doctrine of ethics, nor in Grotius's *Prolegomena*, and even less in rhetorical treatments or in oratorical "disputationes in utramque partem". We can say that Valla and Hobbes are among the very few Renaissance and early modern philosophers who connect criticism of the mediocrity thesis to a strong demand for "reason" and "science" in the discourse of moral theory. Hobbes's "science of Virtues and Vices" or science "of Good and Evil" is to be seen as the direct heir to Valla's attack on the Aristotelian doctrine of virtue.

(b) *The grounds of moral psychology*. The affinity does not regard just this formal requirement of scientificity, since the similarity touches on the fundamental grounds of moral psychology as they are developed in both authors. Despite the refinements of Hobbes's thought, it remains true that for him as for the Epicurean tradition all moral notions can be ultimately translated into terms of desiring and fleeing, appetite and aversion, and that these impulses result in seeking pleasure and avoiding pain. Neither Valla nor Hobbes start with virtue and law; their commencement is with something more fundamental, which is the ideal of living well, avoiding pain and seeking pleasurable states of mind.

That is not to say that there are no dissimilarities between the two authors. In fact, Valla maintains the traditional notion of *summum bonum*, even if he considers it to be about pleasure and happiness (*voluptas*, *beatitudo*), not virtue. Furthermore, he tries to reconcile Epicureanism and Christianity, presenting in the last book the

[38] Ibid., 969.

[39] Ibid., 970.

[40] Ibid.

supreme good as a sort of heavenly beatitude.[41] Hobbes, on the contrary, subverts this tradition, abolishing the idea of "summum bonum" and replacing it by the idea of "supreme evil" ("summum malum"), that is destruction of the subject. According to him, the supreme good, if it ever existed, would be something like the end of desire, which is also the end of human life. There could be nothing worse to flee.

Despite these contrasts, however, the affinities between Valla and Hobbes are more significant, especially if we consider the first book of *De voluptate*. Here, Valla develops (through Antonius's interventions) a purely Epicurean view, which is not yet compromised with Christianity as in the third book. I said purely Epicurean, but it would be better to say neo-Epicurean, because Antonius's discourse "for the Epicureans and nature, and against the Stoics" ("Antonius pro Epicureis et pro natura contra Stoicos") is quite original with regard to the ancient sources. Far from repeating Epicurean orthodoxy, Antonius[42] leaves in the background the value of *ataraxia*, *tranquillitas animi*,[43] stressing instead active fruition of *bona externa* and *bona corporis*, and focusing on pleasures both of the mind and body.[44] He claims not to have slavishly followed the texts, but to have proceeded by looking at "nature itself that teaches and leads" ("ipsa natura magistra et duce").[45]

(c) *The topic of self-preservation*. In this rejuvenation of an old tradition, Hobbes's most striking novelty is the emphasis put on the topic of self-preservation, considered not just as a factual necessity, but as the cornerstone of moral life. The topic of self-preservation and the consequent fear of death did not play any significant role in Epicurean morals; on the contrary, the Epicurean wise man is supposed to be freed from the fear of death. "Death is nothing to us", claims the perfect disciple of the *Kepos*.[46]

By contrast, the neo-Epicurean Antonius establishes a close connection between pleasure (*voluptas*), self-reference and self-preservation. Pleasure (*voluptas*), and not *honestas*, rises to the rank of good *secundum naturam*, because it "maintains life" and promotes one's own good. Needless to say, whereas the former statement

---

[41] See Ebbersmeyer (2010), 238–250. Obviously, Valla's general ontology and consequently his theology are quite different from those of Hobbes, at least because the former splits the general category of *res* or substance into two species (corporeal substance and the incorporeal one, like God, *spiritus*, and angels), whereas for the latter there is in principle no real substance but *corpus* (body). Moreover, the later Hobbes developed an explicitly materialistic theology, while Valla's reform of the tree of Porphyry immediately reveals its religious motivations, setting God and spiritual beings apart from corporeal substances. On this point, see Nauta (2009), 27–29. Specifically on Valla's theology, cf. Monfasani (2000). The companion where this chapter is included contains another chapter, by M. W. F. Stone, which has been denounced as an extensive plagiarism, along with many others (at least 40!) of the same author. See Dougherty et al. (2009). This piece has not, at present, been retracted by the editor and publisher.

[42] Cf. Valla (1540), 905ff.: "Antonius pro Epicureis, et pro natura contra Stoicos".

[43] Valla has recourse to the *Leitmotiv* of "tranquillitas mentis" as an instrumental function of "voluptas". Cf. Valla (1540), 956.

[44] Cf. ibid., 913ff.

[45] Ibid., 926.

[46] On this crucial difference between ancient Epicureanism and modern neo-Epicureanism, see Paganini (2004).

(pleasure as the goal of life) is truly Epicurean, the latter (the focus on maintenance of life) is by no means originally Epicurean: for an ancient Epicurean it is not the conservation of life in itself that constitutes a value, but rather its quality in terms of pleasure and above all in accordance with the supreme pleasure which is tranquillity of mind. In Antonius's opinion, on the other hand, the main function of 'voluptas" is to "preserve life":

> Cuius rei ut fundamentum iaciam, nihil est generi animantium a natura tributum, quam ut se, vitam corpusque tueatur, declinetque ea quae nocitura videantur. Nunc autem quid magis vitam conservat quam voluptas ? ut in gustu, visu, auditu, odoratu, tactu, sine quibus vivere non possumus, sine honestate possumus. Ita si quis in se quod natura praescribit, violare audeat, contra suam utilitatem fecerit, quandoquidem omnia facienda sunt cuique sua causa, nihil aliorum.[47]

This focus on self-preservation is one of the major novelties put forward by the modern neo-Epicurean discourse[48] and, as is well known, Hobbes is the philosopher of self-preservation *par excellence*.

Valla, Hobbes, and the Politics of the Moderns

What is even more important for our topic is that Valla's updating of the Epicurean theory in the wake of strong and modern realism is supplemented by a similar treatment of civil and political philosophy.[49] Values such as honour, glory, and patriotism are under attack in *De voluptate*, from the point of view of pleasure understood as self-interest. Selfish interests are opposed to the so-called public virtues.[50] All this has in Antonius's mouth an evident anti-Stoic flavour, yet, once again, Valla goes farther than his ancient sources, developing topics typical of modern society that will be at the heart of Hobbes's thought too. In his criticism of glory[51] Antonius demolishes it as a kind of vanity in which an individual attempts to affirm superiority over others:

> Omnis gloriae cupido ex vanitate, tumore atque ambitione descendit. Quae quid aliud est nisi velle, vel se inter alios eminentem, vel alios minores cernere, quod quasi discordiarum, odiorum, invidiarum seminarium est. Communitas autem et paritas inter homines parens est pacis et benevolentiae.[52]

---

[47] Valla (1540), 926.

[48] On the specificity of this topic for defining modern neo-Epicurean political discourse, see Mulsow/Schmitz (2004); Paganini (2004); Paganini (2015) and (2016).

[49] For a more extensive comparison between Valla's and Hobbes's ethics and politics, cf. Paganini (1999), esp. 30–53.

[50] Cf. Valla (1540), 929–934. Antonius opposes the Stoic principle of good of the collectivity ("melius est bonum plurimorum quam unius") to the more realistic rule: "Maius bonum est vita mea quam universorum hominum in ipso". See Valla (1540), 929.

[51] For the detailed critique of the value of glory see *De voluptate*. Cf. chapter xv "Gloriam nihil ad mortuos pertinere"; chapter xvi "Gloriam non esse cum voluptate comparandam"; chapter xx "Numquam Stoicos potuisse consequi gloriam." Valla (1540), 935–937, 937–938, 939–940.

[52] Ibid., 935, from chapter xiv (book 2), bearing the title: "Qui propter gloriam aliquid facit, recedit ab honesto".

Peace and benevolence, on the contrary, rest on acknowledging equality among partners ("communitas et paritas inter homines"). In this play of crossed oppositions between glory and equality, vanity and self-preservation, one could easily recognize the same basic notions and relations that are at the basis of Hobbes's political philosophy. Take into account that in Valla's dialogue all this discourse against the value of glory connects to the above-mentioned topics of the fear of death and the striving for self-preservation, and you will have the same basic moral antithesis that Leo Strauss had described with regard to Hobbes: the moral antithesis between vanity, as the root of the natural appetite, on one hand, and on the other fear of violent death that induces man to reasoning, dismissing vanity and admitting human equality.[53]

Just as it remains rather problematic and at the same time absolutely essential, according to Hobbes, to distinguish between true glory and vainglory, between actual and merely imagined superiority (indeed even the cleverest and strongest is still exposed to the offense that might be perpetrated by the weaker and less intelligent),[54] so Valla had already questioned the distinction between true and false glory. "De vera et falsa gloria"[55] is the title of the chapter in *De voluptate* in which he argues that the "regula honestatis" (moralistic rule) is utterly incapable of providing a valid criterion for making a distinction, since "nobody has ever made a choice between the means to employ when what is at stake is reaching glory".[56] Anyone maintaining the contrary would be behaving "as though historians were to write on Stoics and Cynics rather than on kings and generals" like Alexander or Hannibal.[57] Ultimately, even the alternative between *gloria* and *infamia* is placed within practical parameters, dependent on the seeking of pleasure and the avoidance of discomfort, similarly to the approach adopted by Hobbes in the seventeenth century.[58]

With all his bent for emphasizing the importance for Hobbes of Aristotelianism, sometimes described as "new humanistic Aristotelianism", Strauss was honest enough to avow that he did not find in Aristotle this typically modern cluster of moral notions (emphasis on self-preservation and the equality of mankind, rejection of glory as a basic moral value) that one can discover in Hobbes. On Aristotle Strauss was absolutely right; but he was wrong not to search elsewhere, in another

[53] Cf. Strauss (1952).

[54] Cf. for instance Hobbes (1889), 70–71 (*Elements of Law*, I, xiv, 1–4).

[55] Valla (1540), 938.

[56] Ibid.: "Si quis vero curiosior discrimen mihi faciat inter veram gloram et non veram, hic nihil attulerit. Determinat enim gloriam regula honestatis, quam honestatem nihil esse docuimus". "Neminem enim video unquam delectum habuisse, quoquo modo forent, dummodo forent gloriosi. Quanquam quisquis aliquid spe huius praemii facit, utique ab honesto recedit, sive in tolerando et fortiter faciendo, ut dixi, sive quod proximum est, aliquid libertatis exercendo."

[57] Ibid.

[58] On the distinction between "glory" e "false glory" see Hobbes (1889), 36–37 (*Elements of Law* I, ix, 1). Cf. ibid, 70–71 (*Elements of Law* I, xiv, 3) for the clash between "moderate" e "vainly glorious", which degenerates into a state of "general diffidence in mankind, and mutual fear one of another", because the ambitious are incapable of acknowledging the effective "equality" in the state of nature.

illustrious and ancient tradition, rejuvenated as neo-Epicureanism in the long span between the Quattrocento with Valla and the seventeenth century with Gassendi and Hobbes.[59]

Another major mistake that is pervasive in scholarship on Hobbes consists in referring the basis of his notion of natural law to the *koiné* of continental Stoicism, via Grotius or Lipsius: whereas it is clear that Hobbes's system of law is rooted in an anti-Stoic and neo-Epicurean anthropology that makes virtue depend on selfish reference to pleasure and self-preservation. If labels were useful, one might say that Hobbes's morals, and also to a great extent politics, result in a sort of creative anti-anti-Epicureanism, in the sense that he rescued Epicurean moral psychology from an age-old damnation,[60] not simply reinstating it against the traditional criticism, but also supplementing it with a law-based approach that was lacking or weak in ancient and modern Epicureanism, with the sole exception of his contemporary and friend Gassendi. In this case finding the proper context is at once a move toward a more historically grounded understanding and a sounder philosophical interpretation of Hobbes's work. Considering the neo-Epicurean and Renaissance background is not merely a way to satisfy some sort of antiquarian curiosity; it is absolutely necessary to understand Hobbes's position and to avoid serious misinterpretations, such as that of framing his morals into a more or less Aristotelian theory of virtue ethics.

This sort of interpretation plays down Hobbes's polemics against the mediocrity thesis, ignores the neo-Epicurean background of his anthropology, neglects the first appearance in Valla's work of crucial modern topics like self-preservation, equality, the dismissal of "glory" as valuable political notion, and the critique of the mediocrity-theory of virtue. To this latter point it might be objected that this is not exclusive to Valla, and that other Renaissance authors criticised the doctrine of virtue as a midpoint between two extremes,[61] and that therefore Hobbes could have turned to these other authors, or to one of his contemporary opponents of Aristotelian philosophy, such as Pierre Gassendi. Nevertheless, as I have shown, it was in Valla that Hobbes could have read a broader collection of the topics of greatest interest to him, within the framework of an anthropology and moral investigation, which must have been congenial to him for its realism and critical power. It must have been the

[59] In fact Strauss dealt with Hobbes in relation to Epicurean thought in an early work of 1933: see Strauss (2001). In this work Hobbes is basically viewed as a representative of the Epicurean tradition because of his materialism, his profound humanism, the rejection of the fear of death as well as of the gods, etc. (see esp. 315–319, 322). But the comparison is based on ancient Epicureanism, not the neo-Epicureanism of the moderns, from Valla to Gassendi. On this, see Paganini (2016).

[60] See Wilson (2008), 179–193. This book says very little about Valla's wide influence in the early modern period (see 243–244).

[61] See Kraye (1988), esp. 339–342, who examines the critiques above all by Valla, Vives and Sanches of the Aristotelian doctrine of the *via media*. To these authors should be added at least Jean Bodin, author of a short work bearing the telling title: *Paradoxon quod nec virtus ulla in mediocritate nec summum hominis bonum in virtutis actione consistere possit*, first published in Latin in 1596 and then in French (*Paradoxe de Jean Bodin qu'il n'y a pas de vertu en médiocrité ni au milieu de deux vices*, publ. 1598).

cluster of these topics and not only the critique of the Aristotelian doctrine that made Hobbes turn to Lorenzo Valla, who thanks to the mediation of Erasmus was still a great modern classic in the seventeenth century.[62]

Bibliography

Boonin-Vail, David. 1994. *Thomas Hobbes and the Science of Moral Virtue*. Cambridge: Cambridge University Press.

Curley, Edwin. 1998. Religion and Morality in Hobbes. In *Rational Commitment and Social Justice*, ed. Jules Coleman and Christopher Morris, 90–121. Cambridge: Cambridge University Press.

Dougherty, M.V., Pernille Harsting, and Russell L. Friedman. 2009. 40 Cases of Plagiarism. *Bulletin de Philosophie Médiévale* 51: 350–391.

Ebbersmeyer, Sabrina. 2010. *Homo agens. Studien zur Genese und Struktur frühhumanistischer Moralphilosophie*. Berlin: De Gruyter.

Ewin, R.E. 1991. *Virtues and Rights. The Moral Philosophy of Thomas Hobbes*. Boulder: Westview Press.

Grotius, Hugo. 1712. *De Jure belli ac pacis libri tres* [...] *Cum annotatis Auctoris* [...] *Joann. Frid. Gronovii Notae*. Amsterdam: Apud Janssonio-Waesbergios.

Hobbes, Thomas. 1839–1845a [abbreviated in this essay as OL]. *Thomae Hobbes Malmesburiensis Opera Philosophica Quae Latine Scripsit Omnia in Unum Corpus Nunc Primum Collecta studio et labore Gulielmi Molesworth*, 5 vols. London: Bohn; reprint Aalen: Scientia, 1962–1966.

Hobbes, Thomas 1839–1845b [abbreviated EW]. *The English Works of Thomas Hobbes of Malmesbury. Now First Collected and Edited by Sir William Molesworth*, 11 vols. London: John Bohn; reprint Aalen: Scientia, 1962–1966.

Hobbes, Thomas. 1889. *The Elements of Law Natural and Politic*, ed. with a preface and critical notes by Ferdinand Tönnies. London: Smipkin, Marshall, and Co.

Hobbes, Thomas. 1983. *De cive: The English Version*, ed. Howard Warrender. Oxford: Oxford University Press.

Hobbes, Thomas. 1990. *Behemoth, or the Long Parliament*, ed. Ferdinand Tönnies. Chicago: University of Chicago Press.

Hobbes, Thomas. 1994. *Leviathan with Selected Variants from the Latin Edition of 1668*, ed. Edwin Curley. Indianapolis-Cambridge: Hackett. I have collated this edition with the new critical edition by Noel Malcolm (Oxford: Clarendon Press, 2012).

Hobbes, Thomas. 1999. *De corpore. Elementorum Philosophiae Sectio Prima*. Edition critique, notes, appendice et index par Karl Schuhmann. Paris: Vrin.

Hobbes, Thomas. 2008. *Historia Ecclesiastica*, critical edition, including text, translation, introduction, commentary and notes by Patricia Springborg, Patricia Stablein and Paul Wilson. Paris: Honoré Champion.

Hobbes, Thomas. 2012. *Leviathan*, critical edition by Noel Malcolm. Oxford: Clarendon Press.

Kavka, Gregory. 1986. *Hobbesian Moral and Political Theory*. Princeton: Princeton University Press.

Kraye, Jill. 1988. Moral Philosophy. In *The Cambridge History of Renaissance Philosophy*, ed. Charles B. Schmitt and Quentin Skinner, 303–386. Cambridge: Cambridge University Press.

Laffranchi, Marco. 1999. *Dialettica e filosofia in Lorenzo Valla*. Milan: Vita e Pensiero.

[62] Valla's influence on authors such as Vives, Bacon, Hobbes, Spinoza, and up to Hume, is pointed out by Ebbersmeyer (2010), 255.

Lloyd, S.A. 1992. *Ideals as Interests in Hobbes's Leviathan.* The Power of Mind over Matter. Cambridge: Cambridge University Press.

Monfasani, John. 2000. The Theology of Lorenzo Valla. In *Humanism and Early Modern Philosophy*, ed. Jill Kraye and Martin W. F. Stone, 1–23. London: Routledge.

Mulsow, Martin, and Claudia Schmitz. 2004. Eigennutz, Statuserhaltung und Naturzustand: Tradierungen des ethisch-politischen Epikureismus vom 15. bis zum 17. Jarhundert. In *Der Garten und die Moderne. Epikureische Moral und Politik vom Humanismus bis zur Aufklärung*, ed. Gianni Paganini and Edoardo Tortarolo, 47–86. Stuttgart-Bad Cannstatt: Frommann-Holzboog.

Nauta, Lodi. 2009. *In Defense of Common Sense: Lorenzo Valla's Humanist Critique of Scholastic Philosophy*. Cambridge: Harvard University Press.

Pacchi, Arrigo. 1987. Hobbes and the Passions. *Topoi* 6: 111–119.

Paganini, Gianni. 1999. Thomas Hobbes e Lorenzo Valla. Critica umanistica e filosofia moderna. *Rinascimento* 39: 515–568.

Paganini, Gianni. 2003. Hobbes, Valla and the Trinity. *British Journal for the History of Philosophy* 11: 183–218.

Paganini, Gianni. 2004. Hobbes, Gassendi and the Tradition of Political Epicureanism. In *Der Garten und die Moderne. Epikureische Moral und Politik vom Humanismus bis zur Aufklärung*, ed. Paganini Gianni and Tortarolo Edoardo, 113–137. Stuttgart-Bad Cannstatt: Frommann-Holzboog.

Paganini, Gianni. 2007. Hobbes's Critique of the Doctrine of Essences and Its Sources. In *The Cambridge Companion to Hobbes's Leviathan*, ed. Patricia Springborg, 337–357. Cambridge/New York: Cambridge University Press.

Paganini, Gianni. 2013a. How Did Hobbes Think of the Existence and Nature of God?: *De Motu, Loco et Tempore* as a Turning Point in Hobbes's Philosophical Career. In *The Bloomsbury Companion to Hobbes*, ed. Sharon A. Lloyd, 286–303, 321–325 (notes). London: Bloomsbury.

Paganini, Gianni. 2013b. Hobbes e Valla. Teologia, linguaggio e riforma della filosofia prima. In *La diffusione europea del pensiero del Valla*, vol. 1, ed. Mariangela Regoliosi and Clementina Marsico, Atti del Comitato Nazionale VI Centenario della nascita di Lorenzo Valla, 327–344. Firenze: Edizioni Polistampa.

Paganini, Gianni. 2015. Hobbes's Galilean Project and Its Theological Implications. *Oxford Studies in Early Modern Philosophy* 7: 1–46.

Paganini, Gianni. 2016. The Political Neo-Epicureanism of the 17th Century: Gassendi's Dialogue with Hobbes. In *Oxford Handbook of Ancient and Modern Epicureanism*, ed. Phillip Mitsis. Oxford: Oxford University Press (in print).

Sarasohn, Lisa T. 1996. *Gassendi's Ethics: Freedom in a Mechanistic Universe*. Ithaca: Cornell University Press.

Skinner, Quentin. 1996. *Reason and Rhetoric in the Philosophy of Hobbes*. Cambridge: Cambridge University Press.

Springborg, Patricia. 2003. Hobbes's Theory of Civil Religion. In *Pluralismo e religione civile*, ed. Gianni Paganini and Edoardo Tortarolo, 61–98. Milan: Bruno Mondadori.

Springborg, Patricia. 2004. Hobbes and Epicurean Religion. In *Der Garten und die Moderne: Epikureische Moral und Politik vom Humanismus bis zur Aufklärung*, ed. Gianni Paganini and Edoardo Tortarolo, 161–214. Stuttgart-Bad Cannstatt: Frommann-Holzboog.

Strauss, Leo. 1952. *The Political Philosophy of Hobbes: Its Basis and Its Genesis*, 2nd ed. Chicago/London: University of Chicago Press.

Strauss, Leo. 2001. Die Religionskritik des Hobbes. Ein Beitrag zum Verständnis der Aufklärung (1933–1934). In Leo Strauss, *Gesammelte Schriften*, vol. 3, ed. Meier Heinrich, and Meier Wiebke. Stuttgart: J. B. Metzlersche Verlagbuchhandlung und Carl Ernst Poeschel.

Tuck, Richard. 1993. *Philosophy and Government 1572–1651*. Cambridge: Cambridge University Press.

Valla, Lorenzo. 1540. *Opera omnia.* Basel: H. Petrus; 1962 repr. with extra vol., introd. by Eugenio Garin. Turin: Bottega d'Erasmo.

Wilson, Catherine. 2008. *Epicureanism at the Origins of Modernity*. Oxford: Clarendon Press.

Wolfers, Benedikt. 1991. *"Geschwärtzige Philosophie". Thomas Hobbes' Kritik an Aristoteles*. Würzburg: Königshausen & Neumann.

Chapter 14
Debating "Greatness" from Machiavelli to Burton

Sara Miglietti

Abstract From early humanist treatises on city government in Italy to Rousseau's *Social Contract*, "greatness" (*grandezza*, *grandeza*, *grandeur*) was often presented as both the aim that political communities should pursue and the touchstone to measure their relative success. But what exactly should be understood by "greatness", and how could it be achieved? Although most authors agreed that it took more than a large territory for a state to be truly "great", they all seemed to prioritise different things: political liberty, military strength, material wealth, absence of strife, a solid social and political order, or the happiness and overall wellbeing of the citizens. In an age of state- and empire-building, the debate on the nature of political "greatness" raised critical questions and contributed to shaping the agenda and the self-representation of European powers. By concentrating on a few selected thinkers (Machiavelli, Bodin, Botero, Bacon, Burton) whose works form a complex network of mutual influences, this chapter seeks to investigate an exemplary case of unceasing dialogue between the Renaissance and the early modern period.

Machiavelli, Italian Humanism, and the Ideology of Greatness

> Truly it is a marvelous thing to consider to what greatness (*grandezza*) Athens came in the space of a hundred years after she freed herself from the tyranny of Pisistratus. But above all, it is very marvelous to observe what greatness (*grandezza*) Rome came to after she freed herself from her kings.[1]

Notwithstanding his well-justified "marvel" at the progress made by Athens and Rome under a "free government" (*vivere libero*), the Florentine writer Niccolò Machiavelli (1469–1527) was probably aware that his words would be of little surprise to the readers of his *Discourses on Livy* (composed around 1513–1517 but first published in 1531). Many of these, indeed, would have easily recognised that the

[1] Machiavelli (1965), II.2, 329; Machiavelli (1999), II.2, 297.

S. Miglietti (✉)
Department of German and Romance Languages and Literatures,
Johns Hopkins University, Baltimore, MD, USA
e-mail: smiglie1@jhu.edu

C. Muratori, G. Paganini (eds.), *Early Modern Philosophers and the Renaissance Legacy*, International Archives of the History of Ideas Archives internationales d'histoire des idées 220, DOI 10.1007/978-3-319-32604-7_14

Florentine was subscribing to a discourse whose roots could be traced back to Roman moralists and historians such as Sallust, and which had been revived in Italy since at least two centuries.[2] The very word *grandezza* had made its first appearance in thirteenth-century vernacular writers such as Matteo de' Libri, Giovanni da Vignano, Guido Faba and Filippo Ceffi, who coined this term to signify the highest end to which political communities could aspire.[3] Although the concept was typically Roman,[4] the word itself was not. As noted by Skinner, classical Latin lacked an expression "at once denoting grandeur and magnitude",[5] and those "pre-humanist" Italian authors who wrote in Latin usually turned to vaguer synonyms (such as "incrementum")[6] or created curious *pastiches* by weaving the Italian word *grandezza* into Latin texts.[7] For these authors, *grandezza* was an essential component of a healthy civic life (*bon stato*),[8] alongside domestic peace (*riposo*), dignity (*honore*) and, crucially, political freedom.[9] Towards the beginning of the fifteenth century, this notion of self-government as the only way to achieve *bon stato* and *grandezza*,[10] originally inspired by classical authors such as Aristotle, Cicero and Sallust, was bound to become the keystone of Florentine self-representations,[11] and was frequently used to sustain and justify the aggressive "republican imperialism" of the Tuscan city-state.[12]

Machiavelli's *Discourses*, written more than a hundred years later and under profoundly changed historical circumstances, have been rightly identified as the ultimate and possibly highest expression of this tradition coupling liberty and empire in the pursuit of political *grandezza*.[13] For the late medieval and early humanist authors who wrote before Machiavelli, *grandezza* had at once a spatial, a material, and an ethical-political meaning, denoting not only the magnitude of a state's territory, but also its wealth, standing, power, and most importantly its *vivere civile* (a well-ordered, republican way of life).[14] The same holds true for Machiavelli, who describes *grandezza* in terms of both territorial extension (*dominio* and *corpo*)[15]

---

[2] Hörnqvist (2004), 74.

[3] Skinner, quot. in Springborg (1992), 198–199; Skinner (2003), 92–93.

[4] Springborg (1992), 197–198.

[5] Skinner (2003), 92.

[6] Ibid., 92.

[7] See for instance Giovanni da Viterbo's *Liber de regimine civitatum*, quot. in Skinner (2003), 93.

[8] Skinner (1995), 108.

[9] Skinner (2003), 92–93; Hörnqvist (2004), 39–40.

[10] Skinner (1995), 104.

[11] The most famous example is Leonardo Bruni's *Laudatio Florentinae Urbis*, written around 1403–1404, but see Hörnqvist (2004), 55–70, for further examples.

[12] Hörnqvist (2004), 42, building on previous work by Hankins (1995, 2000), Najemy (2000) and Brown (2004).

[13] Springborg (1992), 197; Skinner (2003), 103–106; Hörnqvist (2004), 72–74.

[14] Skinner (2003), 91, correcting Springborg's somewhat unilateral view of *grandezza* as something "literally spatial" (1992, 198).

[15] Machiavelli (1999), II.2, 297, 301; Machiavelli (1965), II.2, 329, 333.

and wealth (*ricchezza*),[16] equates it with military power,[17] and establishes a direct link between self-government (*vivere libero*) and greatness thus conceived. The polysemy of Machiavellian *grandezza* did not escape early translators such as Johann Nikolaus Stupanus, whose Latin version of *Discourses* (1588) formed the basis for most subsequent Latin editions of the work[18]: in the absence of an exact Latin equivalent for the Italian word *grandezza*, Stupanus chose to translate the latter in more than one way, thus brilliantly capturing its multiple meanings.[19]

While interpreting *grandezza* in accordance with a well-established tradition, Machiavelli moved a step forward in identifying the factors that could affect its achievement. Although a constitutional form ensuring political liberty was just as decisive for him as it had been for his forebears, the Florentine saw that other aspects needed to be taken into account as well. First and foremost among these was a consideration of the size and features of the local population. According to Stangeland, Machiavelli was "perhaps the first to express ideas of modern tone on population"[20]; at the very least, he was among the earliest authors to acknowledge the centrality of demographic issues for political theory, and advocated population growth in a time when the dominant ideal was still that of a fixed and relatively small population size. Machiavelli's main point in favour of population growth was that no great empire can be established without a sizeable population, given the high demographic costs of expansionism.[21] Not by chance, he argued, the Romans sought to increase their population by all possible means, and particularly through their citizenship policies.[22]

In sum, just as the achievement of *grandezza* requires territorial expansion, the latter demands population growth. Machiavelli did not regard expansionism and demographic increase as good in themselves, but only as means to a higher end – namely greatness. He was also acutely aware of their problematic nature: while territorial expansion was among the causes that ultimately led the Roman Empire to collapse,[23] a large population can turn from a strength into a handicap if the country is not sufficiently wide, or sufficiently rich, to sustain it.[24] Territory and population

[16] Machiavelli (1999), II.2, 297, 300; Machiavelli (1965), II.2, 329, 332.

[17] See for instance Machiavelli (1965), II.2, 332: "The Roman Empire with her arms (*arme*) and her greatness (*grandezza*) wiped out all the republics and all the self-governing communities (*e viveri civili*)"; Machiavelli (1999), II.2, 299–300.

[18] See Ruffo Fiore (1990), 40.

[19] Machiavelli (1588), 270–271, 276: "Admiratione summa dignum est Athenienses Pisistrati tyrannide liberatos, centum annorum spacio, *vires* atque *potentiam* suam tantopere augere potuisse. Et longe magis stupendum adhuc, Romanorum Rempublicam, expulsis Regibus, ad tantum *fastigium* pervenire potuisse [...] equidem Romani imperii *magnitudo*, non parum etiam ad id fecit, cum per universum orbem liberas Respublicas sub suum iugum subiecerit" (my emphasis).

[20] Stangeland (1904), 92.

[21] Machiavelli (1999), I.6, 77; II.3, 301–302.

[22] Ibid., II.3, 302–303.

[23] In keeping with Sallust's account in *Bellum Catilinae* (X.1-2), which became the standard view of Roman decadence in the Renaissance: see Armitage (2002).

[24] Cf. Stangeland (1904), 93.

are thus subtly related by Machiavelli so as to emphasise their crucial, and often delicate, balance. It is not simply a matter of size: the qualitative features of the terrain influence the people who live on it,[25] just as the latter constantly alter and reshape the former through industry, architecture and agriculture.[26] We shall see in the next sections how Machiavelli's perceptive insights into the relationship between territory and population inspired and stimulated later authors grappling with the question of political greatness.

"The Greatness of Cities": Botero and Bodin

The year 1588 constituted a milestone in the discussion on *grandezza*. It was in this year that Giovanni Botero (1544–1617) published in Rome his *Causes of the Greatness and Magnificence of Cities* (*Cause della grandezza e magnificenza delle città*). The work, comprised of three books and written in Botero's elegant Italian, was an immediate success. It was reprinted the following year as an appendix to Botero's even more influential *Reason of State* (*Ragion di Stato*); and while the two treatises were frequently published together, the *Greatness of Cities* also enjoyed a separate fortune, with new editions and translations all over Europe until the mid-seventeenth century.[27] The tract was especially successful in England, where it was printed twice in two distinct English translations (by Robert Peterson in 1606, and by Thomas Hawkins in 1635) and, as we shall see in section 3, received the close attention of many leading authors of the time.

Botero's treatise was seminal in many respects. As noted by Symcox, it departed from a century-long tradition of urban literature – from Leon Battista Alberti to Filarete and Francesco di Giorgio Martini – by shifting "the focus of enquiry from the *forma urbis* and the geometry of fortifications to economics, demography, and the political factors that foster urban development, causing some cities to prosper while others do not".[28] Among such causes of urban development, Botero identified three main topographical factors ("the commodity of the site, the fertility of the soil and easiness of conduct", according to Peterson's translation),[29] and a number of economic, social and political aspects that he carefully reviewed in book 2. Some of these were clearly inspired by a reading of republican Rome through the lens of Machiavelli's *Discourses*, as is especially evident in the sections on citizenship and colonies.[30] Recent studies have shown how the influence of Machiavelli was crucial

---

[25] Machiavelli (1965), I.1, 193–194; Machiavelli (1999), I.1, 62–63. See Glacken (1976), 433–434.

[26] See Stangeland (1904), 93.

[27] See Botero (2012), xiii–xiv.

[28] Ibid., vi.

[29] Botero (1956a), I.7, 234 and (1990a), I, 318.

[30] Botero (1956a), II.1-2, 244–247 and (1990a), II, 330–334.

for the development of Botero's thought[31]; yet at the same time it must be noted that Botero was critical of Machiavelli for reasons that go well beyond the Florentine's alleged atheism and amorality, and which involve their divergent understandings of what political greatness is and how it can (or should) be achieved.

"The greatness of a city is said to be, not the largeness of the site or the circuit of the walls, but the multitude and number of the inhabitants and their power", wrote Botero in the opening lines of his treatise,[32] thus revealing at once his dependence on, and departure from, the author of *Discourses*. Indeed, by saying that the "greatness" of a city should be measured by the number (*moltitudine*) and power (*possanza*) of its population rather than by its spatial extension, Botero was directly targeting Machiavelli's equation of territory and population, as well as his emphasis on territorial expansion. On the other hand, the two agreed that an expansionist programme necessarily requires the implementation of policies encouraging population growth.[33] In his *Universal Relations* (*Relationi Universali*, a geographical and political description of several world countries) of 1596, Botero gave numerous examples of how engaging in wars of conquest can prove devastating for an underpopulated country. He argued with remarkable foresight that Spain and Portugal were destined to a rapid decline, for so many young men were lost every year in the colonial effort that "the homeland remains almost lifeless and devoid of vigour".[34] The two Iberian countries were following "a reason of state exactly opposed to that which made Rome great and powerful", having failed to augment their population in preparation for their expeditions and thus exhausting themselves in an undertaking which would bring them ruin rather than *grandezza*.[35] As a counterexample Botero cited the case of the Chinese, who – he said – wisely abandoned their aggressive expansionism as soon as they realised that the war against India could only be won at the expense of immense human and material losses, and now happily prospered within fixed borders.[36]

"What good is it to struggle to expand, when expansion not only does not bring about well-being (*bene essere*), but destroys being (*essere*)?" wonders Botero.[37] For him, the primary goal of a good ruler should be to protect and preserve the state, not to increase it. He is aware, of course, that this viewpoint contradicts Machiavelli's famous argument that rulers do not really have a choice between preservation and

---

[31] See in particular Descendre (2009).

[32] Botero (1956a), I.1, 227 and (1990a), I, 309.

[33] See for instance Botero (1596), II, "Proemio" (unpaged); Botero (1956b), VII.12, 144–7 and (1990b), VII, 201–205.

[34] Botero (1596), I.1 ("Cina"), 17 (all translations from *Universal Relations* are mine). Cf. Botero (1956b), VII.12, 145 and (1990b), VII, 203–204. Botero was expressing concerns that would spread in Spain several years later, particularly in the late reign of Philip III: cf. Elliott (1977) and Spedding in Bacon (1900a), 127–129, footnote 2.

[35] Further distinctions between ancient Roman policies and the short-sighted colonialism of the Spanish and the Portuguese are made in Botero (1956b), VIII.5, 156–157 and (1990b), VIII, 219–220.

[36] Botero (1596), I.1 ("Cina"), 125–126.

[37] Ibid., II.1 ("Moscovia"), 41.

expansion, since states are generally forced to expand in order to survive.[38] But Botero is unconvinced: in fact, middle-sized states have better chances of surviving than large empires,[39] and by rechanneling their energies from wasteful military undertakings into the systematic exploitation and improvement of their own territory, they are also often wealthier, healthier, and more densely populated – which, for Botero, is itself a measure of "greatness".[40]

Besides, military conquest is not the only way in which states can expand. Although war remains unquestionably an important component of Botero's mental universe,[41] the former Jesuit tends to see trade as a more benign and cost-effective way of securing one's power.[42] Among the many examples of relatively small countries that have thrived by choosing commerce over war, he mentions the Flanders: here, where a substantial amount of money and energy has been invested into the construction of artificial channels that would make the transportation of goods easier and faster, a significant boost "to merchandising and to the traffic of other nations" has ensued.[43]

Botero was among the first to grasp the economic principle according to which the wealth of a state depends essentially on its having a favourable balance of trade – a principle that would later be formulated by authors such as Thomas Mun.[44] But clearly commerce is only a viable option where there is a surplus available for export. No wonder then that Botero would express his admiration for all those countries that not only exploit their territory to the fullest and do not let anything go to waste,[45] but seek to produce more than they need for internal consumption and also engage in manufacturing (*arteficii*)[46] – an activity that enables them to export transformed goods rather than less lucrative raw materials.[47] Botero's hints at the added value of labour have been seen by some as an anticipation of later theories

---

[38] Machiavelli (1999), I.6, 78–79 and (1965), I.6, 210–211. Botero admits that his preference for preservation over expansion "would not be approved by modern wisdom (*prudenza moderna*)" (1596, II.1, 41).

[39] Botero (1956b), I.6, 7–9 and (1990b), I, 7–10.

[40] For examples of "improvement" (*miglioramenti*), see Botero (1956b), III.2, 76 and (1990b), III, 106; Botero (1956b), VIII.2-3, 148–153 and (1990b), VIII, 207–214. Slack (2014) briefly discusses the place of Botero within the early modern tradition of "improvement".

[41] Descendre (2009) has stressed the Machiavellian inspiration behind Botero's discourse on "force", "power", and military prowess (see particularly chapter 4, 140–142).

[42] See Hartman and Weststeijn (2013), 14. According to Hörnqvist (2004), Machiavelli would not have seen trade as "a tenable alternative" to military conquest, since "in his zero-sum world, commerce is also a form of imperialism" (74, n. 110).

[43] Botero (1956a), I.10, 238 and (1990a), I, 323.

[44] Botero (1956b), VII.10, 142–143 and (1990b), VII, 199. Mun's *England's Treasure by Foreign Trade* was written around 1630 but only published in 1664.

[45] China is again proposed as a positive example: see Botero (1596), I.2 ("Cina"), 125; Botero (1956a), II.11, 266–9 and (1990a), II, 359–60; Botero (1956a), II.11, 268 and (1990a), II, 359.

[46] See again Botero's praise of Flanders in Botero (1596), II.2 ("Cina"), 65.

[47] Botero (1956b), VIII.3, 151 and (1990b), VIII, 211.

(particularly by William Petty and Richard Cantillon),[48] while his pioneering insights into the strategic importance of manufacturing are likely to have inspired part two of Antoine de Montchrétien's *Traicté d'oeconomie politique* (1615), one of the key economic works of the seventeenth century.[49]

Botero thus added a crucial component to the multifaceted understanding of *grandezza* that had been framed by the Italian humanists and redefined by Machiavelli. A flourishing economy – based on a range of extractive, productive, and commercial activities ("Agricoltura, Arte e traffichi")[50] – was for him just as important as, if not more important than, a strong army for securing political greatness.[51] Both Machiavelli and Botero advocated population growth; but while the former did so mainly for military reasons, Botero, though certainly not insensitive to the strategic advantages of a large population, also appreciated the fact that a well-inhabited country is usually better tilled, more thoroughly exploited, and ultimately more productive than an under-populated one.[52]

Botero was also acutely aware that demographic expansion was only good to the extent that it remained manageable. Machiavelli had already stressed the importance of keeping a sustainable balance between the size of a state's territory and that of its population: besides restating his point,[53] Botero adds that a large population demands a careful management of urban spaces so as to minimise conflict, and public policies that guarantee high standards of hygiene.[54] Botero's interest in public health is not merely motivated by the obvious consideration that large concentrations of people are ideal cradles of epidemic diseases; the former Jesuit is also concerned with providing each citizen with a good quality of life, which includes a safe and clean environment and an adequate education.[55] Overall, he argues, the greatness of a state is not merely a matter of quantity. Quality counts too, and a wise prince should seek to "increase his forces intensively" as well as "extensively"[56] – striving to improve the "character" of his subjects rather than just focusing on numbers.

Botero was not alone in his effort to rethink Machiavellian *grandezza* in the service of early modern governmental practices. The Flemish Justus Lipsius (1547–1606) – whose *Politica* came out in Leiden in the same year as Botero's *Reason of State* (1589) and was translated into English 5 years later – was also moving in the

48 See Descendre (2009), 192–194.

49 See Reinert (2011), 5.

50 Botero (1956b), VII.10, 143 and (1990b), VII, 199–200.

51 See Weber (2003), 328.

52 Botero (1956b), VII.12, 144–147 and (1990b), VII, 201–205. Spain is again presented as a negative example.

53 See Botero (1956a), III.2, 276–277 and (1990a), III, 371–372. Cf. Botero (1956b), VIII.4, 155 and (1990b), VIII, 217.

54 See Botero (1956a), V.7, 109–110 and (1990a), V, 152 (on public order); Botero (1956b), VIII.4, 155 and (1990b), VIII, 217 (on public health).

55 Botero (1956b), VIII.4, 154 and (1990b), VIII, 215. Cf. Weber (2003), 330.

56 Botero (1956b), IX.1, 168 (modified translation) and (1990b), IX, 234.

same direction, drawing attention to economy and taxation as fundamental aspects of statecraft and thus marking a significant step in the discussion on political greatness.[57] Behind Botero's and Lipsius's reframing of *grandezza* lurks the influence of Jean Bodin (1529/30–1596), the French jurist and political writer who crucially revisited the concept of sovereignty in his *Six Books of the Commonwealth* (*Six livres de la République*) – first published in French in 1576, largely reworked for the Latin edition of 1586, and translated into English by Richard Knolles in 1606.

Surely, the word *grandeur* (French for "greatness") retains in Bodin the same polysemy that it originally had in the Italian tradition before Botero. Nothing indicates that the Frenchman ever tried to circumscribe its meaning and uses in the same way that Botero would a few years after him. When the term appears in his vernacular works, it is often to indicate very literally the size of something, as when the Frenchman speaks of the "grandeur" of buildings and human bodies.[58] When used alone, *grandeur* usually signifies social standing,[59] dignity (particularly royal dignity),[60] or actual power,[61] although on a few occurrences it stands for liberality and magnificence, in an interesting recovering of the Hellenistic notion of *megalopsychia* – another distant relative of humanist *grandezza* according to Springborg.[62] Finally, the word is also employed to denote sheer territorial extension, as when Bodin refers to the "grandeur" of the Turkish Empire.[63] Bodin's understanding of greatness is in sum quite flexible and, overall, not particularly original. Besides, the Frenchman does not seem to view *grandeur* (as he conceives of it) as a primary objective of good statecraft: stable government, absence of strife, and the happiness and well-being of the citizens (their *vivre heureusement*) are indeed the parameters by which he thinks that a state's success should be measured, with all other considerations coming second.

Yet, even though Bodin does not specifically contribute to the discussion on "greatness" in any significant way, his work contains highly original insights into the conditions of political development that would prove a fundamental source of inspiration for Botero. On the one hand, he acted as a positive mediator of Machiavellian ideas, for instance by echoing the Florentine's enthusiasm for demographic growth ("there is no wealth nor strength but in men", he famously argued in his *Six Books*),[64] while at the same time voicing similar concerns about the delicate balance between territory and population.[65] On the other hand, he provided the first substantial critique of indiscriminate territorial expansionism, carefully

---

[57] See Weber (2003), 327–379.

[58] E.g. Bodin (1593), IV.2, 544; V.1, 667, 671, 672, 673, 688.

[59] E.g. ibid., III.3, 393; VI.6, 1020.

[60] E.g. ibid., I.9, 162, 204; I.10, 215; III.1, 355, 359; IV.6, 618; V.4, 748 (on Alexander the Great).

[61] E.g. ibid., I.1, 7; V.6, 794, 799, 800; VI.6, 1051.

[62] Springborg (1992), 197. See Bodin (1593), IV.6, 631; V.4, 748 ("grandeur & liberalité"); VI.2, 906.

[63] Bodin (1593), V.1, 672. Also cf. ibid., V.5, 780.

[64] Ibid., V.2, 705–706: "Il n'y a de richesse, ny de force, que d'hommes".

[65] Ibid., I.1, 6.

distinguishing between states that *must* expand, states that *can* expand, and states that *should not* or *cannot* expand. The difference lies not only in the constitutional form of each state (monarchies, for instance, are well suited for expansion, whereas aristocracies are not),[66] but also in the "natural character of the people" (*naturel du peuple*) who inhabit it: while Northerners and mountain-dwellers are naturally drawn to warfare on account of the harsh, toughening climate, those who have been favoured with a bountiful environment make very poor soldiers.[67] Yet excelling at combat is not all that it takes to build an empire: other qualities are needed to maintain what has been acquired, and for this reason the best suited for long-term expansion are not the valiant Northern peoples, but those of the middle region (the so-called temperate zone), who stand out for their political wisdom.[68]

Bodin's point – possibly inspired by Aristotle[69] – is that a universally valid recipe for success does not exist. Each state must be ruled according to specific conditions – its constitutional regime, the character of its inhabitants – and what works for one country could prove catastrophic for another. Yet, while this is true, the functioning of a state is essentially the same in all cases. Indeed, Bodin's treatise was ground-breaking for its systematic attention to structural elements of statecraft such as public welfare, monetary and fiscal policies, trade, and the regulation of human movements across borders. Although Botero's dependence on Bodin's economic ideas might occasionally have been overblown,[70] there is little doubt that the Frenchman's novel emphasis on the interconnections between economy and politics, as well as between politics and geography,[71] was an important source of inspiration for the author of the *Greatness of Cities*. So was it for so many writers of the following generations, who could read the *Six Books of the Commonwealth* in any of the numerous editions and translations which appeared from the late sixteenth century onwards. Just as Botero inevitably read Machiavelli through the lens of Bodin, and vice versa, so many readers of Bodin interpreted his works through those of Botero (again, the reverse is also true). The *joint* reception of these three authors in the seventeenth century is an extremely complex and fascinating topic which has not yet received the attention it warrants. In the next section, I shall limit myself to a few remarks on the role played by their works in stimulating the English debate on political greatness.[72]

---

[66] Ibid., V.5, 763–767.

[67] Ibid., V.5, 763 and V.1, 671–674. Cf. Glacken (1976), 435–447.

[68] Bodin (1593), V.5, 671.

[69] Cf. Aristotle, *Politics*, VII.7 (1327b19-35).

[70] See Descendre (2009), 194–5, for a critical assessment.

[71] Botero famously devotes an entire chapter of *Reason of State* (II.5) to a discussion of environmental influences that owes much to Bodin. See Chabod (1967), 340.

[72] On reception of Machiavelli in England see Raab (1965); Pocock (2003); Arienzo and Petrina (2013). On reception of Bodin see Burgess (2013); Miglietti (2013). On reception of Botero see Fitzmaurice (2007).

Conflicting Ideals: Bacon, Burton, and the Anatomy of England

The first decades of the seventeenth century saw the rise of England's maritime power, the intensification of its colonial attempts in Ireland and America, as well as the *de facto* "regnal union" of England and Scotland under James I (1603–1625). It was a time of exciting change and growth,[73] but also one of shifting self-representations, conflicting agendas, and lively debates about the direction that the country should take. With Spain and the Netherlands as powerful competitors, England's strive for international predominance could only be successful by candidly assessing the country's strengths and weaknesses, and then by acting to maximise the former and minimise the latter. Such a task became even more pressing in the final years of James I's reign, when the country was threatened by economic instability, scarcity of money and commercial decline, partly in connection with the outbreak of the Thirty Years' War.[74] It was precisely in those years that two of the most renowned intellectuals of the time, the multitalented Francis Bacon (1561–1626) and the Oxonian scholar Robert Burton (1577–1640), advanced fundamental and largely opposing views on "the true Greatness of Kingdomes and Estates, and the Meanes thereof"[75] – both by engaging with the works of Machiavelli, Bodin and Botero.

Bacon's essay *Of the true Greatness of Kingdomes and Estates*, composed around 1622 but only published 3 years later, was in fact the culmination of almost two decades of meditation on this theme.[76] A shorter version of the same piece had appeared in the 1612 edition of *Essays*; even earlier, in 1608, Bacon had taken a firm stance on the issue in an unfinished letter to king James I, where he claimed that "true greatness" should not be identified with "largeness of territory", "treasure or riches", "fruitfulness of the soil or affluence of commodities", nor finally with "the strength and fortifications of towns or holds"; instead, "true greatness consisteth essentially in population and breed of men", and specifically in their "valour and military disposition".[77] Asserting the "predominancy of valour above treasure", Bacon quoted Machiavelli's famous line that men, not money, are the true "sinews of war"[78] – a statement that he had already recalled in an earlier speech to the House of Commons to explain why the "communication of naturalisation" to "the whole

[73] Bowden (1990), 60–61.

[74] See Gould (1955).

[75] Bacon (1985), XXIX, 90.

[76] The composition of the essay can be placed around this date on the basis of internal evidence, namely Bacon's reference to a "Pragmaticall Sanction, now published" which was issued by Philip IV of Spain in 1622: cf. Bacon (1985), XXIX, 95, and Spedding in Bacon (1900a), 127.

[77] Bacon (1900c), 233. The letter was published posthumously in 1634.

[78] Ibid., 243–244, explicitly referencing Machiavelli (1999), II.10, 317; Machiavelli (1965), II.10, 350. Cf. Bacon (1985), XXIX, 91: "The Principal Point of Greatnesse in any State is to have a Race of Military Men. Neither is Money the Sinewes of Warre, (as it is trivially said) where the Sinewes of Mens Armes, in Base and Effeminate People, are failing".

Scotch nation" would lead to England's "greatness and power": by incorporating a nation known for its brave and warlike character, he argued, "this kingdom of England" would quickly become "one of the greatest monarchies, in forces truly esteemed, that hath been in the world".[79]

England, besides, was already blessed with a number of natural advantages: the "fit situation of the region",[80] with "no intermixture or interposition of any foreign land, but only of the sea", of which the English nation was "absolutely master"[81]; a geographically compact, yet extremely varied territory, each of whose provinces performed different but equally "profitable" functions[82]; and, most importantly, the "prowess and valour" of the English people, not yet made "slothful and effeminate" nor "insolent and arrogant" by "excess of riches".[83] Despite this final point, Bacon was not blind to the strategic advantages of wealth. He appreciated that affluent nations are able "to aid and defray great charges for wars" more promptly and for a longer time than poor ones[84]; he also thought that they are less exposed to a dangerous type of domestic seditions – the "Rebellions of the Belly", as he tellingly called them – which have their first cause in "Want and Necessity".[85] In his revised version of the essay *Of Seditions and Troubles*, published in 1625,[86] Bacon suggested a series of measures intended to prevent popular revolts by promoting economic growth, full employment, and a positive balance of trade.[87] As he was keen to stress, however, wealth was only beneficial if it made the whole country rich, and not just a small portion of it; if it was used for the profit of all, rather than selfishly squandered by a minority. Wealth should be "dispersed" among the active classes – "merchants, burghers, tradesmen, freeholders, farmers",[88] who "live lower, and gather more"[89]—rather than "ingrossed" in the wasteful "hands of the nobility and

[79] Speech to the House of Commons of 17th February 1606/1607, quot. by Spedding in Bacon (1900c), 221–222. In "The true greatness of Britain" Bacon similarly argues that the granting of the rights of citizenship to the Latins and the Gauls was what allowed Rome to "grow great": Bacon (1900c), 238. Also cf. Bacon (1900b), 378 ("States liberall of naturalization, are capable of greatnesse"); Bacon (1985), XXIX, 93–94.

[80] Bacon (1900c), 252.

[81] Ibid., 242.

[82] Ibid., 240–241.

[83] Ibid., 248.

[84] Ibid., 250.

[85] Bacon (1985), XV, 45.

[86] An earlier version of this essay was published in the second edition of 1612, but most of the passages quoted here were introduced in the revised version of 1625.

[87] Bacon (1985), XV, 47–48.

[88] Bacon (1900c), 250–251.

[89] Bacon (1985), XV, 47.

gentlemen",[90] who "spend more, and earne less".[91] Then, and only then, would wealth "add true greatness and strength to a state".[92]

Bacon was convinced that the displacement of social and economic power from the unproductive classes to the labourers would also exert positive effects on the "Breed and disposition of the people"[93]: "Let States that aime at Greatnesse, take heed how their Nobility and Gentlemen doe multiply too fast," he wrote in 1625, rephrasing a concept already expressed in 1612, "for that maketh the Common Subject grow to be a Peasant, and Base Swaine, driven out of Heart".[94] Speaking out against the advocates of indiscriminate population growth, Bacon argued that it was pointless, not to say dangerous, for a state to have "Great Population and Little Strength"[95]: an overpopulated country full of hungry and degraded subjects makes a poor candidate for "Empire and Greatnesse"[96] and an ideal cradle for civic discord. Such a proportion must therefore be kept between "the Population of a Kingdome" and "the stock of the Kingdome which should maintain them", "as may breed a Subject to live in Convenient Plenty, and no Servile Condition".[97]

Bacon's emphasis on human qualities rather than numbers,[98] as well as his continuous insistence on the "essential difference between the scale of miles and the scale of forces",[99] have induced some scholars to speculate that the English philosopher might have developed his ideas on greatness in opposition to Botero.[100] As shown above, however, Botero too allowed that sheer quantity (of territory, riches, population) was only a feeble indicator of a state's "greatness". Like Bacon, he agreed with Machiavelli that valour, not money, was the true "sinews of war" (*nervo della guerra*), and quoted Bodin's famous motto that "there is no wealth nor strength but in men".[101] Far from being an uncritical advocate of territorial expansion, he was alert to what Bacon called the "weakness of states possessed of large territories",[102] and recommended a more intensive exploitation of existing resources and the improvement of living conditions for all as key components of genuine political development. As readers of Machiavelli and Bodin,[103] Bacon and Botero also shared a common vision of the Romans as a clever empire-building nation who adopted

[90] Bacon (1900c), 250–251.

[91] Bacon (1985), XV, 47.

[92] Bacon (1900c), 246.

[93] Bacon (1985), XXIX, 91.

[94] Ibid., XXIX, 92; cf. Bacon (1900b), 377.

[95] Bacon (1985), XXIX, 93.

[96] Ibid., XXIX, 95.

[97] Ibid., XXIX, 93 and XV, 47.

[98] Cf. ibid., XXIX, 90–91.

[99] Bacon (1900c), 236.

[100] Weber (2003), 338, and Slack (2014), 46, notably argue that Botero was the main intended target of Bacon's critique of "largeness of territory".

[101] Botero (1956b), VII.10, 141 and (1990b), VII, 197.

[102] Bacon (1900c), 234.

[103] A full study of Bacon's relationship with Bodin has not yet been done; but cf. Yates (1999), 145.

policies such as the "communication of naturalisation" and the foundation of settler colonies to increase their population and thus their "forces".[104]

At the same time, Bacon was less rigidly opposed to territorial expansion than is often thought. It suffices to look at his Latin self-translation of *The true Greatness of Kingdomes and Estates*, which appeared in the eight book of *De augmentis scientiarum* (1623), to realise that Bacon's "greatness" (indifferently translated with *magnitudo* and *amplitudo*) had a lot to do with "pushing the state's borders farther", as is literally stated in the Latin title of the essay ("De proferendis finibus imperii"): indeed, "augmenting the state and extending its borders" was described here as a task of equal importance as "preserving the state" and "making it happy and prosperous".[105] Such emphasis on territorial expansion was not necessarily in contradiction with the mistrust for large empires so vocally expressed in Bacon's 1608 letter to James I. While claiming that "largeness of territory is so far from being a thing inseparable from greatness of power, as it is many times contrariant and incompatible with the same", Bacon had also specified a series of conditions under which "greatness of territory" does in fact "add strength".[106] First and foremost among such conditions was that "the territories be compacted, and not dispersed"[107]: while the "defence" of "provinces dispersed [...] doth commonly consume and decay and sometimes ruin the rest of the estate",[108] it is much easier to keep a hold on colonies and dominions that are within close and easy reach from the mother-country.[109] Here again Bacon was following Botero, who had similarly drawn attention to distance as a discriminating factor in choosing the seat for a colony, and preferred the Roman way of "sending colonies near the mother-country" over the frail transatlantic empire of the Spanish and the Portuguese.[110]

Where Botero and Bacon crucially disagreed was in assessing England's "forces" and its chances of future greatness. In this regard, Bacon's 1608 letter to James I may be read as a sort of counterpoint to Botero's preamble to the second book of *Universal Relations*, which only a few years before (1601) had been translated into English.[111] In this section, dedicated to the question of how to "extend one's empire" (*estendere il dominio*), Botero carefully discussed all the factors that would later reappear in Bacon's letter on greatness: abundance of people, military valour,

---

[104] See Bacon (1985), XXIX, 94.

[105] Bacon (1900a), 120 (my translation).

[106] Bacon (1900c), 236–237.

[107] Ibid., 237.

[108] Ibid., 238.

[109] Ibid., 237–238.

[110] Botero (1956b), VIII.5, 157 and (1990b), VIII, 219. Yet Botero was better disposed in principle towards "dispersed states" (*stati disuniti*), especially if all of their provinces could easily be defended by sea: cf. Botero (1956b), I.7, 11 and (1990b), I, 13, referring again (this time approvingly) to Spain and Portugal.

[111] This was the second, expanded edition of Robert Johnson's translation (printed in London by John Jaggard as *The worlde, or an historicall description of the most famous kingdomes and common-weales*). An earlier edition, also published by Jaggard in 1601 but entitled *The Travellers Breviat*, did not contain the preamble to book 2. See Shackleton (1948), 405–406.

money and wealth, nature of the site and so forth. Among the countries favoured with an advantageous location Botero mentioned England, whose vast plains surrounded on all sides by the sea were open to all sorts of commercial exchanges yet easy to protect from the enemy. "How come then," he asked, "that islands endowed with such a site have never had a great empire (*dominio grande*)?" His answer was that just as "land forces are superior to maritime forces", so maritime empires, however powerful, are always weaker than land empires: "no empire can be great unless it extends itself inland".[112] Bacon's reaction was prompt: "To be Master of the Sea, is an Abridgement of a Monarchy," he wrote in 1622,[113] adding that "hee that Commands the Sea, is at great liberty, and may take as much, and as little of the Warre, as he will. Whereas those, that be strongest by land, are many times nevertheless in great Straights".[114] Consequently, Bacon's picture of England's future greatness, framed in opposition to Botero's preamble, envisaged a relatively small land empire (limited to the British Isles) with a firm hold on key commercial bases in Europe and maritime trade routes across the globe.[115]

Other readers of Botero, however, came to different conclusions. As noted by Fitzmaurice, Botero's theory of greatness "attracted many adherents who were at the forefront of colonization".[116] For men such as Robert Johnson and Walter Raleigh, both personally involved in England's colonial efforts in North America,[117] Botero's ideas on empire were an important source of guidance. Raleigh, whose *Observations concerning the Causes of the Magnificency and Opulence of cities* (probably composed in the early seventeenth century, but published posthumously in 1651) are actually an abridged translation of Botero's *Greatness of Cities*,[118] was "one of the first to see in Botero's writing a means of emulating and therefore rivaling Spanish *grandezza*"[119] – quite ironically, in fact, given "Botero's pro-Spanish

[112] Botero (1596), II, "Proemio" (unpaged).

[113] Bacon (1985), XXIX, 97. Cf. Bacon (1900a), 132: "Maris Dominium monarchiae quaedam epitome est". Cf. Bacon (1900c), 233: "true greatness […] consisteth in the commandment of the sea".

[114] Bacon (1985), XXIX, 98. In fact, Botero too conceded that maritime forces, while they might not add "strength" (*nervo*), do usefully contribute "greater mobility" (*agilità*), making it possible to "attack the enemy unexpectedly in many places and to keep him continually undecided": Botero (1956b), X.7, 217–218 and (1990b), X, 301.

[115] In his speech at the House of Commons of 17th February 1606/1607, Bacon dreamed of having "Scotland united, Ireland reduced, the sea provinces of the Low Countries contracted, and shipping maintained": quot. by Spedding in Bacon (1900c), 222.

[116] Fitzmaurice (2007), 798.

[117] Johnson (dates unknown) was deeply involved with the Virginia Company and in 1609 published a promotional text entitled *Nova Britannia, Offering Most Excellent Fruites by Planting in Virginia* (see Fitzmaurice (2007); Paul and Meschkat (2013)). Raleigh (ca. 1554–1618) took personally part in the exploration and colonisation of Virginia, where he founded the unsuccessful colony of Roanoke Island.

[118] See Symcox in Botero (2012), xiv.

[119] Fitzmaurice (2007), 798.

sentiments".[120] Competition with Spain was also Johnson's primary motivation for producing no less than six English translations of Botero's *Universal Relations*, whose textual layers – closely examined in a recent study by Paul and Meshkat – reveal Johnson's "attempts to intervene in British international policy-making by putting forward a specific vision of global order as maintained by a British empire balancing that of the Spanish".[121] Behind such efforts was a reading of Botero that took seriously the latter's calls for inland expansion, the establishment of colonial settlements, and – in Raleigh's case – urban growth as vital measures towards a strong empire.

Despite their disagreements, Bacon, Johnson and Raleigh all shared a relatively optimistic view of England as a world power with legitimate aspirations to greatness; the Oxonian scholar Robert Burton, on the other hand, saw things from a radically different perspective. The first edition of his monumental masterpiece, *The Anatomy of Melancholy*, came out in 1621, only a few months into the economic crisis that shattered the final years of James I; Burton, who already had a clear sense of its gravity,[122] devoted large part of his general preface to analysing its causes in detail, in close intertextual dialogue with Botero's writings,[123] as well as with those of Machiavelli, Bodin and others. "Our land is fertile wee may not deny, full of good thinges, and why doth it not then abound with Citties, as well as Italy, France, Germany, the Low countries?" he wondered in his preface.[124] The search for an answer led him to paint a dreadful image of England as a country overwhelmed by melancholy,[125] "a Paradise turned to a wilderness"[126]:

> so many thousand acres of our fens lye drowned, our Cities thin, and those vile, poore, and ugly to behold […] our trades decayed, our still running rivers stopped, and that beneficiall use of transportation, wholly neglected, so many Havens void of ships and Townes, so many Parkes and Forrests for pleasure, barren Heaths, so many villages depopulated.[127]

While the Low-Countries are described as an "epitome of China by reason of their industry and commerce",[128] England stands as a painful example of self-inflicted decadence. For Burton, the chief causes of the country's decline were

---

[120] Paul and Meshkat (2013), 127.

[121] Ibid., 110. Johnson's translations came out in 1601 (two editions), 1603, 1608, 1611, and 1616. A new translation expanding on the 1616 edition came out in 1630, but it was no longer the work of Johnson.

[122] See Oliveira and Grignon (2003).

[123] See Gowland (2006), 234. Burton quotes from Draud's Latin translation of *Reason of State* and *The Greatness of Cities*, published in 1602 (Ursellis: Apud Cornelium Sutorium, impensis Lazari Zetzneri) as *Tractatus duo: De Illustrium statu & politia libris X; De origine urbium… libris III*).

[124] Burton (1989), 76. Since this edition is based on the 1635 print, I have collated its text with that of the first edition of 1621 to ascertain the absence of variants in the passages quoted here.

[125] Burton (1989), 66–67. Cf. Oliveira and Grignon (2003).

[126] Ibid., 74.

[127] Ibid., 75.

[128] Ibid., 77.

indeed bad government[129] and the natural laziness of the English people: "Idlenesse is the MALUS GENIUS of our Nation. For as *Boterus* justly agrees, fertility of a Country is not enough, except Art and Industry be joined unto it".[130] Industry is what makes "a barren soyle to be fertile and good": it is "a load-stone to draw all good things",[131] the one infallible means "to the ornament and enriching of a kingdom".[132] Without industry, the richest country decays; where industry thrives, on the other hand, "you shall see the people civill [...] peaceable and quiet, rich, fortunate, and flourish, to live in peace, in unity and concord, a Country well tilled, many faire built and populous Citties".[133]

While they both drew on Botero's writings to develop their theories, Burton's view of England was in many respects at odds with that of Bacon. Idleness, which the former saw as the root of all ills, was for Bacon a distinctive trait of "all Warlike People", thus a quality to encourage, rather than curb, "if they shall be preserved in vigour".[134] Clearly Burton did not share Bacon's dreams of "Martiall Greatnesse"; his plan for the regeneration of England focused on intensive development of the country's lands rather than on struggles for territorial expansion. "The lesser the Territory is, commonly the richer it is", he wrote in fact, offering a number of examples from Botero's writings.[135] Not by chance, that "UTOPIA of mine own, a new ATLANTIS",[136] which he sketched in the final pages of the preface, was an exemplary experiment in land management that would inspire generations of British thinkers and "improvers" after him.[137]

Conclusion

All concepts have a history, and the hundred years that separate Machiavelli's *Discourses* from the final edition of Bacon's *Essays* were arguably a golden age in the century-long history of "greatness". For Fitzmaurice, *grandezza* is *the* essential concept in the history of early modern state- and empire-building, as its pursuit "facilitated the transformation of Europe's numerous feudal political units, cities, and principalities into the beginnings of the modern state system"; for Pocock, the long shadow cast by Machiavelli's *grandezza* marked the entire politico-intellectual

129 Ibid., 74 and 76.

130 Ibid., 76. The reference is to Botero (1956a), I.9, 235 and (1990a), I, 320.

131 Burton (1989), 77.

132 Ibid., 78,

133 Ibid., 66.

134 Bacon (1985), XXIX, 95.

135 Burton (1989), 78. The reference is in particular to Botero (1956b), VIII.3, 150–153 and (1990b), VIII, 210–214.

136 Burton (1989), 85.

137 Cf. ibid., 88–89; on improvement see Slack (2014).

life of the seventeenth century.[138] What has not been sufficiently emphasised thus far, however, is that in the course of these hundred years Machiavelli's *grandezza* changed shape and took on new meanings as it was translated into different languages, applied to new contexts and adapted to the political agendas that it helped create.

This chapter has sought to show how the debate over the meaning of "greatness" and the proper ways to achieve it brought together thinkers from the four corners of Europe to engage in a lively intertextual conversation which was not without consequences at the level of actual policy-making. Machiavelli, Bodin and Botero quickly came to represent a canon of authors to whom any discussion of "greatness" should necessarily refer. In England, where the reception of their works was particularly intense, opposing understandings of what "true greatness" is stimulated writers such as Bacon, Johnson, Raleigh and Burton to develop different schemes for the advancement (or the regeneration) of the country.

But the history of "greatness" does not end here: the concept was still to have a fascinating afterlife, for instance in France – Rousseau's chapter on "The People" in the second book of *Social Contract* is in part a meditation on what a state's "true greatness" (*véritable grandeur*) is, and how it can be measured and achieved[139] – and Scotland, if Istvan Hont is right in suggesting that "the positive core of Machiavellian ideas of *grandezza*, namely that flourishing political communities had to be able to grow" inspired Adam Smith's thesis that "the most flourishing political communities were not the richest, but the ones that grew the fastest".[140] While a complete genealogy of "greatness" will have to wait for future studies, following its trajectories from the first decades of the Cinquecento to the early seventeenth century has enabled us to shed some light on an exemplary case of unceasing dialogue between the Renaissance and the early modern period.

Acknowledgements I would like to thank David Lines for comments on an earlier draft of this chapter.

Bibliography

Arienzo, Alessandro, and Alessandra Petrina (eds.). 2013. *Machiavellian Encounters in Tudor and Stuart England. Literary and Political Influences from the Reformation to the Restoration*. Aldershot: Ashgate.

Armitage, David. 2002. Empire and Liberty: A Republican Dilemma. In *Republicanism: A Shared European Heritage*, vol. 2, ed. Martin van Gelderen and Quentin Skinner, 29–46. Cambridge: Cambridge University Press.

[138] Pocock (2003), part 3.

[139] Rousseau (1992), II.10, 73–74; cf. Rousseau (1997), 75–77.

[140] Hont (2008), 307.

Bacon, Francis. 1900a. De proferendis finibus imperii (De augmentis scientiarum, VIII.3). In *The works of Francis Bacon*, vol. 3, ed. James Spedding et al., 120–135. Boston: Houghton, Mifflin and Company.

Bacon, Francis. 1900b. Of the Greatnesse of Kingdomes. In *The Works of Francis Bacon*, vol. 12, ed. James Spedding et al., 376–378. Boston: Houghton, Mifflin and Company.

Bacon, Francis. 1900c. On the True Greatness of Britain. In *The Works of Francis Bacon*, vol. 13, ed. James Spedding et al., 219–255. Boston: Houghton, Mifflin and Company.

Bacon, Francis. 1985. *The Essayes or Counsels, Civill and Morall*, ed. Michael Kiernan. Oxford: Clarendon Press.

Bodin, Jean. 1593. *Les Six Livres de la Republique*. Lyon: Barthelemy Vincent.

Botero, Giovanni. 1990a. Cause della grandezza e magnificenza delle città. In *Della ragion di Stato, e Cause della grandezza e magnificenza delle città*, ed. Luigi Firpo. Bologna: Forni. English translations: Botero, Giovanni. 1956a. The Greatness of Cities. In *The Reason of State & the Greatness of Cities*, trans. Robert Peterson, ed. D.P. Waley. London: Routledge and Kegan Paul. Botero, Giovanni. 2012. *On the Causes of the Greatness and Magnificence of Cities*, trans. Geoffrey Symcox. Toronto: University of Toronto Press.

Botero, Giovanni. 1990b. Della ragion di Stato. In *Della ragion di Stato, e Cause della grandezza e magnificenza delle città*, ed. Luigi Firpo. Bologna: Forni. English translation: Botero, Giovanni. 1956b. The Reason of State. In *The Reason of State & the Greatness of Cities*, trans. P.J. and D.P. Waley, ed. D.P. Waley. London: Routledge and Kegan Paul.

Bowden, Peter J. 1990. Agricultural Prices, Farm Profits and Rents 1500–1640. In Id. *Economic Change: Wages, Profits and Rents, 1500–1750. Chapters from the Agrarian History of England and Wales,* vol. 1, Cambridge: Cambridge University Press.

Brown, Alison. 2004. The Language of Empire. In *Florentine Tuscany. Structures and Practices of power*, ed. William J. Connell and Andrea Zorzi, 32–47. Cambridge: Cambridge University Press.

Burgess, Glenn. 2013. Bodin in the English Revolution. In *The Reception of Bodin*, ed. Howell A. Lloyd, 387–407. Leiden: Brill.

Burton, Robert. 1989. *The Anatomy of Melancholy*, vol. 1, ed. Thomas C. Faulkner et al. Oxford: Clarendon Press.

Chabod, Federico. 1967. Giovanni Botero. In Id. *Scritti sul Rinascimento*. Turin: Einaudi.

Descendre, Romain. 2009. *L'état du monde. Giovanni Botero entre raison d'État et géopolitique*. Genève: Droz.

Elliott, J.H. 1977. Self-Perception and Decline in Early Seventeenth-Century Spain. *Past and Present* 74(1): 41–61.

Fitzmaurice, Andrew. 2007. The Commercial Ideology of Colonization in Jacobean England: Robert Johnson, Giovanni Botero, and the Pursuit of Greatness. *The William and Mary Quarterly* 64(4): 791–820.

Glacken, Clarence. 1976. *Traces on the Rhodian shore: Nature and Culture in Western Thought from Ancient Times to the End of the Eighteenth Century*. Berkeley: University of California Press.

Gould, J.D. 1955. The Trade Crisis of the Early 1620s and English Economic Thought. *Journal of Economic History* 15: 121–133.

Gowland, Angus. 2006. *The Worlds of Renaissance Melancholy: Robert Burton in Context*. Cambridge: Cambridge University Press.

Hankins, James. 1995. The "Baron Thesis" After Forty Years and Some Recent Studies of Leonardo Bruni. *Journal of the History of Ideas* 56: 309–338.

Hankins, James. 2000. Rhetoric, History, and Ideology: The Civic Panegyrics of Leonardo Bruni. In *Renaissance Civic Humanism Reconsidered*, ed. James Hankins, 143–178. Cambridge: Cambridge University Press.

Hartman, Jan, and Arthur Weststeijn. 2013. An Empire of Trade. In *The Political Economy of Empire in the Early Modern World*, ed. Sophus Reinert and Pernille Røge, 11–31. Basingstoke/New York: Palgrave Macmillan.

Hont, Istvan. 2008. The "Rich Country- Poor Country" Debate in Scottish Classical Political Economy. In *David Hume's Political Economy*, ed. Carl Wennerlind and Margaret Schabas, 243–323. Oxford/New York: Routledge.

Hörnqvist, Mikael. 2004. *Machiavelli and Empire*. Cambridge: Cambridge University Press.

Machiavelli, Niccolò. 1999. *Discorsi sopra la prima deca di Tito Livio*, ed. Giorgio Inglese. Milan: BUR. Latin translation: Machiavelli, Niccolò. 1588. *Disputationum de republica libri*, trans. Johann Nikolaus Stupanus. Mompelgarti (Montbéliard): Per Iacobum Folietum Typographum. English translation: Machiavelli, Niccolò. 1965. *Discourses on the First Decade of Titus Livius*. In *The Chief Works and Others*, vol. 1 trans. Allan Gilbert. Durham: Duke University Press.

Miglietti, Sara. 2013. Reading from the Margins: Some Insights into the Reception of Bodin's *Methodus* in England. In *The Reception of Bodin*, ed. Howell A. Lloyd, 193–217. Leiden: Brill.

Najemy, John. 2000. Civic Humanism and Florentine Politics. In *Renaissance Civic Humanism Reconsidered*, ed. James Hankins, 75–104. Cambridge: Cambridge University Press.

Oliveira, De, and Claire Grignon. 2003. Mercantilisme et utopie dans la Préface de l'Anatomie de la Mélancolie de Robert Burton. *Revue de Métaphysique et de Morale* 3: 345–363.

Paul, Joanne, and Kurosh Meschkat. 2013. Johnson's Relations: Visions of Global Order, 1601–1630. *The Journal of Intellectual History and Political Thought* 2(1): 108–140.

Pocock, John. 2003. *The Machiavellian Moment: Florentine Political Thought and the Atlantic Republican Tradition*. Princeton: Princeton University Press.

Raab, Felix. 1965. *The English Face of Machiavelli: A Changing Interpretation, 1500–1700*. London: Routledge and Kegan Paul.

Reinert, Sophus A. 2011. *Translating Empire. Emulation and the Origins of Political Economy*. Cambridge: Harvard University Press.

Rousseau, Jean-Jacques. 1992. *Du contrat social*, ed. Pierre Burgelin. Paris: Garnier Flammarion. English translation: Rousseau, Jean-Jacques. 1997. *The Social Contract and Other Later Political Writings*, trans. Victor Gourevitch. Cambridge: Cambridge University Press.

Ruffo Fiore, Silvia. 1990. *Niccolò Machiavelli: An Annotated Bibliography of Modern Criticism and Scholarship*. Portsmouth: Greenwood Publishing Group.

Shackleton, Robert. 1948. Botero, Bodin and Robert Johnson. *The Modern Language Review* 43(3): 405–409.

Skinner, Quentin. 1995. The State. In *Political Innovation and Conceptual Change*, ed. Terence Ball et al., 90–131. Cambridge: Cambridge University Press.

Skinner, Quentin. 2003. The Vocabulary of Renaissance Republicanism: A Cultural *longue durée*? In *Language and Images of Renaissance Italy*, ed. Alison Brown, 87–110. Oxford: Clarendon.

Slack, Paul. 2014. *The Invention of Improvement. Information and Material Progress in Seventeenth-Century England*. Oxford: Oxford University Press.

Springborg, Patricia. 1992. *Western Republicanism and the Oriental Prince*. Cambridge: Polity Press.

Stangeland, Charles Emil. 1904. *Pre-Malthusian Doctrines of Population: A Study in the History of Economic Theory*. New York: Columbia University Press.

Weber, Dominique. 2003. Grandeur civique et économie dans la pensée politique de Francis Bacon. *Revue de Métaphysique et de Morale* 39: 323–344.

Yates, Frances. 1999. The Mystery of Jean Bodin. In *Ideas and Ideals in the North European Renaissance: Collected Essays*, 139–152. London/New York: Routledge and Kegan Paul.

Chapter 15
John Upton from Political Liberty to Critical Liberty: The Moral and Political Implications of Ancient and Renaissance Studies in the Enlightenment

John Christian Laursen

Abstract John Upton (1707–1760) was a humanist scholar in the tradition of Shaftesburian moral philosophy who began as a robust supporter of political liberty but ended up withdrawn to the position of a promoter of critical liberty. His earlier work on ancient moral philosophy went hand in hand with criticism of the political establishment, but his later Renaissance studies were associated with a withdrawal from political commitments. He exemplified a pattern according to which literary criticism which begins in the service of a moral and political ideal ends up narrowing its purposes to goals and ideals internal to literary criticism.

As the last chapter of a book on seventeenth-century appropriations of Renaissance philosophy, this chapter broadens the perspective in three dimensions. It takes the story into the eighteenth century. It demonstrates that in addition to drawing on Renaissance philosophers in the narrower sense, early modern moral philosophers could also draw on Renaissance authors such as Shakespeare, Ben Jonson, and Edmund Spenser. And it expands the study of the influence of early modern Renaissance studies in early modern philosophy by looking at the implications for political and moral philosophy.

This is an essay on how Renaissance studies served to depoliticize a scholar who had begun his career politicized by ancient studies. John Upton (1707–1760) was a humanist scholar in the tradition of Shaftesburian moral philosophy who began as a robust supporter of political liberty but ended up withdrawn to the position of a promoter of critical liberty. He exemplified a pattern according to which literary criticism which begins in the service of a political ideal ends up narrowing its purposes to goals and ideals internal to literary criticism.

J.C. Laursen (✉)
Department of Political Science, University of California at Riverside, Riverside, CA, USA
e-mail: johnl@ucr.edu

C. Muratori, G. Paganini (eds.), *Early Modern Philosophers and the Renaissance Legacy*, International Archives of the History of Ideas Archives internationales d'histoire des idées 220, DOI 10.1007/978-3-319-32604-7_15

Coming as the last chapter of a book on seventeenth-century appropriations of Renaissance philosophy, this chapter broadens the perspective in three dimensions. One, it moves into the eighteenth century. Two, it demonstrates that in addition to Renaissance philosophers, early modern philosophers could also draw on Renaissance authors such as Shakespeare, Ben Jonson, and Edmund Spenser. And three, it expands the study of the influence of early modern Renaissance studies in early modern philosophy into looking at the implications for political and moral philosophy.

John Upton as a Shaftesburian Humanist

"Shaftesburian" here means following in the footsteps of Anthony Ashley Cooper, third Earl of Shaftesbury (1671–1713), a moral philosopher who drew heavily on the tools of literary criticism in formulating his philosophy. John Upton was an accomplished Shaftesburian humanist who provided us with a scholarly edition of Epictetus in the Greek, together with a translation into Latin, in 1739–41.[1] After his heroic edition of Epictetus, Upton concentrated most of his work on English Renaissance writers, issuing *Critical Observations on Shakespeare* in 1746, *Remarks on three plays of Benjamin Jonson* in 1749, and *Spenser's Faerie Queene* in 1758. Upton was part of a number of networks of elite philosophers, humanists, and translators in the England of his day. For the edition of Epictetus, he drew on annotations provided by Shaftesbury, and had the assistance of James Harris.[2] Thus his work is symptomatic of larger currents of the moral philosophy and sophisticated humanistic scholarship of his time.

We do not always think of the work of scholars of philosophical and literary history in terms of political intentions, but that is precisely what I shall be looking for in this essay. In the earliest of his works Upton transmitted Epictetus's Stoicism, including Stoicized Cynicism, together with its political implications. In the first of his writings on Renaissance authors he retailed the political language of Whiggish and Shaftesburian republicanism: the vocabulary and tropes that insisted that Britain should model itself on the ancient Greek and Roman republics. And in his last work he wrote of "critical liberty". I will suggest that by the time of that work he seems to have realized that his work was only going to have political implications of the most indirect and attenuated sort, and that the real driver in his life was the claims of critical scholarship.

Upton's clues about his political sympathies provide us with a set of questions about the relationship between elite philosophy, humanistic scholarship, and politics. Do philosophy and humanistic scholarship have special affinities with certain kinds of politics? What makes a closeted scholarly type with no experience in real political life think he has special insight into political matters, and that his readers

[1] Uptonus [Upton] (1739–1741).

[2] Oldfather (1925), vol. 1, xxxii–xxxiii.

will want to hear his thoughts on them? Is there any reason to trust the insights of such intellectuals? These questions apply widely, to any intellectuals who opine about politics,[3] but we will limit our inquiry here to Upton.

I conclude that Upton's elite scholarship dovetailed very well with the elitism of Shaftesburian Whiggish politics. Rather like Shaftesbury himself, the cultural politics of Upton combined a justification of his own withdrawal from actual politics for the life of the scholar with a subversive critique of the political system which was, paradoxically, just as elitist. The republican politics that he idealized was not an egalitarian, leveling sort of politics, but rather a republicanism for elites. Then as now, most intellectuals really favor rule by mandarins.

By way of background, we may begin with a quote from the first sentence of the Preface to his edition of Spenser's *Faerie Queene*. "As every original work, whether of the poet, philosopher, or historian, represents, mirrour-like, the sentiments, ideas and opinions of the writer, so the knowledge of what relates to the life, family, and friendships of such an author, must in many instances illustrate his writings; and his writings again reflect the image of the inward man."[4] If we can assume that criticism can be original, and falls somewhere under poetry, philosophy, or history, then we should look for such contexts for understanding our author.

So it is worth knowing that John Upton was born in 1707 in Taunton, the second son of schoolmaster and classicist James Upton (1670–1749). His father edited Gulston's *Poetics of Aristotle* (1696), translated Dionysius of Halicarnassus (1702), and published *A Selection of Passages from Greek Authors* (1726). John was educated by his father and matriculated at Merton College, Oxford, in 1724. He was a fellow of Exeter College from 1728 to 1736/7. Part of the meaning of what I have been calling "elite" was the access to the many years of education that it took to master Greek and Latin. Having a classical educator for a father helped, and so did networks of patronage that made further study and ecclesiastical benefices possible. After receiving the rectory of Seavington (1732) he was admitted prebendary of Rochester (1736/7), and eventually held several other benefices. His duties must not have been very time-consuming because, judging from his output, his literary efforts must have been all-consuming.[5]

Upton did not work alone. He belonged to intellectual circles that included his father, James Harris, and even the German composer Handel.[6] As we shall see, the work of some of these other scholars helps us understand Upton's.

It is worth noting that Upton's father depended on patronage from two Royalist families, the Pouletts and the Sydenhams. Yet Upton's politics were clearly Whig

---

[3] For distrustful answers to these questions, see Lilla (2001) and Fish (1995).

[4] Upton (1758), vol. 1, v.

[5] Marchant (1921–22), vol. 20, 39; Marchant and Skedd (2004), vol. 55, 932. In these works, paragraphs on John are part of the entry on his father, James. It is regrettable that there is no article on Upton in Todd (2004). The best summary of the life and career of Upton, based in part on original archival research, is Radcliffe (1987), v–xx.

[6] See Smith (1995). Upton failed to persuade Handel to write an oratorio based on Milton's *Paradise Lost*: Smith (1997), 102. See also Probyn (1991), 72–3.

and anti-court. It is a pity that we do not have any details about how they may have felt when Upton's published work expressed his political leanings. We can speculate that there were at least two possible responses. One was that they felt betrayed by this son of a recipient of their generosity, but that he felt independent enough from them to not worry about the loss of their patronage. The other is that the world of provincial patronage allowed the expression of such differences of political opinions as part of normal social life, given that there was little or no fear that it would lead to any real political change. The establishment was so well established that stereotypical radical language about freedom from tyranny and praise for ancient republicans could be seen as no threat whatsoever.

Epictetus

The Epictetus translation is a *tour de force*, a careful collation of Greek manuscripts, commentary from several sources, and an accurate and elegant Latin translation. His work was used extensively in later editions and translations including Elizabeth Carter's translation into English of 1758, Schweighäuser's major scholarly edition of 1799 which picked up Upton's Latin translation entire, and Oldfather's translation for the Loeb Classical Library in 1925.[7] Why translate Epictetus? Career incentives were surely one reason. Epictetus's *Enchiridion* was widely recommended for use by students and clergymen[8] and had even been translated into English,[9] but Arrian's version of Epictetus's *Discourses* was not well known. Upton may have had opening up this market in mind. Money could be made from the classics: Elisabeth Carter made a thousand pounds sterling from her translation of Epictetus into English in 1758.[10] And it was also a path to preferment to higher office in the church. In a state-run church that sometimes seemed to approach David Hume's apparent ideal of government by the Confucian literati,[11] many men were appointed to bishoprics and other benefices largely on the basis of their classical scholarship.[12]

---

[7] Epictetus (1759); Epictetus (1799–1800); Epictetus (1925). In Epictetus (1799–1800) Schweighäuser also reprinted "Ioannis Vptoni Praefatio ad lectorem", v–xii. Volume 2 of Epictetus (–1800) consisted of "Notae et emendationes Hieron. Wolfii, Io. Vptoni aliorumque virorum doctorum".

[8] Clarke (1959), 169, 222.

[9] Epictetus (1694). Stanhope's preface justifies the reading of a pagan philosopher on the ground that it would provoke Christians into behaving better in order to vindicate their religion (p. 9 of unnumbered Preface).

[10] Clarke (1945), 5, 115. Clarke is aware that Upton "Edited Epictetus, 1739-41" (p. 233), but does not include him in his "List of translations from Greek authors" (p. 242); conceivably because he means "translations into English"?

[11] Hume (1987), 78.

[12] Clarke (1945), 3–4.

Why translate Epictetus from Greek into Latin, and not into English? Even if the market for translations into Latin may have been narrower than that for translations into English, the Latin translations had a near-monopoly on the school text market. In the Grammar Schools of Upton's day, editions of Greek authors were commonly furnished with Latin translations and the Greek was taught by way of the Latin.[13] Upton also may have translated into Latin in order to sell into the European market, where they were unlikely to read English. As we have just seen, at the end of the century Schweighäuser borrowed the translation and Upton's notes. But that point also calls for caution in assuming that Upton wanted to sell to the Europeans: he must have known, in the days before enforceable copyright, that Europeans might reprint his work without sending him any royalties. There remains the explanation of Upton's purposes as a personal matter: an opportunity to exercise and demonstrate his virtuosity, and gain fame. Products of the kind of education he received from his father could get to the point where they thought and wrote Latin better than English.[14] He may have thought of Latin as the appropriate language for any serious edition of a classical work.

As already mentioned, Upton did not have to take on Epictetus alone. Rather, his work was part of a larger project promoted by Shaftesbury and other friends of Upton such as James Harris to redeem English intellectual life by reappropriating the treasury of classical thought. As we shall see, they even thought this would have beneficial political effects. Just as a reminder of the elite status of many of these men, James Harris and his wife "patronized the best artists of the day at their house in the cathedral town of Salisbury", as his great-grandson, the Third Earl of Malmesbury, put it.[15]

Harris provided Upton with Shaftesbury's detailed annotations in a copy of Epictetus's *Enchiridion*.[16] These show up in Upton's work as many notes identified as "Com. Shaftsburiensis". Shaftesbury's annotations, in turn, were not merely the random product of an idle passtime. Rather, they were an important part of his philosophical project. Epictetus was by far the most-cited author in Shaftesbury's manuscript "Askemata", published by Benjamin Rand as his "Philosophical Regimen".[17] And he was also quoted or paraphrased often in Shaftesbury's published *Characteristics of Men, Manners, Opinions, Times* (first ed. 1711; 4th ed. 1732).

Epictetus, it is well known, is one of our chief sources for understanding Stoicism. Shaftesbury's interest in Stoicism was part and parcel of his overall philosophy, many of the elements of which were promoted by followers such as Harris and Upton. Chief among these were the rejection of religious superstition and enthusiasm, the defense of political liberty understood in ancient republican terms, and a belief that classical scholarship could help support such liberty. Shaftesbury's grandfather, the First Earl of Shaftesbury, was a key figure behind the restatement of

[13] Clarke (1959), 51–2; Clarke (1945), 15.

[14] Clarke (1959), 58.

[15] Malmesbury (1870), vol. 1, vii.

[16] Probyn (1991), 58.

[17] Rand (1900).

the old antithesis of Court and Country in England for the new conditions of ministerial corruption of Parliament, and the Third Earl generally followed his grandfather's Whig political inclinations.[18]

The Third Earl was also an innovator and consolidator of the growing "culture of politeness".[19] This was the culture of coffee houses and town homes in London where the landed elite had been mixing with writers and pundits since the seventeenth century to create a new public philosophical sphere.[20] It depended on the circulation of a wide variety of literary materials ranging from newspapers and moral weeklies down through an underground of clandestine manuscripts, pornography, and deist and atheist tracts, and up to grand scholarly editions such as Upton's Epictetus. Intellectuals and writers could rub elbows with the landed elite if they knew how to meet the standards of politeness in dress, languages, and behavior. John Upton and his circle in the West of England adapted this culture for the smaller towns. Frequent visits of city people out to the country and of country people in to the city made this circulation of texts and ideas possible. As James Harris's great grandson put it somewhat elegiacally a century later, in those days "the Provincial Gentry filled and enlivened, during a portion of the year, our now deserted and mournful cathedral Cities".[21]

The Third Earl of Shaftesbury set the groundwork for understanding classical studies as a form of political action. In his *Characteristics* he repeatedly associated literary criticism with political liberty. Taste and judgment are a product of criticism, he wrote, and without taste and judgment you cannot have good politics.[22] In turn, criticism will always be in repute when "Antient Authors… are in request" (3.149). Liberty is the "native privilege of criticism", and vice versa (3.165). In a late sketch added for the first time to the 1732 edition of his *Characteristics*, "A Letter Concerning the Art, or Science of Design", Shaftesbury makes the connection between criticism and political liberty over and over (3.247 ff.).

Shaftesbury promoted classical studies as a way of forming one's own philosophical judgment and learning how to think for oneself. It allows one to stand back from the prejudices of one's time so that one will not fall for the narratives that the Church and other powerful institutions will try to sell. The political implications were obvious to him: this kind of training would make one independent and enable one to take advantage of political liberty. Shaftesbury never had to spell out where one should do his thinking for himself and exercise his political liberty because everyone knew he had held a seat in Parliament. Upton's friend James Harris later held a seat in Parliament, and served as Lord of the Treasury and Secretary and Comptroller of the Queen's household. Nevertheless, many writers and intellectuals like Upton would never have such seats unless their gentry friends put them up

[18] Pocock (1975), 406–420.

[19] Klein (1994).

[20] See, for example, Laursen (2002).

[21] Malmesbury (1844), vol. 1, vi.

[22] Shaftesbury (2001), vol. 3, 101.

for them and they held them as a matter of patronage. But all of them could imaginatively superimpose the images of republican Greek and Roman politics from their classical educations upon their own roles in contemporary political life, no matter how self-deceptive this might be.

Since ancient times, Stoicism and neo-Stoicism had always been the basis of a politics for elites. Cicero, Marcus Aurelius (much cited by Shaftesbury), and other Roman elites had taken over the Greek Stoics for Roman republican and Roman imperial purposes. In early modern Europe neo-Stoicism had provided the elements for a political language for elites, especially appealing to soldiers.[23] Shaftesbury's contribution was to bring Stoicism into "polite" circles of the gentry and nobility in the newly arising conditions of intellectuals who moved between London, the universities, and their country seats or livings.

This Shaftesburian Whiggism was endorsed by many in Upton's circles, including James Harris.[24] In letters to Upton, Harris wrote of the need for a "rational and philosophical" commentary on Stoics such as Epictetus.[25] Harris also collated Upton's version of Epictetus with several previous editions including Salmasius's and "an old MS collation, which I lately purchased, written in the margin of the old Venice edition of 1528... [with] the commentary of Simplicius upon it".[26] And he also annotated the proofs of Upton's edition.[27]

Harris made his purposes in helping his friend with Epictetus clear in *Three Treatises* (1744), *Hermes: or, A Philosophical Inquiry concerning Language and Universal Grammar* (1751), and *Upon the Rise and Progress of Criticism* (1752).[28] Classical philosophical studies would prevent the decline of English culture that came from forgetting the past. They were a therapy for the insular and ahistorical culture of the period, and a remedy for the materialism and cultural complacency he saw around him. The study of the past was a political act that would give strength to political liberty in the present.[29]

There is not much in Upton's notes to Epictetus from which to make out any political intentions, although there is an occasional declaration of sympathy. So, for example, as a comment on Epictetus's report that Cynics were not upset by exile, Upton quotes from Shakespeare's *Cymbeline*: "There's living out of Britain".[30] For a Cynic, exile was not the worst thing in the world; nationalism and patriotism are not important. This is not the sort of critical apparatus that would have been strictly

---

23 See Oestreich (1982).

24 Probyn (1991), 63–4.

25 Quoted in Probyn (1991), 58. Oddly enough, Probyn cites three different dates for Upton's edition of Epictetus: 1769 (p. 58), 1739–1740 (p. 71), and 1738 (p. 98).

26 Probyn (1991), 58, 71.

27 Ibid., 71.

28 In *Rise and Progress* Harris lists Upton's Epictetus among outstanding examples of modern criticism (cited in Probyn (1991), 70).

29 This summarizes material in Probyn (1991), 79, 105, 159–60.

30 Uptonus [Upton] (1739–1741), vol. 2, 206.

necessary in a scholarly commentary on Epictetus, and rather shows both Upton's interest in Shakespeare and perhaps some sympathy with Cynic self-sufficiency and moralism.

Epictetus's discourse *Peri Kunismou* (On the Calling of the Cynic)[31] represents a "cleaning up" of the ancient Cynic tradition for Stoic and early modern moralism. As Robert Voitle put it in his book on Shaftesbury, Epictetus was "the most Cynic of the great teachers of [Stoicism]".[32] We can turn Voitle's observation around and add that Epictetus's and Shaftesbury's Cynicism was a very Stoicized Cynicism. As a matter of fact, Shaftesbury died in Naples, "living out of Britain".

In Epictetus, Cynics are self-denying moral paragons, using their outsider status to chide the establishment and its beneficiaries, and Cynic nihilism or destructive criticism is played down. Thus, Upton transmitted a "civilized" and moderated Cynicism. It was not the kind of libertine, "in your face" and anti-social Cynicism that soon got Johann Friedrich Struensee in trouble in 1770s Denmark.[33] Rather, it was closer to the Cynicism also transmitted by Pedro de Valencia in the late sixteenth and early seventeenth centuries, which centered on retirement and self-control.[34] Upton's version contributed to the European-wide rediscovery of this aspect of ancient Cynicism in the eighteenth century.[35] Serious moral philosophers like Thomas Reid may have read Epictetus in Upton's edition.[36]

Right after his translation of Epictetus, Upton began to transfer his interests to English literature of the Renaissance. In 1740 he wrote remarks in the margins of a 1640 edition of Benjamin Jonson's plays, which he published later in 1749.[37] He made it very clear that his interests in this English writer revolved around Jonson's classical allusions. In Upton's opinion, Jonson's "good genius seems to have forsaken him, whenever he forsook the guides of antiquity, and trusted to his own natural strength".[38] His text consists largely of quotes from Jonson followed by the identification of his sources in Horace, Pindar, Lucian, Plautus, Euripides, Apollonius, Juvenal, Ovid, and others. "Never was there a poet so fond of introducing Roman and Greek modes of speech into the English language, as Jonson".[39] This was evidently something of which he highly approved.

31 Ibid., vol. 1, book 3, ch. 22.

32 Voitle (1984), 149.

33 Struensee (1763a) and (1763b); Laursen (2011).

34 Valencia (n.d.a) and Valencia (n.d.b). Sánchez Suárez de León (1997), 119–126, 140–157, 175, 189–192, 273–276; Laursen (2009), 139–158.

35 Unfortunately, this is not noted in Niehues-Pröbsting (1996) and Niehues-Pröbsting (1988).

36 Reid (), 308.

37 Upton (1749). Some catalogs wrongly attribute this work to Upton's father, James. See Radcliffe (1987), xvi.

38 Upton (1749), Preface, 1–2.

39 Ibid., 100.

Shakespeare and Republican Politics

The reference to Shakespeare in Upton's commentary on Epictetus foreshadows Upton's next work, a full-length critical commentary on the great playwright. This one makes his political leanings much clearer. If anyone ever thought that French café society of the twentieth century was the first example in history of erudite and sophisticated intellectuals using the latest in philosophy, philology, and historiography to support elitist and radical politics (Maoist, Marxist), this is an example to prove them wrong. Our scholar-translator mixed appeals to radical Whig and republican politics with his scholarly labors. We will survey those appeals, and then draw the obvious connections between the elitism implicit in scholarly criticism and radical politics. I cannot claim that Upton was a particularly influential writer, nor particularly representative. What I can do is present him as an example of the connections that can emerge between the elitism of the most advanced scholarly circles and that of radical politics.

Upton endorses Shaftesbury's position on criticism and political liberty, as we have reviewed it above. The notes to his *Critical Observations upon Shakespeare* (1746, expanded and reissued in 1748) contained what amounts to a manifesto of the critical philosophical project.[40] Criticism "is not so foreign to [Socrates's] grand design of bettering mankind".[41] Upton declares that he will be answering the question as to whether there is in nature a standard of taste for the critic. His answer, many pages later, is that right and wrong in criticism "have their standard in nature"; criticism is "fixed by nature, not by whim or caprice".[42] This echoes Shaftesbury's insistence that there is a standard of morals in nature. Upton would not have liked the answer that David Hume gave to the question in "Of a Standard of Taste" of 1757. Hume argued that the only standard that we have is the judgment of qualified critics, reinforced over time by generations of such critics.[43]

The manifesto of criticism rapidly became a manifesto of Upton's political sympathies. These were largely and almost stereotypically anti-court and Whig. "Hypocrisy, nonsense, and superstitious fanaticism" came to Britain after the time of Shakespeare, Upton notes; they came with Charles II, a "frenchified King", which should be a "shame to freeborn Englishmen"![44] Over and over, he reports that the French are "an enslaved nation" (16), "slavish" (123); in France, "genius has to flatter" (124). This is both anti-monarchist and nationalist: Upton probably enjoyed characterizing Rabelais as a "facetious Frenchman" (235).

To such a Whig, politics affects philosophy and literature. "Popular governments" are the "only nurse of genius" (123). There is a "mutual connection" between civil

---

[40] It is a pity that Lynch (2003) does not know Upton's work. Lynch knows John's father James Upton's edition of Ascham, and refers to James as "John Upton" at one point (p. 34).

[41] Upton (1748), 294.

[42] Ibid., 295, 385.

[43] Hume (1987), 226–249.

[44] Upton (1748), 6.

liberty and polite literature; corruption of taste (by court society) leads to corruption of morals and civil servitude (18).

The "stoics were all Commonwealthmen", Upton declares (74): this must have been in the back of his mind when he translated the Stoic, Epictetus. At another place, he quotes Epictetus on preferring the whole to the part, the city to the citizen, and the importance of the general good, confirming his interpretation as a republican (315–6).

Commonwealthman politics could be radical, and violent. Milton was a commonwealthman, too, Upton observes (141). He is the object of Upton's most frequent comparisons, suggesting Upton's sympathies. *Samson Agonistes* is a transparent parallel to Milton's own life and "the republican party after the restoration, afflicted and persecuted", in Upton's view, waiting for deliverance from oppression (144).[45] The crucial point about Commonwealthmen is that they do not hesitate to kill usurping kings. Milton was notorious for his defense of the regicide of Charles I. For Upton, historians who criticize Brutus for assassinating Julius Caesar are court-flatterers (75). Where Samuel Johnson hotly rejected Milton's republican politics,[46] Upton embraced them –at least rhetorically.

Observe, however, that these remarks are in a book attributed on the title page to "John Upton, Prebendary of Rochester". We can presume he would have published the book anonymously if he thought it might bring retaliation from the court or the Church. He must have thought that what could have been taken as inflammatory rhetoric in other times and places was evidently acceptable in both provincial gentry circles and in London in the 1740s. It is probably safe to say that the once-incendiary republicanism of the Commonwealth had become so well naturalized by the 1740s that it was not perceived as radical. In fact, it may have been understood as both patriotic and Protestant. Britain was now often described as a "republican monarchy", and much of the criticism of monarchy of republican language had now been internalized as normal opposition politics, no longer understood to apply to prevailing arrangements as a whole.[47]

One could get away with anti-clericalism even if one was a member of the clergy if it was understood that this was only anti-Catholic. Shaftesbury and the other Whigs were convinced anti-Catholics. Upton was very much a Protestant in his sympathies, ready to heap scorn on priests and monks. He would, no doubt, have been horrified by the claims of Stephen Greenblatt and others that Shakespeare was actually a crypto-Catholic.[48]

Considering that it was published in 1746, Upton's book can be read as a direct response to the Jacobite rebellion known as "The '45". The grandson of Charles II landed in Scotland with 6,000 French soldiers to spark a rebellion that would return the Catholic Stuarts to the British throne. It was brutally suppressed, but the outcome could not be taken for granted at the time. This was, naturally, an issue of

[45] Also noted by Smith (1995), 215.

[46] See, e.g., Lynch (2003), 151–152.

[47] See Blom, Laursen, and Simonutti (2007).

[48] Greenblatt (2004) and Asquith (2005).

some importance in Upton's circles. Harris's great-grandson published a copy of the Bishop of Salisbury's letter to the Dean and Chapter calling on the people "to defend their Religion and Liberties against Popery and Slavery".[49] Upton's similar language was probably taken as patriotism rather than subversion.

Part of Upton's republican vision is a certain measure of sexism. Shakespeare is properly "masculine and nervous" (15). Plato and Cicero favored censoring music and the stage, which now is left in the hands of women, to Upton's obvious disgust (16–17). Theater declined after the Restoration readmitted women to the stage (73). At a later point, undermining his own claims about Shakespeare's manliness, Upton has to admit that even Shakespeare engaged in flattery (125).

Nothing in Upton's commentary was meant to support a democratic, leveling politics. Even though he had translated a Greek author, there is no reason to think he represented what has been called "The Greek Tradition in Republican Thought", according to which agrarian laws or the abolition of private property were necessary to ensure political equality.[50] His republicanism was surely as elitist as his criticism was. Such a republicanism supports the overthrow of tyrants, but only by the *pars maiores*, the older, wealthier, better-educated, and wiser.

A commentary on Shakespeare may not seem like the most likely place to find a manifesto of mid-eighteenth-century Whiggism. But to a politically-engaged scholar, it may have been the logical way to express his opinions. He had the model of Shaftesbury and the support of friends who thought that philosophy and literary criticism could carry over into framing the conditions for moral philosophy and republican politics.

Spenser and "Critical Liberty"

In his last major work, *Spenser's Faerie Queene* (1758), Upton used the phrase "critical liberty" at least twice toward the beginning of his commentary. I think this is a clue to Upton's final synthesis of criticism and politics. It is a pull-back from the more obvious political message of the commentary on Shakespeare to a clearer recognition that his liberty and rebelliousness was much more philosophical and literary than political. If Upton still saw himself as a kind of political actor, it was only by transfer of concepts from the political dimension to the literary-critical one.

Although a Renaissance author from two centuries before, Spenser could be read as a good author for promoting the republicanism of Upton's most radical days. One commentator has written that Upton "championed British liberty" in an imitation entitled *A New Canto of Spenser's Fairy Queen* of 1747/8, setting this work in the context of the Shaftesburian Whiggism we have reviewed above.[51] But although this

[49] Malmesbury (1870), vol. 1, 5.

[50] Nelson (2004).

[51] Radcliffe (1996). I date the work to 1747/8 because at pp. 67 and 68 David Hill Radcliffe dates it to 1747, and at pp. 35 and 51 he dates it to 1748.

anonymous work is ascribed to Upton, even if Upton was a republican sympathizer in 1747/8, much of that seems to have dropped away by the time of his 1758 edition.

Upton's edition was a scholarly tour de force, and our commentator observes that "Upton's notes broached most of the philological and critical issues that would preoccupy Spenser scholars for the next two centuries".[52] Upton's commentary consisted largely of comparisons to texts of Virgil, Homer, Horace, Ovid, Dante, and especially to Shakespeare and Milton, and more. But he did allow himself reflection on his role as critic. He announced that he was assuming a "critical liberty" to denounce the errors of the poet, which could include everything from wrong word choice to the "errors" of "sophistical and polemical divinity; cabalistical and scholastic learning, &c".[53] He would "speak freely, and with critical liberty".[54]

Upton is still very much an admirer of the ancients. Perhaps he had forgotten that he had previously credited Jonson with being the English writer most influenced by the classics. "No poet borrows so much from the learned languages as Spenser", he writes.[55] There is "no writer that has so many latin idioms in his poem as Spenser".[56] And he does draw the republican political implications from Spenser's text. On the one hand, he is still anti-clerical: "'twas owing to blind devotion that abbies, monkeries, &c. were built".[57] On the other, he is still anti-court: Persian pomp is servile,[58] and "Spenser looks askew at the Court Ladies: his poem is to be considered always with more than one meaning",[59] and at least one of them will be subversive.

But the Whiggishness is not as pervasive as in his commentary on Shakespeare. He does not take advantage of his point that Spenser's Fairy Land is a metaphor for England to make republican points.[60] I cannot prove that this was deliberate. Commentaries can contain more or less of an author's political preferences in accord with other dynamics and other demands on their time and interests. But it is also possible that Upton realized, more or less self-consciously, that his scholarly endeavors were not likely to be appreciated as political commentary, nor to have the political effects that he might have intended at an earlier point.

[52] Radcliffe (1996) 52. Nevertheless, all that remains of Upton in Alpers (2001) and the rest of the *Cambridge Companion to Spenser* is the observation that "Upton's notes, with their precise interpretive questions and their resourceful answers remain valuable to this day" and one further mention (259, 261).

[53] Upton (1758), vol. 2, 337, 344. There is a good summary of Upton's notes in Wurtsbaugh (1936), 74–102. Radcliffe (1987) is a modern edition of the notes only.

[54] Upton (1758), vol. 2, 338.

[55] Ibid., vol. 2, 354.

[56] Ibid., vol. 2, 353.

[57] Ibid., vol. 2, 363.

[58] Ibid., vol. 2, 368.

[59] Ibid., vol. 2, 369.

[60] Ibid., vol. 1, xxvii. "Upton" is mentioned at p. 130 [not in the index] of Lynch (2003), but there is no discussion of Upton on Spenser.

Diminished republicanism may have also been the product of changes in politics in England, which was moving beyond the Shaftesburian Whiggishness of earlier decades. It is hard to believe that Upton would have been sympathetic to the slogans of "Wilkes and Liberty" that were to grow from the antics of John Wilkes, first elected to Parliament in 1757.[61] This sort of proto-democratic, or at least demagogic, politics in which the artisans and tradesmen of London were replacing the elites and intellectuals of the coffee houses and salons would not have been to Upton's taste. The James Harris correspondence suggests some of the discomfort and dismay at this sort of thing among Upton's elite scholarly friends.[62]

Caroline Robbins has observed that the predicament of many eighteenth century commonwealthmen in the reign of George II was that although they still had a rhetoric, they no longer had a program. Whiggish radicalism had become respectable, at least as long as it led to no particular action.[63] This may be a good frame within which to understand Upton. He still had the rhetoric, but by his last writing he may not have seen any point in harping on it.

If this is so, then "critical liberty" was now mostly a matter of the liberty of a chastened previously-engaged would-be political actor to continue with his scholarly activities without much attention to possible political effects. Ancient and Renaissance provocations to republican politics had lost their political edge. It also might have been the effect of age. Perhaps Upton was simply not prepared to change to a new idiom for politics, and withdrew to the fortifications of his critical efforts.

Bibliography

Alpers, Paul. 2001. Spencer's Influence. In *Cambridge Companion to Spenser*, ed. Andrew Hadfield. Cambridge: Cambridge University Press.

Asquith, Clare. 2005. *Shadowplay: The Hidden Beliefs and Coded Politics of William Shakespeare*. New York: Public Affairs.

Blom, Hans, J.C. Laursen, and Luisa Simonutti (eds.). 2007. *Monarchisms in the Age of Enlightenment: Liberty, Patriotism, and the Common Good*. Toronto: University of Toronto Press.

Clarke, M.L. 1945. *Greek Studies in England 1700–1830*. Cambridge: Cambridge University Press.

Clarke, M.L. 1959. *Classical Education in Britain, 1500–1900*. Cambridge: Cambridge University Press.

Epictetus. 1694. Reprinted 1700. *Epictetus, His Morals, with Simplicius, His Comment*, ed. George Stanhope. London: Richard Sare.

Epictetus. 1759. *All the Works of Epictetus*, ed. and trans. Elizabeth Carter. London: Richardson.

Epictetus. 1799–1800. *Epicteteae philosophiae Monumenta*, ed. Johannes Schweighäuser. Leipzig: Weidmann.

[61] The classic survey is Rude (1962). See Thomas (1996). Also see the remarks on Upton's friend James Harris's encounters with Wilkesite politics in Probyn (1991), 182–200.

[62] Malmesbury (1870), vol. 1, 100 ff.

[63] Robbins (1959), 272–288.

Epictetus. 1925. *The Discourses as Reported by Arrian*, ed. W. A. Oldfather. Cambridge: Harvard University Press.

Fish, Stanley. 1995. *Professional Correctness: Literary Studies and Political Change*. Oxford: Oxford University Press.

Greenblatt, Stephen. 2004. *Will in the World: How Shakespeare Became Shakespeare*. New York: Norton.

Hume, David. 1987. Of Superstition and Enthusiasm. In *Essays Moral, Political, and Literary*, 2nd ed, ed. Eugene F. Miller. Indianapolis: Liberty Fund.

Klein, Lawrence E. 1994. *Shaftesbury and the Culture of Politeness*. Cambridge: Cambridge University Press.

Laursen, John Christian. 2002. Abel Boyer, the Translation of Libertines, and the Politics of French Politeness in Britain. In *La Vie intellectuelle aux refuges protestants, II: Huguenots traducteurs*, ed. Jens Häseler and Antony McKenna. Paris: Champion.

Laursen, John Christian. 2009. Skepticism and Cynicism in the Work of Pedro de Valencia. In *Skepticism in the Modern Age*, ed. Jose Maia Neto, Gianni Paganini, and J.C. Laursen. Leiden: Brill.

Laursen, John Christian. 2011. Cynicism as an Ideology Behind Freedom of Expression in Denmark-Norway. In *Freedom of Speech: The History of an Idea*, ed. Elizabeth Powers, 45–60. Lewisburg: Bucknell University Press.

Lilla, Mark. 2001. *The Reckless Mind: Intellectuals in Politics*. New York: New York Review Books.

Lynch, Jack. 2003. *The Age of Elizabeth in the Age of Johnson*. Princeton: Princeton University Press.

Malmesbury, Third Earl of. 1844. Introductory Memoir. In *Diaries and Correspondence of James Harris, First Earl of Malmesbury*, 4 vols, ed. Third Earl of Malmesbury. London: Richard Bentley.

Malmesbury, [Third] Earl of. 1870. *A Series of Letters of the First Earl of Malmesbury, His Family and Friends from 1745 to 1820*. London: Richard Bentley.

Marchant, E.C. 1921–1922. "Upton, James", *The Dictionary of National Biography*, vol. 20. Oxford: Oxford University Press.

Marchant, E.C., and S.J. Skedd. 2004. Upton, James. In *Oxford Dictionary of National Biography*, vol. 55, 932. Oxford: Oxford University Press.

Nelson, Eric. 2004. *The Greek Tradition in Republican Thought*. Cambridge: Cambridge University Press.

Niehues-Pröbsting, Heinrich. 1988. *Der Kynismus des Diogenes und der Begriff des Zynismus*, 2nd ed. Suhrkamp: Frankfurt am Main.

Niehues-Pröbsting, Heinrich. 1996. The Modern Reception of Cynicism: Diogenes in the Enlightenment. In *The Cynics: The Cynic Movement in Antiquity and Its Legacy*, ed. R. Bracht Branham and Marie-Odile Goulet-Cazé. Berkeley: University of California Press.

Oestreich, Gerhard. 1982. *Neo-Stoicism and the Early Modern State*. Cambridge: Cambridge University Press.

Oldfather, W.A. 1925. Introduction. In Epictetus, *The Discourses as Reported by Arrian*. Cambridge: Harvard University Press.

Pocock, J.G.A. 1975. *The Machiavellian Moment*. Princeton: Princeton University Press.

Probyn, Clive T. 1991. *The Sociable Humanist: The Life and Works of James Harris 1709–1780*. Oxford: Clarendon Press.

Radcliffe, John G. 1987. Introduction. In *John Upton: Notes on the Fairy Queen*, 2 vols., ed. John G. Radcliffe, v–xx. New York: Garland.

Radcliffe, David Hill. 1996. *Edmund Spenser: A Reception History*. Columbia: Camden House.

Rand, Benjamin (ed.). 1900. *The Life, Unpublished Letters and Philosophical Regimen of Anthony, Earl of Shaftesbury*. London: Sonnenschein; reprinted London: Routledge/Thoemmes, 1992.

Reid, Thomas. 1990. In *Practical Ethics*, ed. Knud Haakonssen. Princeton: Princeton University Press.

Robbins, Caroline. 1959. *The Eighteenth Century Commonwealthman*. Reprinted New York: Athenaeum, 1968.

Rude, George. 1962. *Wilkes and Liberty*. Oxford: Clarendon.

Sánchez Suárez de León, Juan Luis. 1997. *El Pensamiento de Pedro de Valencia: Escepticismo y Modernidad en el Humanismo Español*. Badajoz: Diputación de Badajoz.

Shaftesbury, Anthony Ashley Cooper, Third Earl of. 2001. *Characteristicks of Men, Manners, Opinions, Times*, ed. Douglas Den Uyl. Indianapolis: Liberty Fund [orig. 1732]).

Smith, Ruth. 1995. *Handel's Oratorios and Eighteenth-Century Thought*. Cambridge: Cambridge University Press.

Smith, Ruth. 1997. Handel's English Librettists. In *The Cambridge Companion to Handel*, ed. Burrows Donald, 102. Cambridge: Cambridge University Press.

Stanhope, George. 1694; reprinted 1700. *Epictetus, His Morals, with Simplicius, His Comment*. London: Richard Sare.

Struensee, Johann Friedrich. 1763a. Nachrichten vom Diogenes. *Monatsschrift zum Nutzen und Vergnügen* 1:57–67.

Struensee, Johann Friedrich. 1763b. Lobrede auf die Hunde. *Monatsschrift zum Nutzen und Vergnügen* 3:233–253.

Thomas, Peter D.G. 1996. *John Wilkes, a Friend to Liberty*. Oxford: Oxford University Press.

Todd, Robert B. (ed.). 2004. *Dictionary of British Classicists*. Bristol: Thoemmes.

Uptonus, Joannes [Upton, John]. 1739–1741. *Epicteti quae supersunt dissertationes ab Arriano collectae: nec non Enchiridion*, 2 vols. London: Thomas Woodward.

Upton, John. 1748. *Critical Observations on Shakespeare*, 2nd ed. London: Hawkins. Reprinted New York: AMS Press, 1973.

Upton, John. 1749. *Remarks on Three Plays of Benjamin Jonson*. London: Hawkins

Upton, John. 1758. *Spenser's Faerie Queene. A New Edition with a Glossary, and Notes Explanatory and Critical*. London: Tonson.

Valencia, Pedro de. n.d.a. Oración, o discurso de Dion Chrysostomo, que se intitula Perianachoreseos, esto es, del Retiramiento. Traducida del Griego, Biblioteca Nacional, Madrid, Manuscript 5586, 19rr–34r.

Valencia, Pedro de. n.d.b. Discurso fundado creo que en el Epicteto de Arriano sobre los que pretenden vivir con quietud, Biblioteca Nacional, Madrid, Manuscript 11160, 72r–76r.

Voitle, Robert. 1984. *The Third Earl of Shaftesbury: 1671–1713*. Baton Rouge: Louisiana State University Press.

Wurtsbaugh, Jewel. 1936. *Two Centuries of Spenserian Scholarship (1609–1805)*. Reprinted New York: AMS Press, 1970

Part V
Epilogue

Chapter 16
A Story in the History of Scholarship: The Rediscovery of Tommaso Campanella

Germana Ernst

Abstract The attempt at reconciling Renaissance natural philosophy with the new foundations for the entire encyclopaedia of knowledge and a radical reform of society may be considered as the most original aspect of the philosophical project of Tommaso Campanella (1568–1639). This contribution offers a panorama of recent editions prepared and published by the author over the last few years. Ernst starts her narrative with the discovery of the manuscript Italian version of the *Ateismo trionfato*, which had long been thought lost, and proceeds to organise the other recent editions in three thematic nuclei. The first part focuses on autobiographical elements which, though interspersed across Campanella's entire corpus, are particularly present in the *De libris propriis syntagma*, which he wrote at the request of Gabriel Naudé, and in the *Lettere*, which bear witness to his links with scholars of the time such as Galileo, Mersenne, Fabri de Peiresc and Gassendi. Natural philosophy constitutes the second thematic nucleus. Campanella outlines an image of nature as a unified body endowed with life and sensation, a 'great and perfect animal'. Although this image is very different from Galileo's view of nature as a book written in mathematical characters, this did not stop Campanella from writing the courageous *Apologia pro Galileo* (1616). In this work the Calabrian philosopher denounces the uncalled for dogmatic value accorded to Aristotelian philosophy and lucidly redefines the relations between theology, philosophy and science, with a view of defending the scientist's right to read the book of nature free from any principle of authority. The last part of the chapter discusses Campanella's ethical and political thought, the richest aspect of his philosophical reflection which is articulated in many of his diverse works.

G. Ernst†

C. Muratori, G. Paganini (eds.), *Early Modern Philosophers and the Renaissance Legacy*, International Archives of the History of Ideas Archives internationales d'histoire des idées 220, DOI 10.1007/978-3-319-32604-7_16

The Encounter with Campanella

It all began in Milan, apparently by chance – although chance is probably nothing more than a more subtle and mysterious variant of the chain of causes. At university I had enrolled in the Faculty of Philosophy and the first exam I took was for a course on Galileo. This was a rather unusual subject to be taught by Mario Dal Pra, whose lucid and rigorous lectures on the philosophy of Hume, Hegel and Marx I was later to follow with great interest and admiration. As I studied the debate surrounding the difficult beginning of the new science, I came across a text that was unknown to me until then: *Apologia pro Galileo* by Tommaso Campanella, who I vaguely remembered as the author of *The City of the Sun*. The *Apologia* would become the subject of my thesis, and its author the starting point of an itinerary which is still ongoing till this day. The work struck me for both its density and depth. Unlike the apparent simplicity of the solarian utopia, this text is dense with quotations cleverly employed to create a bulwark in defence of Galileo against his critics, thus marking the definite divorce between Aristotelian physics and theology. It also claims legitimacy for the *libertas philosophandi*, the reading of the book of nature, placing it on the same divine level as the reading of Scriptures. As I studied this work, I started realizing that flowing into those few pages, almost like stratified concentric circles, were important themes regarding the complex relations between the legacy of the Renaissance, problems of the Counter-Reformations, and the origins of the new science and modern philosophy.[1] I immediately felt a strong desire to immerse myself into that dense forest of multifarious thoughts, which deserved to be explored more adequately. It was multifaceted and versatile but lacked neither coherence nor unity; something which I immediately thought was quite extraordinary for its originality and depth.

It was also thanks to Mario Dal Pra's lectures that I learnt to understand and appreciate the critical role of the close study of an author's texts, and therefore of the important preliminary work of producing reliable editions. This soon developed into a real passion for philological work, even though I was never formally trained in this scholarly field. When I was still very young, I started studying Campanella's *Monarchia di Spagna*, and I remember very well my great surprise at noticing significant divergences between some of the extant manuscripts and the editions and translations produced later, all of which were interspersed with long passages lifted from Giovanni Botero's *Ragion di stato*. After exploring the large number of manuscripts, years later I had the pleasure of tracing and publishing an unknown first

[1] The *Apologia* was written during the first months of 1616, at the time of the denunciation before the Inquisition of the Copernican doctrines canvassed by Galileo. Campanella states that he sent it to Rome from the Neapolitan prison in which he was detained, in the hope of avoiding a condemnation which he viewed as extremely harmful for both science and theology. For an English translation see Campanella (1994); the critical Latin edition with Italian translation is in Campanella (2006).

draft of the text written by Campanella in his youth. I eventually also published the genuine original version, purged of Botero's interpolations.[2]

Over the years, I published many other writings by Campanella. These have included unpublished texts such as the *Articuli prophetales* and the *Compendium physiologiae*[3]; texts that were thought to be lost but which I eventually managed to trace, such as the letter written by Campanella to Fabri de Peiresc on 19 June 1636 (which Firpo had sought for many years)[4]; and, with even greater emotion, texts that were completely unknown or the existence of which had been completely ignored, such as Campanella's *Apologeticum* defending the *Atheismus triumphatus* addressed to Cardinal Bellarmino in 1621 (just before the latter's death), five new sonnets, and a small treatise entitled *Chiroscopia* which Campanella dedicated to Cardinal Richelieu, who had expressly requested it.[5]

The Discovery

At this point of recounting the most important stages of my Campanellian itinerary, I cannot hold back from sharing one of the greatest emotions of my life. It is a story I have already narrated elsewhere, but I am always happy to relive those extraordinary moments of an unrepeatable day. It was a day in April of many years ago, I was at the Vatican Library, and I was very happy. As I leafed through the ms. Chigi, F VI 137, hidden within the four texts contained in the volume I first came across a copy of Campanella's early *Dialogo contro i Luterani* (this copy was not included in Firpo's *Bibliografia*[6]). Then as I continued browsing the codex out of sheer curiosity, I found an untitled pamphlet which bore no sign of the author's name. I read the very first page and my heart started beating faster. Some expressions, some words – a word in particular: *recognoscimento* – started to make me think that it could be a text by Campanella. My hypothesis was confirmed after a closer reading: there I was standing before a short text written during the philosopher's sojourn in Rome. With understandable enthusiasm I frantically started transcribing the text – a truly beautiful handful of pages – in which Campanella expresses his deep disappointment at his transfer from the Neapolitan castles to the city of the Pope: his newly found and much longed-for freedom had brought no improvement in his condition and, more importantly, it had done nothing to quell the hostility towards

[2] Campanella (1989b, 1997b).

[3] Campanella (1977, 1999).

[4] Ernst and Canone (1994); the letter is in Campanella (2010), 454–455; cf. Firpo (1956), in part. 544–545.

[5] Ernst (1992, 1995a, b); see also n. 7 below.

[6] Firpo (1940), 158–159.

him and his ideas. This situation led him to reflect, once again and with greater bitterness, on the inevitable clash between politicians and prophets.[7]

It soon became quite clear that on that long-gone day of spring the stars had formed an exceptionally favourable design in the skies. After a few hours of frenzied transcribing, I allowed myself a short break. As I paced around the catalogue room, I nonchalantly started leafing through a particular volume of the *fondo Barberini*: I opened a page at random and there, before my incredulous eyes, appeared the title – clear and unmistakable – of the original Italian draft of the *Ateismo trionfato*, which had long been thought lost. There was, of course, no visible trace of the author's name. In those days the Vatican Library did not allow researchers to request and consult manuscripts in the afternoon, so I left the library with my heart in my mouth and spent the rest of the evening battling a mix of emotions of doubt, hope and fear. My night was sleepless. I kept asking myself whether it could really be Campanella's original work. And if it was, could it be consulted at the library or had it been lost, mislaid or otherwise made unavailable? The next morning, at 8:30 am, I was handed the small codex Barb. lat. 4458. I could hardly hold back my tears, and perhaps I did not quite manage. It was indeed the first Italian draft, handwritten in its entirety, of the *Ateismo*. I suspect that every scholar is familiar with the strong feeling stirred by the sight of the handwriting of an author: it creates an almost physical, unmediated and – I would dare say – intimate bond between the two.

This discovery was even more exceptional since that very first draft in Italian marked the beginning of a story, which would last 30 years, of a text which would have to follow an extremely arduous path before eventually seeing the light of day. It would have to overcome, only thanks to the tenacious stubbornness of its author, obstacles of all sorts: criticisms, censorships, an inquisitorial trial, the request for further corrections once it had been printed, and the removal from circulation of the editions published in Rome in 1630 and 1631 – until it emerged, finally, in the 1636 Paris edition, which the Papal Nuncio in France had not managed, to his great regret, to stop from being published.[8]

Due to what some call the 'irony of fate', the original draft of Campanella's most persecuted text reached us precisely thanks to a confiscation. On 9 April 1615, the Nuncio in Naples, Deodato Gentile, wrote to Cardinal Scipione Borghese to inform him that some days earlier he had confiscated some handwritten papers ('written in Campanella's own hand') during a search in the prison cell where the Dominican was held under arrest. Gentile hurried to send the papers over to Rome after judging them to be 'full of his old errors and atheisms, albeit masked by a pious and religious appearance'.[9] After lying buried and forgotten in the *fondo Barberini*, the codex came to light as a confirmation that – as Campanella had stated in the opening

---

[7] I published the text under the heading *Politici e cortigiani contro filosofi e profeti* in Ernst (1996), 104–152; and Ernst (2002), 143–179.

[8] Firpo (1950, 1951); Ernst (1991), 73–104 ; Campanella (2004), VII-LV; Campanella (2013b), XI-XXX.

[9] Campanella (2004), vol. 1, 236.

lines of the *Praefatio* to his first published work, the *Philosophia sensibus demonstrata* – truth may be rejected, occulted and unjustly persecuted, but in the end it emerges from the shadows to return to its full splendour, illuminated by the divine light.[10] This idea is expressed visually on the frontispiece of the *Philosophia sensibus demonstrata*: a young monk, immersed in water, is shown struggling to swim towards a large sphere, which, though threatened by hostile winds blowing from the four angles of the sky, keeps afloat. The wording framing this image is at once an explanation and a commentary: 'The truth which I keep searching for spontaneously can never be drowned'.[11]

Autobiography: The *Syntagma* and the *Lettere*

The *Ateismo trionfato* was published by the Scuola Normale di Pisa as the inaugural volume of the series entitled *Opere di Tommaso Campanella*. In this present contribution I shall limit myself to giving a brief indication of the most recent editions of works by Campanella, grouping them into three thematic groups: autobiography, natural philosophy and ethical-political works.

Autobiographical allusions pervade all Campanella's works, but the *Syntagma de libris propriis* and the *Lettere* stand out as shining examples of self-narration. After living in Rome as a free man following the long years of imprisonment in the 'cyclops' cavern' – the 'Polyphemus' cave' of the Neapolitan prisons[12] –, between 1630 and 1632 Campanella held regular meetings with the Parisian physician Gabriel Naudé, who insisted that the philosopher should dictate to him an autobiography and a commented list of his works. These conversations developed into *Vita Campanellae* (which unfortunately remains lost) and *Syntagma de libris propriis*. Dictated rather impulsively (*stans pede in uno*, or *alla peggio*) and entrusted to his young friend, the *Syntagma* would only be published in 1642, 10 years after its dictation and 3 years after the death of its author.[13] In this brief treatise Campanella narrates the circumstances in which he wrote his works, recalling the painful loss of many manuscripts following thefts, searches and confiscations. He also formulates a list of requirements and rules for conducting one's studies with profit. As a general principle, true knowledge needs to be based on a fertile, critical doubt through which, while abandoning false dogmatic certainties, one seeks to establish an ongoing comparison between the books of men and the infinite book of nature. The most difficult obstacle one comes across along the path of the search for truth is "that because of which one thinks to have found that which is sought in that which one

---

[10] Campanella (1992), 3.

[11] 'Verum quod sponte recepto submergi haud potuit'. The frontispiece is reproduced in Ernst (2010, [x]); old editions, with their respective frontispieces, are reproduced in digital format in Archivio Tommaso Campanella (2010–2012).

[12] Campanella (1998), 285, 634.

[13] Campanella (1642). On autobiographical themes in Campanella's works, see Ernst (2007).

likes". If undertaken in this manner, one's explorations are bound to fail since "of every minor thing, of an ant or of a nail, there is an almost infinite science."[14]

The truly wise person should then aim at communicating the truth by means of clear and effective language, feeling free to coin new words for new concepts without caring too much about the external elegance sought by courtesans and by those who, not having anything new or original to say, worry about embellishing ideas taken from others. Campanella views the excessive care for the merely formal aspect of discourse as one of the signs indicating the moral and political decadence of a state. He has no qualms with including great humanists such as Poliziano, Valla and Bembo among those he condemns for delighting themselves in the fancy of grammar and rhetoric as ends rather than means; they communicate nothing new and profound, but limit themselves to embellishing with formal elegance what is otherwise vacuous in content.[15] In the fourth and last part of the *Syntagma* Campanella gives pointed opinions about books and authors. Just to cite a few examples, I should like to recall his vibrant take on Savonarola ("even though he was mistaken on many accounts, he is nonetheless effective as a preacher who ignites the spirit of those who listen, forcing the seeds of virtue to bear fruit, forcefully rescuing the soul from futile things, laying bare vices and burning them"), Justin Martyr, who "teaches [...] and shows that religion, which others hold to be planted only in their little gardens, is really sown in the entire human race", and Tertullian, whose efficacy is described in a few incisive lines: "in just a couple of words [he] explains more than others do in a thousand arguments; serious, difficult, rigid, effective, succinct, elegant, passionate about philosophical rather than civil grammar, and with his Africanism he attracts readers as though to exotic goods".[16]

Many pages in Campanella's works are interspersed with strong autobiographical allusions. Some of the best examples are found in his letters, which, as he writes in the *Syntagma*, are of different kinds, but the most specific is "that in virtue of which we inform whoever is far away so as to communicate something about us or about them: and this is entirely right, insofar as letters were invented precisely to render almost present that which we cannot express due to absence." [17] In 2010, on the basis of materials left by Luigi Firpo – who had worked on this project for years without seeing it come to fruition – I edited a new edition of Campanella's epistolary, which supplants and enriches considerably Vincenzo Spampanato's 1927 edition. The epistolary spans Campanella's entire lifetime, with a high concentration of letters written in certain years and a complete absence in others. The letters offer a broad spectrum of contents, length and expressive style but are all connected through a profound, hidden unity. Each letter contributes an added stroke to the author's self-portrait, be it a terse note to the viceroy prompting the payment of a maintenance allowance to the prisoner who is "dying of hunger", or the passionate memorandum to the pope and cardinals in which he, a worm "covered in garbage, dirt and

[14] Campanella (2007a), 66.

[15] Ernst (2011).

[16] Campanella (2007b), 98, 110.

[17] Ibid., 102.

stench, with [his] mouth stoppered by madness and terrorized by a hundred armed leopards",[18] has the audacity to address the gods of this earth, the lords of the world, to proclaim the inevitable destiny of the wise and the prophets, who are persecuted and put to death by politicians and powerful men for daring to bring light where there is darkness and reveal their deceit.

I will here limit myself to recalling the six letters Campanella wrote to Galileo between April 1631 and October 1632, in which he senses the storm that was about to break. Campanella's satisfaction upon receiving the copy of Galileo's brilliant "philosophical comedy", the *Dialogo sui due massimi sistemi del mondo*, is short-lived. Barely two weeks later, he hastily puts pen to paper to inform the mathematician of the alarming rise in the threats by people who were even more dangerous and violent on account of their incompetence: "I am afraid of the violence of people who do not know."[19] Campanella promptly offers himself as Galileo's defender, but he is curtly refused on the grounds that he "has written a similar prohibited work [the *Apologia*], and being an offender he cannot take up a defence".[20] The situation precipitated dramatically within a few days, causing Campanella to feel extremely disappointed at the unleashing of passions which cannot be governed by reason and the need for a difficult and resigned acceptance of the situation: "Let us comply with the divine will and let us believe that, if all natural things are done with infinite art and wisdom, so are all moral and political things, even though to us the opposite seems true; and let us be sons of obedience. When the blood cools down, I will tell you more."[21]

In a letter addressed to Grand Duke Ferdinand II in July 1638, less than a year before his death, all the threads of Campanella's life seem to come together in what can be considered a true spiritual testament. He praises the Medici house for having supported the revival of Platonic studies and the emancipation of scholars from the yoke of Aristotle. He also recalls his youthful hopes of settling under the protection of the Grand Duke, which were later frustrated by his adherence to Telesio's natural philosophy. He mentions his plans for the reform of knowledge in the light of the two divine books of nature and Scripture, and the "secret of fate" that led to his flight to France with the purpose of publishing his works, including the early and much-treasured *Civitas Solis*, "an idea of the ideal republic". He recounts the unforgettable encounter with Galileo in Padua during his youth, which had marked the beginning of a continued friendship and admiration notwithstanding their disagreement on certain matters: "intellectual disagreement can coexist with the agreement of the will of both". Campanella's last reflection is on the destiny of prophets who, defeated and persecuted, rise again on the third day or in the third century: "the

[18] Campanella (2010), 75.

[19] Ibid., 342.

[20] Letter by Francesco Niccolini, Tuscan envoy in Rome, to Andrea Cioli, Secretary of State of the Grand Duke, in Galilei (1890–1909), vol. 14, 389.

[21] Campanella (2010), 343.

future century will be our judge, for the present always crucifies its benefactors, but they then rise again on the third day or in the third century." [22]

Natural Philosophy: *Del senso delle cose e della magia*

The second thematic nucleus is that of natural philosophy. One of the most representative texts concerned with this fundamental aspect of Campanella's philosophy is *Del senso delle cose e della magia*, of which I prepared a new edition published by Laterza in 2007.[23]

Using a markedly expressive language, this complex work outlines an image of the world as a "great and perfect animal",[24] a living organism in which each part is endowed with the degree of sensitivity (*sensus*) required for the preservation and propagation of life. The *sensus* consists in the ability to distinguish between what is beneficial and what is harmful, in such a manner that every being may pursue the preservation of its own life and avoid whatever it perceives as destructive: "the escape from, and hatred towards, death and the love for life are found in every being. Therefore, preservation must be the highest good of every being." [25] Sensation and the tendency towards self-preservation are inherent in every natural being, albeit in different ways and to different degrees. Sensation is found in the sky and in all celestial bodies, in air, water, earth – which Pythagoras described as "a great animal" ("and grass and trees are its hair, stones its bones") –, and light, which "expands above the earth so that it may multiply, generate and amplify itself". Plants feel great pleasure in "sprouting, growing, flowering, giving fruit and multiplying". Minerals and metals are also endowed with sense, albeit in a more veiled manner, and even the shadows of our bodies desire to unite themselves with other shadows.[26] Animal organisms are endowed with sense as well as *spiritus*, the warm breath made of extremely attenuated matter and purified by the heat of the sun. Mobile, subtle and receptive, the spirit is located in the brain, from where it flows through the very subtle channels of the nervous system and comes into contact with the external world. Its passions and knowledge are derived from the changes it undergoes on account of exhalations, movements and light deriving from bodies.

It is in the light of the principles of the sense of things, the spirit and the passions that Campanella revisits prophecy and natural magic. Premonitions and predictions can be explained by the fact that air is a sort of common sensor, in communication with the various spirits contained in single individuals, in which it imprints passions

[22] Ibid., 509–510.

[23] The first edition of the Italian text had been published by Antonio Bruers in 1925, while the Latin version (*De sensu rerum et magia*) was widely known in the seventeenth century thanks to editions published in Frankfurt (1620) and Paris (1636, 1637).

[24] Campanella (1998), 37.

[25] Campanella (2007a), 97.

[26] Ibid., 131, 133.

and affections. Those whose spirit is very subtle may perceive lighter movements as well as the motions of the cause of events that are about to take place. A magician is a wise person who, having knowledge of the various voices, correspondences and attractions in nature, and above all of the specific quality of the sense that is present in every being, is capable of causing in the mobile and subtle spirit passions such as pain and happiness, love and hatred, hope and fear.

The alterations experienced by the spirits explain the transformations of those who, after being bitten by a rabid dog, are weakened and scream and in the end bark "thinking that they are dogs"; or of the peasants from Puglia who, when bitten by tarantulas, start dancing and jumping to the sound of various instruments until they succumb to fatigue.[27] In both cases, the acrid spirits introduced through animal bites alter the temperament and imagination of the unfortunate victims, whose organism is taken over by the spirit of the animals to such an extent that they lose their own identity and forget what they were. Apparently prodigious events can be explained on account of the latent forms of the permanence of the sense in beings and in the air. For instance, the corpse of a person who died a violent death, in which an "obtuse" sense lingers, starts bleeding again when it senses the presence of the assassin. Campanella does not refute the efficacy of a much-discussed remedy, often attributed to Paracelsus, of nursing a wound of a person (who might be far away) by applying a magical ointment to the weapon that would have caused it.[28] And it is again the latent sense that endures in things that makes possible an explanation of a famous example cited in every book on magic: due to the revival of an old fear, a drum made of sheep skin breaks into pieces with the rolling of a drum made of wolf skin. This had led a much-feared Bohemian captain to order that a drum made of his skin should be manufactured upon his death so as to terrorize enemies.[29]

Campanella's view of life running through each of nature's fibres is in stark contrast with Galileo's image of nature as a book which has been written, as he puts it in a famous passage of his *Saggiatore*, "in a mathematical language, and the letters are triangles, circles and other geometrical figures, without which it is humanly impossible to understand a single word of it; without them it is like wandering hopelessly through a dark labyrinth."[30] Moreover, Campanella held reservations with regard to heliocentrism on account of difficulties in its compatibility with the principles of Telesio's natural philosophy, according to which the sun – where the principle of heat is found – is light and endowed with motion, while the earth – where the antagonistic principle of cold is found – is heavy and motionless. His *Apologia pro Galileo* can therefore be regarded as all the more admirable and courageous. The Dominican philosopher makes use of his vast theological knowledge to redefine the relation between philosophy, science and theology. The real problem,

[27] Ibid., 189–193.

[28] Ibid., 184, 188.

[29] Ibid., 186–187.

[30] Galilei (1890–1909), vol. 6, 232.

according to him, lay with the unjustified dogmatic value given to Aristotelian philosophy, which, like any other human doctrine, could in reality be modified, corrected or abandoned. Notwithstanding his occasional disagreements with Galileo, Campanella defends in the strongest terms possible the right of the scientist to read the book of nature so as to modify, correct or improve the books of men.

Human Behaviour and Relations: *Ethica*

The third thematic nucleus includes ethics and politics. The Latin *Ethica* (2011) was another important addition to the above-mentioned series of the Scuola Normale di Pisa in which the *Ateismo trionfato* and the *Apologia pro Galileo* had been previously published. Even though it occupies a central place in Campanella's system, this text was hitherto scarcely known since it had never been reprinted following the two seventeenth-century editions contained in the quadripartite *Philosophia realis* (Frankfurt, 1623; Paris, 1637). In the first, complex part of the work, the author places ethical themes within the theoretical framework of his metaphysics founded on the doctrine of the primalities. Here he deals with problems such as the relation between the impulses of the senses and the rational and divine *mens*, between physical inclinations and free choices, founded on freedom, the necessary condition and premise for every moral system. He defines virtue not as *habitus* or *medietas* or science or operation, but rather as the rule imposed by reason on passions. He then moves on to outline a colourful picture of the various virtues, organised according to three kinds of preservation: in oneself, in one's children, and in various social relations. These are in turn framed within a fourth kind of preservation, which is God, giver of being.

By far the most extended discussion in the work is dedicated to the third kind of preservation, namely in friends, fame and society. By way of a simple example, I should like to mention a few considerations on the bond of friendship and on veracity. As is often said, a friend is an *alter ego* and as such amplifies our personality. Insofar as it is a form of mutual love for the benefit of reciprocal utility, honour and pleasure, friendship is nourished by various virtues, such as benevolence (thanks to which we desire for others that which is good for them, a virtue that counters the vices of malignity and envy), or praise (*benedicentia*), which regulates our conversations with others: affability, for instance, teaches us to be helpful and kind towards others, "conversing with amiability even with stableboys and laundrypersons, behaving affectionately, greeting others and wishing them well". The opposing vice, i.e. abstaining from speaking or taciturnity, which is typical of "melancholic" people, is especially detestable when it is a deliberate choice particularly by the powerful and the rulers. When they do not deem it worthy to engage with others either because they consider "themselves divine, and others beasts" or because they want to appear wise without being so and hide their ignorance behind a veil of silence, "they reach a point when they hold back from laughing, especially when they would want to, so as to appear serious and weighty, and indeed they are, but more like lead than like gold".[31]

The opposite of loyalty is betrayal. Campanella, whose condition as a prisoner often made him a victim of plagiarism, recalls the unbecoming behaviour of friends in whose hands he had entrusted his manuscripts and condemns such contemptible acts: "Some reveal secrets, others deliver their country to enemies, while others claim as their own writings and ideas which friends would have entrusted them with, and publish them under their own name, which is even more detestable than sacrificing a son before the eyes of his father". He then concludes dejectedly that "those of sincere heart attract traitors like a magnet".[32]

Highlighting his disappointment with regard to unkept promises by those whom he had considered his friends, Campanella recalls that many of those who had reassured him of their assistance had held back when push came to shove, confirming the abyss that exists between words and actions:

> It is easy to wish the best for your friends and to speak well about them, but the real difficulty arises when it comes to acting in the right manner, at the right moment. Hesitations, the fear of losing one's money, one's life, one's fame, one's pleasures and one's commodity and everything else – this is what causes turmoil. When you request facts, they offer you words and advice, they refer you to others for help, and they declare themselves available to do anything to help, except what you ask of them.[33]

Just to give an example of these dishonest manipulations, a recent discovery by Gianni Paganini revealed that in 1624–25 Marin Mersenne promised to help Campanella publish his *Metaphysica* and received the manuscript, but failed to honour his promise (the work was published only in 1638, thanks to the intervention of Claude de Bullion). To make matters worse, Mersenne plagiarized conspicuous portions of the *Metaphysica* (especially from book I, devoted to a discussion of scepticism) in order to build the character of the "sceptic" for his own *De la Vérité des sciences* that came out in 1625.[34]

A fundamental virtue for every human relationship is veracity, which is articulated in various ways and to which a corresponding number of vices is opposed. Campanella condemns harshly lying and deceit, which betray the truth in the name of some miserable personal advantage. He also condemns hypocrisy insofar as it is a mask that covers the face of the deceiver in order to harm others and acquire advantages for himself. However, he makes a distinction between hypocrisy and 'officious simulations': if their purpose is good, the latter might well be useful and can be considered as a form of "admirable prudence" which is not only necessary but also praiseworthy. Campanella himself admits to having been forced on some occasions to employ deceit and tricks. In the most dramatic moment of his life, in order to avoid death, he had to resort to the simulation of insanity. This is one of the extreme solutions used by wise men in order to escape violence by those who

[31] Campanella (2011), 91–92.

[32] Ibid, 97.

[33] Ibid., 110.

[34] See Paganini (2005, 2009).

exercise power in an unjust manner, as can be seen in the stories of David, Solon and above all Brutus, whose madness had been simulated as a means of defying tyrannical violence.[35]

The *Quaestiones* in Appendix to *Ethica* and *Politica*

The large tome, in folio, of the Paris edition of the *Philosophia realis* contains a series of *Quaestiones* appended to the four main parts constituting the work, including the *Ethica* and the *Politica*. Although they are little known, these questions play an important role in Campanella's thought.[36] Without spending too much time on the intricacy of the subtle arguments contained in these pages, it is worth highlighting some key points common to both ethics and politics. Aristotle is given centre stage as the main antagonist and interlocutor in both sets of questions, and his *Nicomachean Ethics* and *Politics* are amply cited or paraphrased with the intention of exposing those aspects which the Calabrian philosopher considered to be contradictory or mistaken. Yet the target of Campanella's strongest criticisms are the positions held by those he refers to as 'politicians' or 'Machiavellians', who identify virtue and the highest good with power and domination achieved through any means, uphold the doctrine of the 'reason of state' according to which everything is licit in the pursuit of power, and consider religion as nothing more than an *ars dominandi* or an astute *figmentum* invented and used by rulers and the priestly class to obtain and maintain political power.

In his replies to the arguments put forward by classical philosophical traditions, Campanella refuses to identify the highest good and happiness with glory, honour, riches, pleasure or the contemplative life. Instead, he supports the position of the Stoics, for whom the "only true good resides in virtue, the only evil in vice: all other things – riches, poverty, honour, dishonour, the skies, the earth, the homeland, every good of the body or external to it – are indifferent", insofar as their value depends on the good or bad use that can be made of them. For the Stoics, it is only the virtuous person who is happy, for he knows how to put to good use even evils such as misfortunes, death and poverty. Aristotle thinks – stupidly, according to Campanella – that the wise person cannot be happy when he suffers, for he cannot dedicate himself to contemplation. However, he fails to understand that even when the body is lacerated virtue is not annihilated, for it resides in the mind, and the person who suffers in a virtuous manner is happier and deserves more admiration than he who takes pleasure in vice. Moreover, the virtuous person is a lord by right, even if not in fact, insofar as he merits to be lord; he is a teacher to other men and the law to

[35] *Ibid*, 97–99.

[36] The annotated edition of the three ethical questions in Campanella (2011), 211–349; the Latin text with an Italian translation of the first three political questions in Campanella (2013a); the Latin text with an Italian translation of the fourth political question, concerning *Civitas Solis*, in Campanella (1996), 96–173.

himself; he is wise, insofar as he is able to dominate and lay down a rational rule for his own passions.[37]

In his replies to the 'politicians', Campanella claims that even the value of power depends on how it is put to use. Tyrants may appear happy at a first, superficial and vulgar glance that considers only external appearances and ephemeral pleasures such as the crimson dress, lavish food and music. The philosopher, who can see into the inner parts of the soul, understands that evil rulers are like "whitewashed walls" or like a fruit that appears beautiful on the outside but is ravaged by worms on the inside. Recovering the unity between the ethical and political dimensions, which Machiavelli had kept separate, Campanella refuses to consider heroes those who, blinded by their own passions, equate the law with the use of armed force. He shows the intrinsic weakness of power isolated from the other two primalities that together constitute true power, i.e. wisdom and love: this weakness is testified by the unhappy life of the Machiavellian heroes ("in their lifetime they are tormented by hatred, curses, suspicions and fears"), by their often inglorious deaths and by the sudden fall of their fragile political systems.[38]

In their attempt to sustain the legitimacy of their use of force and their right to impose what they consider to be individual excellence, politicians appeal to what happens in nature and in the animal world. Campanella argues that these examples are incorrect. Relations between animals cannot be compared to the mutual relations between human beings. Not only do bees, ants and cranes offer excellent examples of collective organisation, but those animals that attack others do so in order to feed themselves: the 'natural' sin of the sparrowhawk that kills the dove for its own preservation cannot be compared to the 'moral' sin of one man's oppression of another. Should domination be based on and justified through the use of force, man would by nature be a slave to stronger animals such as lions, horses and elephants, yet the opposite is true: "Therefore man rules on account of his virtue, not on account of force". The supporters of such deviant doctrines deserve to be subjected to domination by a stronger animal, just like the Danes who were once subjected to a dog by the Nordics as a sign of contempt.[39]

If the 'Machiavellians' rely on the natural world to legitimize power based on the use of force, Aristotle's reference to nature seems to consent to the inequality between human beings on the basis of the natural relation between the rulers and the ruled. According to the Stagirite, there exist among human beings natural differences related to the body and the mind, in such a way that some are more suitable by birth to rule (e.g. men and those endowed with greater intelligence), while others are by nature subordinate (e.g. women and those who are uncouth) These natural differences serve as the basis and justification for the legitimacy of the domination by nature of the more excellent and the servitude by nature of the others.[40] According to Campanella, it is society that is natural, rather than the relation between servitude

[37] Campanella (2011), 223 ff.

[38] Ibid., 236 ff.

[39] Campanella (2013b), 627–628.

[40] Ibid., 615, 616.

and domination. "[T]he more excellent should rule, not dominate", which is equivalent to manifesting their "own excellence by doing good, which propagates itself". It is true that there exist various forms of legal servitude that were introduced by the law of peoples as a lesser evil, such as when prisoners of war are reduced to slavery rather than killed. However, there is no such thing as slavery by nature: a child, "even if he is more stupid than a sheep, is not a slave who can be dominated but one who can be governed". Uncouth and uneducated people, who have an "obfuscated reason", have to serve "so as to improve their condition, rather than to get used to serfdom as though it were something natural". The role of rulers is similar to that of a physician, aimed at offering remedies for the ills of fellow humans, and "the wise person shall not discourage, but rather raise the dignity of his own species."[41]

But it is above all another aspect of Aristotle's political philosophy which Campanella finds completely disgraceful. Aristotle considered all those who carry out manual work, such as artisans, farmers, herdsmen and merchants, as unworthy of being considered as an integral part of society and as citizens with full rights within the political body since, according to him, they lack the virtues and requirements necessary to attain happiness, which consists in leading a contemplative life and partly in an active life. According to Campanella, those who carry out manual work form an essential part of society, from which only those who are idle and brag about false titles of nobility without contributing to the collective well-being should be excluded. Against Aristotle's position, Campanella quotes the famous passage from St Paul's first Letter to the Corinthians, in which he presents an analogy between the Christian community and the human body with the intention of defending the full and equal dignity of all individual members insofar as they collaborate to build the body as a unified whole.[42] The full dignity of work is recognised even by Solomon, according to whom each craftsman is a king within his own craft: in the exercise of their activity the physician, the cook and the barber command even the king, who is very happy to obey the outstanding sailor when a storm strikes.[43]

It is then completely false to claim that those who work cannot pursue virtue and happiness. Jacob, who is taken by St Ambrose as a model of the happy life, was a shepherd, as were the patriarchs, Moses the lawgiver and kings such as David and Saul, while Noah was a farmer. Going beyond the examples of Jewish figures, Campanella refers to the Roman republic, in which senators were farmers who sometimes derived their names from legumes; yet, as Pliny observed, "for their military and civil gallantry, conquered the world and enlightened it as a second sun". The apostles were artisans and fishermen, and in Florence men of letters such as Gelli and Burchiello were a stocking-maker and a barber respectively. Moreover, those who are culturally unrefined are not necessarily lacking in virtues but might actually possess them in a simpler and purer manner: they respect the laws with a sincere heart, and are often better than the learned who are agitated by opposing positions. Recalling that Cato the Censor had predicted the impending ruin of

[41] Ibid., 631, 632.

[42] Ibid., 642–643.

[43] Ibid., 621–22.

the republic upon noticing the avid curiosity with which the young Romans had listened to Carneades's words in favour of and against justice, Campanella observes sarcastically: "it was then that Caesar, Pompey and Crassus, still youngsters, started thinking about tyranny. It would have been better for the republic had they kept themselves busy with the vegetables in the fields."[44]

Acknowledgement I would like to express my heartfelt thanks to Jean-Paul De Lucca for translating this contribution into English.

Bibliography

Archivio Tommaso Campanella. 2010–2012. http://www.iliesi.cnr.it/Campanella. Directed by Eugenio Canone. Rome: Lessico Intellettuale Europeo e Storia delle Idee, CNR.

Campanella, Tommaso. 1642. *De libris propriis et recta ratione studendi syntagma*. Paris ap. viduam G. Pele.

Campanella, Tommaso. 1950. *Opuscoli inediti*, ed. Luigi Firpo. Florence: Olschki

Campanella, Tommaso. 1977. *Articuli prophetales*, critical edition by Germana Ernst. Florence: La Nuova Italia.

Campanella, Tommaso. 1989a. Cristianesimo e religione naturale. Le censure all'*Atheismus triumphatus* di Tommaso Campanella. *Nouvelles de la République des Lettres*, 1–2: 137–200.

Campanella, Tommaso. 1989b. *Monarchia di Spagna. Prima stesura giovanile*, critical edition by Germana Ernst. Naples: Istituto italiano per gli studi filosofici.

Campanella. Tommaso. 1992. *Philosophia sensibus demonstrata*, ed. Luigi De Franco. Naples: Vivarium.

Campanella. Tommaso. 1994. *A Defense of Galileo*, ed. and trans. Richard J. Blackwell. Notre Dame/London: University of Notre Dame Press.

Campanella, Tommaso. 1996. *La città del Sole – Questione quarta sull'ottima repubblica*, ed. Germana Ernst. Milan: Rizzoli.

Campanella, Tommaso. 1997a. *La città del Sole*, ed. Luigi Firpo. New edition by Germana Ernst and Laura Salvetti Firpo. Afterword by Norberto Bobbio. Bari-Rome: Laterza.

Campanella, Tommaso. 1997b. *Monarchie d'Espagne et Monarchie de France*, edition of the Italian texts by Germana Ernst, French translation by Serge Waldbaum and Nathalie Fabry. Paris: Presses Universitaires de France.

Campanella, Tommaso. 1998. *Le poesie*, ed. Francesco Giancotti. Turin: Einaudi.

Campanella, Tommaso. 1999. *Compendium physiologiae – Compendio di filosofia della natura*, unpublished Latin text ed. by Germana Ernst, with trans. and notes by Paolo Ponzio. Milan: Rusconi.

Campanella, Tommaso. 2004. *L'ateismo trionfato*, 2 vols. (vol. 1, ed. of the text; vol. 2: anastatic copy of ms. Barb. lat. 4458). Pisa: Edizioni della Normale.

Campanella, Tommaso. 2006. *Apologia pro Galileo*, ed. Michel-Pierre Lerner. It. trans. Germana Ernst. Pisa: Edizioni della Normale.

Campanella, Tommaso. 2007a. *Del senso delle cose e della magia*, ed. Germana Ernst. Rome-Bari: Laterza.

Campanella, Tommaso. 2007b. *Sintagma dei miei libri e sul corretto metodo di apprendere/De libris propriis et recta ratione studendi syntagma* (Latin text and Italian translation), ed. Germana Ernst. Pisa-Rome: Fabrizio Serra.

[44] Ibid., 652–653.

Campanella, Tommaso. 2010. *Lettere*, ed. Germana Ernst, based on material prepared by Luigi Firpo, with the collaboration of Laura Salvetti Firpo and Matteo Salvetti. Florence: Olschki.

Campanella, Tommaso. 2011. *Ethica. Quaestiones super Ethicam*, ed. Germana Ernst, with the collaboration of Olivia Catanorchi. Pisa: Edizioni della Normale.

Campanella, Tommaso. 2013a. In *Atheismus triumphatus*. Anastatic copy of the 1631 edition for Germana Ernst. Pisa-Rome: Fabrizio Serra.

Campanella, Tommaso. 2013b. *Tre questioni politiche contro Aristotele*, ed. Germana Ernst. *Bruniana & Campanelliana* 29, 587–697 (and also in *Bruniana & Campanelliana*. Supplementi, XXXVIII, Materiali, 7. Pisa-Rome: Fabrizio Serra).

Campanella, Tommaso. 2015. *Etica*, It. trans. by Germana Ernst. Pisa: Edizioni della Normale.

Ernst, Germana. 1991. *Religione, ragione e natura. Ricerche su Tommaso Campanella e il tardo Rinascimento*. Milan: Franco Angeli.

Ernst, Germana. 1992. Il ritrovato 'Apologeticum' di Campanella al Bellarmino in difesa della religione naturale. *Rivista di Storia della Filosofia* 47(3): 565–586.

Ernst, Germana. 1995a. Cinque sonetti inediti di Campanella. *Bruniana & Campanelliana* 1: 11–20.

Ernst, Germana. 1995b. Note campanelliane. I. L'inedita *Chiroscopia* a Richelieu. *Bruniana & Campanelliana* 1: 83–101.

Ernst, Germana. 1996. L'opacità del male e il disincanto del profeta. Profezia, ragion di stato e provvidenza divina in un testo inedito di Campanella (1627). *Bruniana & Campanelliana* 1: 89–155.

Ernst, Germana. 2002. *Il carcere, il politico, il profeta. Saggi su Tommaso Campanella*. Pisa-Rome: Istituti Editoriali e Poligrafici Internazionali.

Ernst, Germana. 2007. Autobiografia di Campanella. In *Laboratorio Campanella. Biografia Contesti Iniziative in corso. Atti del Convegno della Fondazione Camillo Caetani, Roma, 19–-20 October 2006*, ed. Ernst, Germana, and Fiorani, Caterina, 15–38. Rome: Fondazione Camillo Caetani, «L'Erma» di Bretschneider.

Ernst, Germana. 2011. 'Io vivo come scrivo'. Il diverso modo di parlare di filosofia in Tommaso Campanella. In *"Virtù ascosa e negletta". La Calabria nella modernità*, ed. Germana Ernst and Rosa M. Calcaterra, 13–27. Milan: Franco Angeli.

Ernst, Germana, and Canone, Eugenio. 1994. Una lettera ritrovata: Campanella a Peiresc, 19 giugno 1636. *Rivista di storia della filosofia* 49: 353–366.

Ernst, Germana, and Fiorani, Caterina, ed. 2007. *Laboratorio Campanella. Biografia Contesti Iniziative in corso*. Atti del Convegno della Fondazione Camillo Caetani, Roma,19–20 October 2006. Rome: Fondazione Camillo Caetani, «L'Erma» di Bretschneider.

Firpo, Luigi. 1940. *Bibliografia degli scritti di Tommaso Campanella*. Turin: Vincenzo Bona.

Firpo, Luigi. 1950. Risposte alle censure dell'*Ateismo triunfato*, in Campanella (1950). 9–54.

Firpo, Luigi. 1951. Appunti campanelliani. XXI. Le censure all'*Atheismus triumphatus*. *Giornale critico della filosofia italiana* 30: 509–524.

Firpo, Luigi. 1956. Appunti campanelliani. XXV. Storia di un furto. *Giornale critico della filosofia italiana* 36: 541–549.

Galilei, Galileo. 1890–1909. In *Le Opere di Galileo Galilei*, ed. Antonio Favaro, 20 Vols. Florence: Barbèra (and reprints).

Lerner, Michel-Pierre. 1995. *Tommaso Campanella en France au XVIIe siècle*. Naples: Bibliopolis.

Naudé, Gabriel. 1644. *Panegyricus dictus Urbano VIII Pont. Max. ob beneficia ab ipso in M. Thom. Campanellam collata*. Paris: apud S. et G. Cramoisy.

Paganini, Gianni. 2005. Mersenne plagiaire? Les doutes de Campanella dans la *Vérité des sciences*. *Dix-septième siècle* 57: 747–767.

Paganini, Gianni. 2009. Tommaso Campanella: The Reappraisal and Refutations of Scepticism. In *Renaissance scepticisms*, ed. Gianni Paganini and José R. Maria Neto, 275–303. Dordrecht: Springer.

Index

A
Abelard, P. (Petrus Abaelardus), 23
Achillini, A., 82
Achillinus. *See* Achillini, A.
Adami, T., 122
Agricola, G. (Georg Pawer), 61, 62
Albert the Great (Albertus Magnus), 27, 146
Alberti, L.B., 136, 242
Alembert (La Rond, J.B.), 5
Alexander of Aphrodisias, 102, 106
Alexander the Great, 234, 246
Alfonso de Cartagena, 62
Althusser, L., 160
Ambrose (Aurelius Ambrosius), 290
Anaximenes, 124
Anscombe, E., 223
Anselm of Canterbury, 25
Antonius from Rho, 230
Apollonius (Rhodius), 266
Aquinas, T., 27, 32, 156, 172
Ariosto, L., 138
Aristotle, 22, 27, 29, 31, 33–36, 48, 49, 61, 63, 67, 70–71, 83–85, 89, 90, 92–93, 99, 100, 103, 104, 106, 108, 111–115, 120–122, 127–129, 136–139, 145, 148, 176, 178, 186, 188, 189, 205, 216, 221–236, 240, 247, 261, 277, 278, 286, 288, 290
Arrian of Nicomedia, 262
Augustine, 27, 30, 52, 54
Averroes, 25, 27, 89–92, 100, 104, 110, 146
Avicenna, 25, 46, 146, 177

B
Bacon, F., 5, 6, 15–16, 21, 24, 27, 55, 82, 84–85, 99, 100, 104–115, 119, 122–128, 130–131, 143, 203–217, 239, 243, 248–255
Bacon, N., 105
Barth, K., 53
Bartholin, T., 179
Basson, S., 127, 128
Bayle, P., 2
Bellarmino, R., 279
Bembo, P., 282
Berkeley, G., 66
Bessarion, B., 27
Bianchi, L., 157, 159
Biel, G., 27
Blair, A., 51
Blumenberg, H., 167
Bodin, J., 235–239, 242–248, 250, 253, 255
Boethius, S., 27, 63
Boonin-Vail, D., 222–225
Borghese, S., 280
Botero, G. (Boterus), 16, 207, 208, 242–255, 278, 279
Bovelles, de, C., 165
Box, I., 206
Boyle, R., 85, 127
Brucker, J.J., 4–8, 12
Bruni, L., 62
Bruno, G., 5, 6, 7, 9–11, 21, 24, 82, 127, 128, 138–142, 147
Brunus. *See* Bruno, G.

C. Muratori, G. Paganini (eds.), *Early Modern Philosophers and the Renaissance Legacy*, International Archives of the History of Ideas Archives internationales d'histoire des idées 220, DOI 10.1007/978-3-319-32604-7

Brutus (Marcus Junius Brutus), 268, 288
Bullion, C. de, 287
Burchiello (Domenico di Giovanni), 290
Burckhardt, J., 3, 5, 23
Burthogge, R., 85
Burton, R., 14, 16, 42–50, 52–55, 248–254
Busson, H., 22

C
Caesar (Gaius Julius Caesar), 268, 291
Cajetan. *See* De Vio, T.
Campanella, T., 1, 9–16, 24, 27, 61, 71, 82, 84, 122, 127, 128, 136, 169, 176–180, 182, 197, 277–291
Campanelle. *See* Campanella, T.
Cantillon, R., 245
Cardans. *See* Cardano, G.
Cardano, F., 81, 91, 92
Cardano, G., 5, 9, 15, 24, 26, 44, 61, 81–95, 99–110, 112–125, 127–128, 185–199
Carneades, 291
Carpenter, N., 122, 127
Carter, E., 262
Cartesius. *See* Descartes, R.
Casaubon, I., 42
Casaubon, M., 42–44, 55
Case, J., 108
Cassirer, E., 23
Castiglione, B., 24, 205, 212, 214
Castriota, G., 120
Cato the Censor (Cato the Elder), 290
Ceffi, F., 240
Ceron, A., 16
Cesalpino, A., 82
Chanet, P., 173, 174, 177, 180, 181
Charbonnel, J.-R., 22
Charles I (king), 268
Charles II (king), 267–268
Charron, P., 8
Chassinus, G., 127
Christ, 31, 32, 42, 43, 53, 86, 136, 213
Chrysostom, G., 32
Cicero (Marcus Tullius Cicero), 27, 32, 62, 64, 66, 68, 69, 71, 76, 149, 206, 207, 211–213, 216, 240, 265, 269
Clerselier, C., 170, 171, 178
Clucas, S., 14
Columella (Lucius Junius Moderatus Columella), 43
Conway, E. (Viscont Conway), 82
Copenhaver, B., 22
Copernicus, N., 128, 140, 141, 143, 144, 146, 148
Corbinelli, J., 207
Corvaglia, L., 186
Cousin, V., 156
Crane, R., 205
Cranz, E., 25, 35
Crassus (Marcus Licinius Crassus), 291
Cremonini, C., 82
Cremoninus. *See* Cremonini, C.
Croce, B., 23
Cureau de la Chambre, M., 169–182
Cusa, Nicholas of, 21, 147
Cusano. *See* Cusa, Nicholas of
Cusanus. *See* Cusa, Nicholas of

D
da Vinci, L., 136–139
Dal Pra, M., 278
Dalgarno, G., 61
Daniel, S., 53
Dante Alighieri, 270
David (Biblical king), 288, 290
De Clave, Etienne, 128
De Sanctis, F., 23
Della Casa, G., 205
Della Porta, G.B., 149
Delle Colombe, L., 141
Democritus, 43, 45, 83, 112, 123
Descartes, R., 1, 2, 5–8, 11–13, 15–16, 22, 23, 25, 28, 55, 61, 69, 75, 81, 83, 84, 88–89, 93–94, 114, 119, 122, 126–127, 128, 144–167, 169–182, 234
Dicaearchus of Messana, 47
Diderot, D., 5
Dilthey, W., 23
Diogenes Laertius, 27, 69, 87
Dionysius the Areopagite, 27, 35, 52
Dionysius of Halicarnassus, 261
Doni, A., 122, 125
Donius. *See* Doni, A.
Duns Scotus J., 27, 60, 156
Dupleix, S., 161
Dzelzainis, M., 207

E
Eglinus, R., 28, 34
Elizabeth of Bohemia (princess), 166
Epictetus, 260, 262–268
Epicurus, 9, 70–71, 100, 193–194, 229–233, 235
Equicola, M., 46, 47
Erasmus, D. (Erasmus of Rotterdam), 24, 43, 67, 76, 210, 212, 236
Euripides, 34, 266

F
Faba, G., 240
Fabbri, N., 15
Fabri de Peiresc, N.-C., 279
Faye, E., 15
Ferdinand II (Grand Duke), 283
Fernel, J., 82
Ficino, M., 5, 9, 24, 27, 29, 35, 41, 44, 45, 47, 48, 51–55, 81, 82, 84, 85
Fiorentino, F., 23
Firpo, L., 279, 282
Fish, S., 206
Fitzmaurice, A., 247, 252
Florio, J., 107
Foot, P., 223
Fracastoro, G., 127
Frey, J.-C., 122, 130

G
Galen, 49, 67, 68, 69, 84, 120, 211
Galilei, G., 21, 136, 143–147, 283, 285
Gallaccini, T., 142
Gallanzoni, G., 145
Garber, D., 15
Garin, E., 12, 22, 157
Gassendi, P., 2, 9, 15, 59, 61, 64, 69–71, 76–77, 122, 127, 131, 229, 235, 277
Gauthier, D., 221
Gelli, G.B., 290
Gentile, D., 280
Gentile, G., 22, 23
George II (king), 271
Gessner, C., 27
Geulincx, A., 169
Giglioni, G., 14, 16
Gilbert, W., 127, 142, 143
Gilson, E., 156, 158
Glisson, F., 110–115
Goclenius, R., 14, 21–37
Godefredus. *See* Godoffroy, P.
Godoffroy, P., 46
Godwin, T., 148
Gontier, T., 160
Goorle, van, D., 127, 128
Gorlaeus. *See* van Goorle, D.
Gouhier, H., 158
Grafton, A., 94
Granvellanus, C.P., 53
Greenblatt, S., 268
Grosseteste, R. (Robert of Lincoln), 62
Grotius, H. (H. de Groot), 227–229, 231, 235
Gualterotti, R., 142
Guazzo, S., 205, 206
Guicciardini, F., 207
Gulston, T., 261

H
Hale, M., 99–115
Handel, G.F., 261
Hankins, J., 3, 4
Harris, J., 260, 261, 263–265, 269, 271
Hamlyin, D.W., 22
Heereboord, A., 127
Hegel, G.F.W., 4, 7–12, 23, 24, 156, 278
Heidegger, M., 156, 160
Heraclitus, 124, 211, 212, 216
Hermes (or Mercurius) Trismegistus, 83
Hero of Alexandria, 107, 109
Hesiod, 91, 215, 216
Hill, N., 127
Hippocrates, 49, 211
Hobbes, T., 5, 15, 16, 22, 23, 59, 61, 64, 71–73, 74–75, 81, 88, 92, 94, 120, 122, 171, 206, 221–236
Holland, G., 131
Homer, 270
Hood, F.C., 222
Horace, 266, 270
Hume, D., 262, 267, 278
Hutton, S., 14

J
Jacob (Biblical figure), 290
James I (king), 248, 251
John (evangelist), 213
Johnson, R., 252, 253
Johnson, S., 268
Jonson, B., 260, 266
Jupiter, 143–146, 215
Justin Martyr, 282
Juvenal, 266

K
Kambouchner, D., 158, 159, 162, 166
Kant, I., 7, 23
Kavka, G., 222, 223
Keckerman B., 108
Kenny, A., 21, 22
Kepler, J., 127, 128, 135, 136, 138, 139, 142–149
Kircher, A., 148–149
Knolles, R., 246
Koyré, A., 82, 83
Kristeller, P.O., 3, 12, 22, 157

L
La Boétie, E. de., 216
La Forge, L. de, 169–182
Lagalla, C., 145
Lassels, R., 85
Leibniz, von, G.W., 5, 23, 59, 61, 64, 74, 75, 77
Leone Ebreo (Judah Leon Abravanel), 24, 41, 44, 47, 53–55
Lessing, G.E., 11
Libri, Matteo de', 240
Liburnio, N., 51, 52, 55
Liceti, F., 82, 145
Licetus. *See* Liceti, F.
Lipsius, J., 9, 43, 44, 137, 207, 208, 235, 245, 246
Livy (Titus Livius), 71, 239
Locke, J., 5, 23, 85, 94
Lucian of Samosata, 148, 266
Lucretius, T.C., 93, 120
Lupini, E., 214
Luther, M., 5, 8, 10–12, 25, 27, 32, 35

M
Machiavelli, N., 8, 16, 22, 82, 84, 203, 205, 207–209, 215, 216, 239–243, 245–248, 250, 253–255, 288, 289
Macrobius, 135, 146, 147
Maestlin, M., 142, 147, 149
Magirus, J., 108
Malebranche, N., 156, 161, 181
Malmesbury, Third Earl of (James Howard Harris), 263, 264, 269, 271
Manzo, S., 15
Marcus Aurelius (emperor), 265
Marion, J.-L., 156
Marlowe, C., 53
Martinich, A.P., 222
Marullo, M., 53
Marx, K., 267, 278
Mauri, A., 142
Maximus Tyrius, 46
Medici, L. de', 145, 147
Melanchthon, P., 26, 50
Mersenne, M., 11, 127, 128, 277, 287
Meschkat, K., 252
Michelet, K.L., 9
Milton, J., 268, 270
Montchrétien, A. de, 245
Montaigne, M. (Michel Eyquem de), 8, 21, 43, 82, 155, 157–161, 165, 166, 167, 207, 216, 217
More, H., 15, 81–95, 114, 175
More, T., 24
Morin, J.-B., 128
Mörlin, J., 27
Mun, T., 244
Muratori, C., 16, 156

N
Naudé, G., 82, 84, 85, 127, 277, 281
Nauta, L., 15
Nelson, J.C., 41, 55
Nifo, A., 82, 108
Niphus. *See* Nifo, A.
Nizolio, M., 61, 74, 75
Noah (Biblical figure), 290

O
Ockham, W., 23, 27
Ovid, 266, 270

P
Paganini, G., 16, 73, 156, 157, 159, 287
Paracelsus (P.A.Theophrastus Bombastus von Hohenheim), 127, 211, 285
Parmenides, 123, 124
Pasnau, R., 60
Patricius. *See* Patrizi, F. of Cherso
Patritius. *See* Patrizi, F. of Cherso
Patrizi, F. of Cherso, 7, 82, 120, 122, 127, 128, 135, 137–140, 142
Patrizi, F. of Siena, 213
Paul, J., 253
Paul of Tarsus, 190
Pausanias, 50
Peltonen, M., 208
Pereira, B., 108
Pererius. *See* Pereira, B.
Persio, A., 122, 137
Petit, P., 177–179, 180
Petrarca, F., 61, 62, 75
Petrarch. *See* Petrarca, F.
Petty, W., 245
Philip II (emperor), 53
Piccolomini, A., 82
Piccolomini, F., 41, 48, 49, 55
Pico della Mirandola, G., 27, 47, 69, 82, 85
Pico della Mirandola, G.F., 27, 69
Picot, C. (abbot), 163
Pindar, 266
Pisistratus (tyrant), 239
Pistorius, J., 53
Platina (Bartolomeo Sacchi), 213, 214

Plato, 9, 14, 22, 27, 29–31, 36, 41–42, 44–55, 81–83, 87, 100, 106, 172, 177, 269, 283
Plautus, 266
Pliny, 290
Plotinus, 46, 53, 54, 81, 84, 147
Plutarch, 46, 91, 138, 142, 144–146, 149, 211, 212, 214, 216
Pocock, J., 254
Polansdorf, von, A., 27
Polanus, A.. *See* Polansdorf, A. von
Poliziano, Agnolo Ambrogini, 76, 282
Pompey (Gnaeus Pompeius Magnus), 291
Pomponacius. *See* Pomponazzi, P.
Pomponazzi, P., 9, 29, 81, 82, 84, 86, 89, 92, 95, 186
Pontano, G., 53
Popkin, R., 71
Prévost, B., 51
Proclus, 81, 135, 138, 142
Pythagoras, 87, 89, 138, 140, 145–146, 211, 284

Q
Quintilian, 62, 64

R
Raleigh, W., 252, 253, 255
Ramée, P. de la, 10, 61, 69, 126–128
Ramus. *See* Ramée, P. de la
Rees, G., 122, 123, 125, 126
Reid, T., 266
Reuchlin, J., 9
Rhodiginus, Caelius. *See* Ricchieri, L.
Ricchieri, L., 147
Riccioli, G.B., 148
Rice, E.F., 165
Richelieu, A.-J. du Plessis de, 176, 279
Rixner, T.A., 6
Robbins, C., 271
Rocco, A., 145
Rochas, Henry de, 128
Rorty, R., 3
Rossi, P., 207
Rousseau, J.-J., 239, 255
Ruggiero, G., 2
Russell, B., 22

S
Sallust (Gaius Sallustius Crispus), 240, 241
Salmasius, C., 265
Sanches F., 59, 61, 67–69, 77, 235
Saracenus, J.C., 53
Sarpi, P., 137, 142
Saul (king), 290
Saumaise, C.. *See* Salmasius, C.
Savonarola, G., 282
Scaliger, J.C., 26, 27, 44, 50, 82–85, 100, 105, 186
Schegk, J., 26
Schelling. F.W.J., 11
Schmitt, C.B., 2, 6, 11, 22
Schuyl, F., 170–171, 182
Schweighäuser, J., 262, 263
Sciaccaluga, N., 125
Sebond, R., 155, 165–167
Secundus, J. (Basia), 53
Sejanus, L.A., 214, 215
Seneca, L.A., 27, 44, 207
Sextus Empiricus, 69, 81
Shaftesbury, Third Earl of (Anthony Ashley Cooper), 259, 260–269
Shakespeare, W., 16, 259, 260, 265–270
Simplicius, 265
Skalić de Lika, P., 27
Skinner, Q., 206, 227, 228, 240
Socrates, 51, 52, 157, 161, 165, 267
Solomon (Biblical king), 54, 191, 290
Solon, 288
Sorel, C., 15, 119, 123, 126–131
Spampanato, V., 282
Spaventa, B., 23
Spenser, E., 16, 53, 259–261, 269–271
Spinoza, B., 22, 23, 25, 75
Springborg, P., 246
Stangeland, C.E., 241, 242
Steno, N., 179
Steuco, A., 82
Steuchus. *See* Steuco, A.
Strauss, L., 224, 234
Struensee, J.F., 266
Stupanus, J.N., 241
Suarez, F., 22, 71, 75, 156

T
Tacitus, 207, 215
Taylor, A.E., 222
Telesio, B., 7, 15, 24, 82, 84, 119–132, 283, 285
Telesius. *See* Telesio, B.
Tennemann, W.G., 6–7
Thales, 124
Thévenot, M., 179
Thylesius. *See* Telesio, B.

Tiberius, 214, 215
Tillen, D., 27
Timon of Athens, 194
Tinto Vicini, G., 213
Turnèbe, A., 43

U
Upton, J., 15, 259–271

V
Vairo, L., 45
Vairus. *See* Vairo, L.
Valencia, P. de, 266
Valerius, C. (Kornelis Wouters), 108
Valesius. *See* Vallès, F.
Valla, L., 15, 16, 21, 24, 59, 61–71, 73, 74, 76, 77, 221, 223, 229, 230–236, 282
Vallès, F., 49, 50, 55
Van Eyck, J., 136
van Goorle, D., 127
Vanini, G.C., 9–10, 16, 81–82, 85–86, 89, 92, 95, 185–187, 192–199
Vasoli, C., 22
Vegius, M., 230
Verulamius. *See* Bacon, F.
Vickers, B., 204–206, 208, 214
Villiers, G., 204, 214–215
Villon, A., 128
Vinta, B., 143
Vives, J.L., 24, 59, 61, 65–71, 75–77, 235
Voitle, R., 266
Voltaire (Arouet, F.-M.), 204

W
Ward, R., 86, 87
Webster, J., 127
Wecker, J.J., 44
Wilkins, J., 61, 135, 142, 143, 147, 148
Wirz, J., 28

Z
Zabarella, J., 26, 31, 108, 110
Zambelli, P., 23
Zanchi, G., 26, 35
Zara, A., 44
Zwinger, T., 27

MIX
Papier aus verantwortungsvollen Quellen
Paper from responsible sources
FSC® C105338

If you have any concerns about our products,
you can contact us on
ProductSafety@springernature.com

In case Publisher is established outside the EU,
the EU authorized representative is:
Springer Nature Customer Service Center GmbH
Europaplatz 3, 69115 Heidelberg, Germany

Printed by Libri Plureos GmbH
in Hamburg, Germany